Corporate Communication

Sixth Edition

Paul A. Argenti
*The Tuck School of Business
at Dartmouth*

McGraw-Hill
Irwin

The McGraw-Hill Companies

McGraw-Hill
Irwin

CORPORATE COMMUNICATION, SIXTH EDITION

Published by McGraw-Hill, a business unit of The McGraw-Hill Companies, Inc., 1221 Avenue
of the Americas, New York, NY, 10020. Copyright © 2013 by The McGraw-Hill Companies, Inc.
All rights reserved. Printed in the United States of America. Previous editions © 2009, 2007, and 2003.
No part of this publication may be reproduced or distributed in any form or by
any means, or stored in a database or retrieval system, without the prior written consent of
The McGraw-Hill Companies, Inc., including, but not limited to, in any network or other electronic
storage or transmission, or broadcast for distance learning.

Some ancillaries, including electronic and print components, may not be available to customers
outside the United States.

This book is printed on acid-free paper.

1 2 3 4 5 6 7 8 9 0 QFR/QFR 1 0 9 8 7 6 5 4 3 2

ISBN 978-0-07-340317-5
MHID 0-07-340317-2

Vice President & General Manager: *Brent Gordon*
Publisher: *Paul Ducham*
Brand Manager: *Anke Braun Weekes*
Marketing Manager: *Michael Gedatus*
Development Editor: *Kelly I. Pekelder*
Project Manager: *Melissa M. Leick*
Cover Designer: *Studio Montage, St. Louis, MO*
Cover Image: *Andrew Salmon*
Buyer: *Susan K. Culbertson*
Media Project Manager: *Prashanthi Nadipalli*
Compositor: *S4Carlisle Publishing Services*
Typeface: *10/12 Palatino*
Printer: *Quad Graphics*

All credits appearing on page or at the end of the book are considered to be an extension of the
copyright page.

Library of Congress Cataloging-in-Publication Data

Argenti, Paul A.
    Corporate communication/Paul A. Argenti.—6th ed.
        p. cm.
    Includes bibliographical references and index.
    ISBN-13: 978-0-07-340317-5
    ISBN-10: 0-07-340317-2
    1. Communication in management. 2. Communication in organizations. I. Title.
    HD30.3.A73 2013
    658.4'57—dc23

                                                                        2012024863

www.mhhe.com

For Jackson

Preface to the Sixth Edition

This book grows out of almost 30 years of work developing the field of study referred to in this book as *corporate communication*. Although the term itself is not new, the notion of it as a functional area of management equal in importance to finance, marketing, human resources (HR), and information technology (IT) is more recent. In the last 30 years, senior managers at a growing number of companies have come to realize the importance of an integrated communication function.

In this introduction, I would like to talk a bit more about my expertise, what this book is all about, and why I think everyone involved in organizations today needs to know about this important discipline.

Author's Expertise

For the last 31 years, I have been a professor of management and corporate communication at the Tuck School of Business at Dartmouth. Prior to that, I taught at the Columbia and Harvard Business Schools.

The tradition of teaching communication has been a long one at Tuck, but as at most schools, the focus was on skills development, including primarily speaking and writing. The first development in the evolution of this field was an interest among businesspeople in how to deal with the media. Because this requirement mostly involved applying oral presentation skills in another setting, the faculty teaching communication were a logical choice for taking on this new task.

So when I began teaching the first management communication course at Tuck in 1981, I was asked to include a component on dealing with the media and handling crises. I became interested in this topic through my study of marketing at Columbia and had already written a case on the subject, which appeared in earlier editions of this book.

Over the years, my interest in the subject grew beyond how companies deal with the media to include how they deal with *all* communication problems. As I wrote more case studies on the subject and worked with managers inside companies, I saw the need for a more integrated function. That's because most companies were conducting communication activities in a highly decentralized way.

For example, the employee communication function at Hewlett-Packard (HP) in the mid-1980s was in the HR department, where it had always been, when I wrote a case on how HP dealt with voluntary severance and early retirement programs. As I looked at other companies, I found similarities to HP. Yet the people in those various HR departments were doing exactly the same thing internally that a communication specialist in the public relations (PR) department was doing for the external audience—sending a specific company message to a specific audience.

The same was true of the investor relations (IR) functions, which typically resided exclusively in the finance department in most companies until the 1990s. Why? Because the chief financial officer was the one who knew the most about the company's financial performance and historically had been responsible for developing the annual report. Communication was seen as a vehicle for getting that information out rather than as a function in itself.

Again, as I worked with companies to develop new identities and images, I found marketing people involved because they had traditionally dealt with brand and image in the context of products and services. Yet those marketing experts didn't always know what was being communicated to the press or to securities analysts by their counterparts in other functional areas.

These experiences led me to believe that corporations and other organizations, from universities to churches to law firms, could do a much better job of communicating if they integrated all communication activities under one umbrella. That was the theory at least, but I could find precious little evidence in practice.

Then, in 1990, I was fortunate enough to be given a consulting assignment that allowed me to put into practice what I had been talking about in theory for many years. I received a call from the chairman and chief executive officer of a major corporation after my picture appeared on the front page of *The New York Times* Sunday business section in an article about how professors were teaching business students about dealing with the media.

Ostensibly, the chairman's call was about how his company could get more credit for the great things it was doing. Specifically, he wanted to know if I had a "silver bullet." My silver bullet, as it turned out, was the development of a new corporate communication function for the company.

This company, like most, had let communications decentralize into a variety of other functional areas over the years, with the predictable result: no integration. The media relations people were saying one thing, the investor relations department was saying another; the marketing team was developing communication strategies for the outside, the human resources department for the inside.

No one except the chairman, who sat at the top of this $30 billion organization, could see the big picture, and none of those intimately involved with the various activities had an inside track on the overall strategy for the firm. Over the next year and a half, the chairman and I came up with the first integrated communication function that had all the different subsets I had tried unsuccessfully to bring together at other companies and even at my own university.

We changed everything—from the company's image with customers to its relationship with securities analysts on Wall Street. Today this company has one totally integrated communication function. This book will explain what all the component parts of that function are all about.

What Is This Book About?

Chapter 1, "The Changing Environment for Business," provides a context for the rest of the book. It describes changes in the environment for business that have taken place over the last 60 years and their implications for corporate communication. Although attitudes about business have never been totally positive, they have reached an all-time low in recent years: Mistrust of and skepticism about corporate entities are high, as are expectations that companies will "give back" to society through philanthropy, community involvement, or environmental protection activities.

In the Google in China case, we see how one company had to compromise its values to do business in one of the fastest growing markets in the world.

Chapter 2, "Communicating Strategically," explains how companies need to use a strategic approach to communications. In the past, most communication activities were dealt with reactively as organizations responded to events in the world around them. With the framework for strategic communication provided in this chapter, companies can proactively craft communications tailored to their constituencies and measure their success based on constituency responses.

In the Galen Healthcare System case, new to this edition, we find an example of a manager who failed to use a strategic approach to communication in a rapidly changing corporate environment.

In Chapter 3, "An Overview of the Corporate Communication Function," we take a look at the evolution of the corporate communication function and some of the different ways it can be structured within organizations. This chapter also describes each of the subfunctions that should be included in the ideal corporate communication department.

The Sweet Leaf Tea case, also new to this edition, provides an excellent example of how a company used its communication function to deal with a difficult situation.

Chapter 4, "Identity, Image, Reputation, and Corporate Advertising," describes the most fundamental function of a corporate communication department: to reflect the reality of the firm itself through visual images and the right choice of words. The study of identity and image has blossomed in recent years as graphic designers have worked with companies to develop the right look for a particular approach to the marketplace. Additionally, corporate reputation is gaining increased attention as consumers and investors take a more holistic view of companies and their activities, such as corporate social responsibility.

Organizations also reflect their image and identity through advertising. We end this chapter by looking at how companies use corporate advertising to sell the organization as a whole, as opposed to just the products or services they offer, to the public. Organizations use corporate advertising for a number of reasons: to enhance or alter their image by developing a corporate brand, to present a point of view on a topic of importance to them, or to attract investment.

The case for this chapter, allows students to look inside Jet Blue's Valentine's Day disaster in 2007.

In Chapter 5, "Corporate Responsibility," we see how companies try to do well by doing good, manage the so-called triple bottom line, and deal with increasing demands from antagonists and pressure groups.

The Starbucks Coffee Company case reveals how one company balanced its responsibilities to its customers with demands from a nongovernmental organization (NGO) to improve its sourcing.

In Chapter 6, "Media Relations," we look at how today's corporate communications function has evolved from the "press release factory" model to a more sophisticated approach of building relationships with both traditional and new media before having a specific story to sell them and targeting the appropriate distribution channel for different kinds of stories.

The Adolph Coors Company serves as our case in point for this chapter. In this classic case, which I wrote for the first edition, we see how this company dealt with the formidable "60 Minutes" when it approached Coors with a controversial story idea.

One of the most important functions within corporate communication deals with an internal rather than an external constituency: employees. In Chapter 7, "Internal Communication," we look at employee communications' migration away from the HR area toward a function that is more connected with senior management and overall company strategy.

The Westwood case explores one company's attempt to deal with voluntary severance and outplacement issues related to layoffs.

In Chapter 8, "Investor Relations," we see how companies use communication strategies to deal with analysts, shareholders, and other important constituencies. In the past, this communication subfunction often was handled by managers with excellent financial skills and mediocre communication skills. Today, as IR professionals interact regularly with the media and need to explain nonfinancial information to investors, strong communication skills are equally critical to a solid financial background.

Our case for this chapter, Steelcase, Inc., examines how an IR function was built at that company.

Chapter 9 covers government relations. The business environment historically has fluctuated between periods of relatively less regulation and relatively more, but government relations is always a consideration for companies, whether at the local, state, federal, or international level.

The Disney case provides an example of how a large corporation dealt with challenges from government and local communities in Virginia as it tried to open an historical theme park.

Organizations inevitably will have to deal with some kind of crisis. In Chapter 10, "Crisis Communications," we look at how companies can prepare for the unexpected and provide examples of both good and poor crisis communications, as well as practical steps to creating and implementing crisis communication plans.

Our case at the end of this chapter focuses on Coca-Cola in India as it attempts to work its way out of a crisis in a case involving accusations of environmental contamination in its products.

What Is New to the Sixth Edition?

The sixth edition of Corporate Communication reflects valuable feedback received from both users and reviewers of the previous editions. In addition to new research findings and new examples to illustrate the latest economic, social, political and corporate trends, changes in this edition include the following:

- New cases and case questions
- Expanded coverage of the history of communication theory
- Additional discussion of the impact and role of social media and digital communications
- Increased emphasis on corporate responsibility issues throughout the book
- Additional recommendations for crisis communication
- Timely analysis of the challenges that companies are facing today in this time of low consumer confidence and anti-corporate sentiment (e.g. the Occupy Wall Street movement) in the wake of the credit crisis

Why Is Corporate Communication So Important Today?

Every functional area, at one time or another, was the newest and most important. But in the twenty-first century, the importance of communication is obvious to virtually everyone. Why?

First, we live in a more sophisticated era in terms of communication. Information travels at lightning speed from one side of the world to another as a result of digital communications and social media.

Second, the general public is more sophisticated in its approach to organizations than it has been in the past. People tend to be more educated about issues and more skeptical of corporate intentions. Thus, companies cannot get by on statements such as, "What's good for General Motors is good for everyone" or "If we build a better mouse trap, customers will beat a path to our door." Maybe not, if they don't know who you are.

Third, information comes to us in more beautiful packages than it did before. We now expect to see glossy annual reports and dazzling websites from major corporations. We don't want to walk into grimy-looking stores even for our discount shopping. Gas stations are modern looking and have been "designed" from top to bottom by high-profile New York design firms. The bar is high for a company's message to stand out in this environment.

Fourth, organizations have become inherently more complex. Companies in earlier times (and the same is true even today for very small organizations) were small enough that they could get by with much less sophisticated communications activities. Often, one person could perform many different functions at one time. But in organizations with thousands of employees throughout the world, it is much more difficult to keep track of all the different pieces that make up a coherent communication strategy.

This book describes not only what is happening in an era of strategic communication but what companies can do to stay one step ahead of the competition. By creating an integrated corporate communication system, organizations will be able to face the next decades with the strategies and tools that few companies in the world have at their fingertips.

I am sure that 20 years from now, managers will come to realize the importance of an integrated, strategic communication function. No doubt, much will have been written about the corporate communication function, and most complex organizations will have a corporate communication department with many of the subsets described in this book. Until then, however, I hope you enjoy reading about this exciting field as much as I have enjoyed chronicling its development.

A Note on the Case Method

Throughout this book you will find cases or examples of company situations that typically relate to material covered in each of the chapters.

What Are Cases?

Cases are much like short stories, in that they present a slice of life. Unlike their fictional counterparts, however, cases are usually about real people, organizations, and problems (even though the names may sometimes be disguised for proprietary reasons). Thus, a reader has an opportunity to participate in the real decisions that managers had to make on a variety of real problems.

The technique of using actual business situations as an educational and analytical instrument began at Harvard in the 1920s, but the use of a "case" as a method of educating students began much earlier. Centuries ago, students learned law by studying past legal cases and medicine through the use of clinical work.

Unlike textbooks and lectures, the case method of instruction does not present a structured body of knowledge. This approach often proves frustrating to students who may be used to more traditional teaching methods. For example, cases are frequently ambiguous and imprecise, which can easily confuse a neophyte. This complexity, however, represents what practitioners usually face when making decisions.

In cases, as in life, problems can be solved in a variety of ways. Sometimes one way seems better than others. Even if a perfect solution exists, however, the company may have difficulty implementing it. You also may find that you have come up with a completely different solution to the problem than another student has. Try to forget the notion of finding an "answer" to the problem. The goal in using this method is not to develop a set of correct approaches or right answers but rather to involve you in the active process of recognizing and solving general management problems.

In class, you will represent the decision maker (usually an executive) in a discussion that is guided by the professor. The professor may suggest ideas from time to time or provide structure to ensure that students cover major issues, but each student's insight and analytical prowess is displayed in this context. Often a professor will play devil's advocate or pursue an unusual line of reasoning to get students to see the complexities of a particular situation. As a teaching device, the case method relies on participation rather than passive learning.

Although cases come in all shapes and sizes, two categories define the scope of most cases: evaluative and problematic. An evaluative case presents the reader with a description of a company's actions. The purpose of an analysis is thus to evaluate what management has done and then to determine whether the actions were well founded.

Problem cases, which are far more common, describe a specific problem a manager faces, such as whether to launch a new corporate advertising program, choose one method of handling the media over another, or even choose one form of communication rather than another. Such problems call for development of alternative strategies, leading to a specific recommendation.

Case Preparation

No matter what type of case you're dealing with, a common approach will help you prepare cases before you have time to develop what will eventually become your own style. In time, you will no doubt find a method that works well and proves more suitable to you. Regardless of the approach, a thorough analysis requires a great deal of effort.

Begin with a quick reading of the case. This read-through gives you a sense of the whole rather than what often can appear as a dazzling array of parts if you start by analyzing each section in detail. You should extract a *sense* of the organization, some impressions of what *could be* the problem, and a working knowledge of the amount and importance of information presented in the case.

A more careful second reading of the case will allow you to begin the critical process of analyzing business problems and solving them. What you should hope to cull from this analysis follows.

Problem Definition

First, you must establish a specific definition of the problem or problems. Although this definition may be clearly stated in the case, usually problem definition is a crucial first step in the analysis. You need to go beyond simple problem definition and look for symptoms as well. For example, as part of the analysis, you might wonder why or how the defined problem has developed in the company. Avoid, however, a repetition of case facts or an historical perspective. Assume that your reader has all the facts that you do and choose reasoning that will serve to strengthen, rather than bloat, your problem definition.

Company Objectives

Second, once you have defined the problem, place it within the context of management's objectives. How does the problem look in this light? Do the objectives make sense given the problems facing management?

In some cases, objectives are defined explicitly, such as "increase stock price by 10 percent this year." If the problem in the case proves to be that the company's investor relations function is a disaster, this objective is probably overly optimistic. Goals can be more general as well: "Change from a centralized to a decentralized communication organization in five years." In this instance, a centralized department with independent managers at the divisional level has a good chance of meeting its objectives.

Data Analysis

Third, you next need to analyze information presented in the case as a way of establishing its significance. Often this material appears in exhibits, but you also will find it stated within the case as fact or opinion. Remember to avoid blind acceptance of the data, no matter where they appear. As in the real world, information presented in the case may not be reliable or relevant, but you may find that if you manipulate or combine the data, they ultimately will prove valuable to your analysis. Given the time constraints you will always be under in case analysis and in business, you should avoid a natural tendency to spend more time than you can really afford analyzing data. Try to find a compromise between little or no data analysis and endless number crunching.

Alternative Strategies and Recommendations

Fourth, after you have defined the problem, identified company objectives, and analyzed relevant data, you are ready to present viable alternative strategies. Be sure the alternatives are realistic for the company under discussion, given management's objectives. In addition, you must consider the implications of each alternative for the company and management.

Once you have developed two or three viable alternative solutions, you are ready to make a recommendation for future action. Naturally, you will want to support the recommendation with relevant information from your analysis. This final step completes your case analysis, but you must then take the next step and explore ways to communicate all the information to your reader or listener.

Cases in the Real World

Here are some further thoughts to help you distinguish a case from a real situation: Despite the hours of research time and reams of information amassed by the case writer, he or she must ultimately *choose* which information to present. Thus, you end up with a package of information in writing. Obviously, information does not come to you in one piece in business. A manager may have garnered the information through discussions, documents, reports, websites, and other means. The timing also will be spread out over a longer period than in a case.

Also, given the necessary selectivity of the case writer, you can be sure a specific teaching objective helped focus the selection of information. In reality, the "case" may have implications for several different areas of a business.

Because a case takes place within a particular period of time, it differs in another important way from management problems. These tend to go on and change as new information comes to light. A manager can solve some of the problems now, search for more information, and decide more carefully later on what is best for a given situation. You, on the other hand, must take one stand now and forever.

Finally, case analyses differ from the realities of management in that students do not have responsibility for implementing decisions. Nor do they suffer the consequences if their decision proves untenable. You should not assume that this characteristic removes you from any responsibility. On the contrary, the class (in a discussion) or your professor will be searching for the kind of critical analysis that makes for excellence in corporate communication.

Acknowledgments

Without the help and support of the Tuck School at Dartmouth, I could not have completed this book. Over the last 31 years, I have been given funds to write cases and conduct research as well as time to work on the material in this book. I am particularly grateful to Dick West for initially investing in my career here at Tuck and encouraging me to develop a new area of study, and to Paul Danos and Bob Hansen for their continued support more recently.

I also must thank my friends and colleagues at Tuck who first made me sit down and finally produce a text after years of collecting materials and thoughts in files and boxes: specifically, the late John Shank and Mary Munter. The International University of Japan also deserves credit for providing me with the contemplative setting I needed to write the first edition of this book.

Many clients helped me to test the ideas I have developed over more than 30 years, but I am particularly indebted to Joseph Antonini, former chairman and chief executive officer of Kmart, for allowing me to think creatively about the possibilities for a unified corporate communication function. I also would like to thank Jim Donahue, head of learning and Andy Sigler, formerly chairman and CEO of Champion International, for allowing me to test new ideas with top managers at their company; Nancy Bekavac, former president of Scripps College, for allowing me to work on Scripps' identity program and for her helpful comments on Chapter 4; and Valerie Haertel of Alliance Capital Management and Jack Macauley of Workwise Communication, for their input and help with Chapter 7. David McCourt, former chairman and CEO of RCN, also allowed me to work more recently on developing a corporate communication function in his company. In addition, I thank my many colleagues at Goldman Sachs, where I was fortunate to work as a consultant for over eight years, and to Peter Verrengia, my dear friend Suzanne Klotz, and all of my colleagues at Fleishman Hillard for their support over the last eight years.

I am indebted as well to the students I have taught, especially at Tuck, but also at Erasmus University, Singapore Management University, Hanoi School of Business, the International University of Japan, the Helsinki School of Economics, Columbia Business School, and Harvard Business School. They have tested these ideas in their fertile minds and given me inspiration for coming up with new ways of thinking about communications.

I'd also like to thank Jim O'Rourke for his permission to use the Google in China case in Chapter 1; Elizabeth Dougall for her permission to use the JetBlue case in Chapter 4; Elizabeth Powell for her permission to use the Disney case in Chapter 9, and my wife, Jennifer Kaye Argenti, for writing the Coke in India case in Chapter 10.

Many research assistants helped me with this project over the years, but I am particularly grateful to Christine Keen and Patricia Gordon, Mary Tatman, Adi Herzberg, Thea Haley Stocker, Kimberley Tait, Abbey Nova, Suzanne Klotz, Courtney Barnes, Alicia Korney, and Cassandra Harrington for their incredible help with previous editions. I would also like to thank my longtime former academic assistant at Tuck, Annette Lyman, for her dedication to the fifth edition. This sixth edition would have been impossible to complete without the help of Alina Everett, Georgia Aarons, Genoa Terheggen, Alexandra Angelo, Katie Rosenberg, Lenore Feder, Jordan Fleet, the amazing Kelly Sennatt, and especially the tireless project manager for this edition, Joanie Taylor. Finally, I want to give thanks to my amazing academic assistant at Tuck, Jessica Osgood. I cannot imagine ever having a better team in place to work on a project like this.

The reviewers who helped with the sixth edition also deserve special thanks for their helpful comments and advice:

Bill Margaritis
FedEx

Cees van Riel
Erasmus University

Don Wright
Boston University

Dr. Sherry Roberts
Middle Tennessee State University

Irv Schenkler
Stern School of Business, New York University

Jon Iwata
IBM

Michele Marie Bresso
Bakersfield College

Robert Mead
Aetna

Sherry Southard
East Carolina University

Stephen Greyser
Harvard Business School

I also wish to thank the reviewers from the previous editions who made this book better through their honesty and input:

Carter A. Daniel
Rutgers University

Charlotte Rosen
Cornell University

Chris Kelly
New York University

Cynthia Buhay-Simon
Bloomsburg University

Don Bates
Columbia University

Elizabeth Powell
University of Virginia

Frank Jaster
Tulane University

Gary Kohut
University of North Carolina–Charlotte

James O'Rourke
University of Notre Dame

Jane Gilligan
Clark University

Jerry Dibble
Georgia State University

JoAnne Yates
Massachusetts Institute of Technology

Joan M. Lally
University of Utah

Joel T. Champion
Colorado Christian University

Jonathan Slater
State University of New York at Plattsburgh

Judith Sereno
Medaille College

J. S. O'Rourke
University of Notre Dame

Karen Gersten
Evelyn T. Stone University College

Linda Lopez
Baruch College

Lynn Russell
Columbia University

Otto Lerbinger
Boston University

Margo Northey
University of Western Ontario

Mary E. Vielhaber
Eastern Michigan University

Michael Putnam
University of Texas–Arlington

Rick Calabrese
Dominican University

Robert Stowers
College of William & Mary

Sherron B. Kenton
Emory University

Suzette Heiman
University of Missouri

Valerie Haertel
Alliance Capital Management

Wayne Moore
Indiana University of Pennsylvania

Yunxia Zhu
UNITEC (New Zealand)

My thanks also go to the staff of McGraw-Hill/Irwin, especially my developmental production editor, Lori Bradshaw; senior project manager, Melissa Leick; copyeditor; and former executive editor at Irwin, Bevan O'Callaghan, who initially signed the book. Their patience allowed me the freedom to develop this material for the six editions over a much longer period of time than I would have guessed it would take at the outset. I would also like to thank Andrew Salmon for his incredible book design for the 6th edition.

Finally, I would like to thank my parents for giving me the raw material in the beginning and the education later on that allowed me to become an academic.

Paul A. Argenti
Hanover, New Hampshire
2012

The author would like any comments or questions as well as corrections to the text. Please write to Professor Paul A. Argenti, The Tuck School of Business, Dartmouth College, Hanover, NH 03755; or e-mail comments to paul.argenti@dartmouth.edu.

Brief Table of Contents

Table of Contents

CHAPTER ONE

The Changing Environment for Business

Most of today's business leaders grew up in a different era from the one they find themselves in now: a typical senior executive grew up during one of the most prosperous and optimistic periods in American history. The difference between the world these people knew in their childhood and the one their grandchildren will face in the twenty-first century is nothing short of staggering.

The public's current expectations of corporations are also different from what they were 50 years ago. To attract customers, employees, and investors, companies need to be progressive leaders about a host of global issues and put their vision in a broader social context. Public scrutiny of business is constant and intense, and in the past decade, disillusionment has grown over excesses in executive pay, questionable accounting practices, drug recalls, and moral laxity on the part of corporations.

In this chapter, we will put our discussion of corporate communication in context by looking at some of the events that have influenced the operating environment for business. We begin by looking at a history of public attitudes toward American business and their reflection in popular culture. Next we turn to the effects of globalization (and the antiglobalization backlash) on business. Finally, we look at how improved corporate communication can help companies compete in this constantly changing environment.

Attitudes toward American Business through the Years

Business has never had a completely positive image in the United States. In the 1860s, the creation of the nation's transcontinental rail systems and the concomitant need for steel created hazardous working conditions for steelworkers and railroad builders alike. Soon thereafter, the Industrial Revolution moved American industry away from a model of small workshops and hand tools to mechanized mass production in factories. This shift had the effect of lowering prices of finished goods, but it also contributed to harsh and dangerous working conditions

1

for laborers, as documented in Upton Sinclair's book, *The Jungle*. The exploitation of young women and children working in factories, highlighted by the deadly Triangle Shirtwaist Factory fire in 1911, only added to negative perceptions of business.

As the patriarchs of big business, the Carnegies, Mellons, and Rockefellers—"robber barons," as they came to be known—were perceived as corrupt businessmen looking out for their own interests rather than the good of all citizens. And yet these negative attitudes toward the first modern corporate businessmen were coupled with envy of their material wealth. Most Americans wanted the lifestyle of these business magnates and came to see the pursuit of wealth and the security it provided as part of the "American Dream." The concept of social mobility, captured in author Horatio Alger's rags-to-riches novels, seemed to many to be a tangible reality in America's cities, and immigrants came to the United States in large numbers.

The 1920s were characterized by a sharply rising stock market following the conclusion of World War I and by increasing disparities in wealth distribution. These disparities—between rich and middle class, between agriculture and industry—made for unstable economic conditions, while speculation in the stock market fueled its growth to unprecedented levels. The stock market "bubble" finally burst in 1929, giving way to the Great Depression, which would last a decade and affect the rest of the industrialized world. It was a dark time for businesses and individuals alike.

By the mid-1940s, however, businesses started rebounding from the Depression as companies geared up for the Second World War. The steel industry, the automotive industry, the military-industrial complex—all of which made the prosperity of the 1950s and 1960s a reality—got their start during World War II.

Perhaps the epitome of this era, considered by many a "golden age," was the "Camelot" years of the Kennedy administration. The economy was booming, and in the aftermath of the Cuban missile crisis, the United States felt it had defused the tensions of the Cold War. Even after Kennedy's death, prosperity continued, and public approval of business soared.

Over a period of 30 years, the marketing consultancy firm Yankelovich asked the question of American citizens: "Does business strike a balance between profit and the public interest?" In 1968, 70 percent of the population answered yes to that question. By the time Richard Nixon was on his way to the White House, however, the nation was torn apart by civil unrest, with the continuation of the civil rights struggle and demonstrations against U.S. involvement in the Vietnam War. Disagreement over the role of the United States in Vietnam marked a serious deterioration in public attitudes toward all institutions, including business. For those who were against the war, the executive branch of government came to stand for all that was wrong with America.

Because it helped to make the war possible and profited from the war, American industry was the target of much of the public's hostility. Dow Chemical's manufacture of Napalm and Agent Orange, which would be used to defoliate Vietnamese jungles, led to student protests on American university campuses. Young people in the United States came to distrust the institutions involved in the war, whether

TABLE 1.1
How Much
Confidence
Do You Have
in These
Institutions?*

Sources: *Yankelovich Monitor*, Harris Poll.

	1966	1971	1989	2011
Large companies	55%	27%	14%	13%
U.S. Congress	42%	19%	10%	6%
Executive branch	41%	23%	27%	19%
Supreme Court	51%	23%	26%	24%

*Answers reflect those answering most positively.

government agencies or businesses. This belief represented a dramatic change from the attitudes Americans had during World War II. Those in power failed to see how the Vietnam War was different because Americans were ambivalent about what the country was fighting for.

Toward the end of the 1960s and coinciding with the war in Vietnam, a rise in radicalism in America marked the beginning of a long deterioration of trust in institutions. The events of the early 1970s also contributed to this shift. For example, Watergate only confirmed what most young Americans had believed all along about the Nixon administration. The aftermath of the oil embargo, imposed by Arab nations after the 1973 Middle East war, had even more of an effect on attitudes toward business in America. Cheap, abundant petroleum—the lubricant of the American way of life—suddenly became scarce and expensive as Saudi Arabia and other Arab producers punished the United States for supporting Israel in the war. The cutoff lasted less than three months, but its effects on consumer attitudes are still with us today.

As a result of Watergate, Vietnam, and the oil embargo, by the mid-1970s American attitudes toward business reached an all-time low. In answer to the same question "Does business strike a fair balance between profit and the public interest?" those answering yes in the Yankelovich poll dropped to 15 percent in 1976 when Jimmy Carter took office. This drop of 55 points in just eight years says more about the changing attitudes toward business than a thousand anecdotes.

An opinion research poll that asked members of the general public to rate their confidence in a number of institutions showed declines in all areas, as shown in Table 1.1.

As you read this, you may be asking yourself whether the 1980s and 1990s, which together constituted the final economic boom of the twentieth century, restored America's faith in business to where it had been in the 1960s. They did not, and in 2011, a Harris Poll asking the same questions found the responses to be as follows: trust in major companies, 13 percent; U.S. Congress, 6 percent; White House, 19 percent; and Supreme Court, 24 percent.[1] These percentages all decreased from 2010, indicating a confidence crisis that has reached critical mass; the 2010 results for the aforementioned institutions were 15 percent, 8 percent, 27 percent, and 31 percent, respectively.[2]

[1] Harris Poll May 2011.
[2] Harris Poll March 2010.

TABLE 1.2
Does Business
Balance Profit
and Public
Interest?*

Source: *Yankelovich
Monitor.*

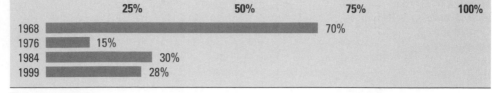

*Percent yes responses.

In response to the question about whether business strikes a fair balance between profit and the public interest, the percentages climbed back to a high of 30 percent answering yes in 1984. And the percentages dropped slightly to 28 percent in 1999 (the last year Yankelovich asked this question). (See Table 1.2.)

The nuances of American distrust of business are further explored by Yankelovich Partners through the following findings:

- 80 percent of surveyed respondents believe that "American business is too concerned about profits, not concerned about responsibilities to workers, consumers and the environment."
- 70 percent believe that "if the opportunity arises, most businesses will take advantage of the public if they feel they are not likely to be found out."
- 61 percent believe that "even long established companies cannot be trusted to make safe, durable products without the government setting industry standards."[3]

A more recent study commissioned by the Public Affairs Council finds that the American public still believes companies are failing to correctly prioritize their constituencies: 83 percent of respondents believe that businesses should put customers, employees, or their communities first; 81 percent of responders feel that companies are instead valuing top executives or shareholders first.[4] The source of the disconnect between upstanding business practices and the current business reality could be one of many factors. First, we must consider the economic instigators.

The 1990s saw the phenomenal rise of the NASDAQ index to 4,000 points by the end of the decade. Individual investors were actively participating in the equity markets and reaping enormous gains as stock prices seemed to be on an unstoppable upward trajectory. Then, in the spring of 2000, the markets came crashing down. By December, the NASDAQ had sunk to less than half its peak level of 5,000, reached at the beginning of the year. And unfortunately for the 100 million individual investors who had poured money into the market during the Internet-fueled boom of the 1990s, it did not stop there in its downward spiral. By early 2002, these individuals had lost $5 trillion since the "Internet bubble" burst, representing 30 percent of their stock wealth.[5]

---

[3] J. Walker Smith, Ann Clurman, and Craig Wood of Yankelovich Partners, Inc., *Point,* February 2005, http://www.RacomBooks.com; results from Yankelovich MONITOR.

[4] Public Affairs Council, *Public Affairs Pulse,* 2011.

[5] Marcia Vickers, Mike McNamee, et al., "The Betrayed Investor," *BusinessWeek,* February 25, 2002, p. 105.

With the bursting of the "dot.com bubble"; the exposure of corporate fraud at large companies such as WorldCom, Adelphia, and Tyco; and the collapse of Enron and its auditor, Arthur Andersen, due to fraudulent accounting, Americans perceived business as actively trying to deceive them. This perception was reflected in the media as well, such as in the *NBC Nightly News* segment entitled "The Fleecing of America."

In the midst of this market turmoil, the actions of unscrupulous financial analysts (see Chapter 8 for more on analysts) and companies like Enron angered the American public further. By February of 2002, some 81 percent of investors polled "did not have much confidence in those running Big Business."[6] This attitude is not surprising when you consider the many highly publicized stories of top executives who sold millions of dollars' worth of shares in their own failing enterprises, further enhancing their wealth as rank-and-file employees lost much of their retirement savings.

The public also has been embittered by the growing pay gap between senior company executives and ordinary workers that reached enormous proportions over recent decades. According to the AFL-CIO, in 1980, CEO pay averaged 40 times the pay of the average American worker, and by 2010, it averaged 343 times the pay of the median worker, for an average of \$US 11.4 million per year.[7] In October 2011, the Congressional Budget Office reported that the middle 60 percent of the American population experienced a growth in household incomes of 40 percent between 1979 and 2007 (after taxes and adjusted for inflation), while the top 1 percent of earners experienced a growth in household incomes of 275 percent. The study also confirmed that the after-tax household income of the top 20 percent of earners was greater than the combined after-tax income of the remaining 80 percent of workers. While top earners enjoy lucrative compensation packages, today 15 percent of Americans live in households receiving food stamps and 48.6 percent live in households receiving some form of government assistance, according to 2010 Census data.[8] Economist and *The New York Times* contributor Paul Krugman refers to this period of increasing income inequality, which he believes started in the late 1970s, as "The Great Divergence." He writes that it is more a product of conservative politics, tax law that is favorable to the wealthy, and inflated executive compensation than it is a product of less personal forces including globalization and technology.[9, 10]

Although executive compensation in general is a controversial subject, in the wake of the 2008 subprime credit crisis, public scrutiny has focused on the outsized annual bonuses doled out on Wall Street. Americans were especially outraged that

---

[6] Ibid., p. 106.

[7] "Executive PayWatch," http://www.aflcio.org/corporatewatch/paywatch/ceopay.cfm.

[8] Sara Murray, "Nearly Half of U.S. Lives in Household Receiving Government Benefits," *The Wall Street Journal*, January 17, 2012.

[9] Paul Krugman, "Introducing This Blog," *The New York Times*, September 18, 2007, http://krugman.blogs.nytimes.com/2007/09/18/introducing-this-blog/.

[10] "Trends in the Distribution of Household Income between 1979 and 2007," Congressional Budget Office, October 2011, http://www.cbo.gov/ftpdocs/124xx/doc12485/10-25-HouseholdIncome.pdf.

financial firms receiving public TARP rescue funds could use the money to pay out executive bonuses. In March 2009, insurance giant A.I.G. earned negative press when it decided to award multimillion-dollar bonuses to its executives despite having just received a $US 100 billion government bailout. In the summer of 2009, then New York attorney general Andrew Cuomo, released a report that detailed compensation at the largest New York–based banks that received public bailout money. The report revealed that Merrill Lynch had paid 149 bonuses greater than $US 3 million and 696 bonuses greater than $US 1 million, despite being in such dire financial straits that it had to merge into Bank of America in early 2009.[11] In July 2010, Kenneth R. Feinberg, who was appointed by President Obama to oversee executive compensation during the bailouts, released a report claiming that nearly 80 percent of the $US 2 billion that banks paid out in 2008 bonuses were unmerited.[12]

Increased tension over growing income inequity combined with relatively high unemployment rates in the United States sparked the Occupy Wall Street movement, a protest against corporate greed and corruption. The largely peaceful Occupy Wall Street movement started in September 2011 in Zuccotti Park in lower Manhattan and quickly spread to other U.S. cities as well as cities around the world, including Paris, London, Berlin, Hong Kong, and Rome.[13] Occupy Wall Street organizers made extensive use of social media and published a daily newspaper to communicate news and marching orders with participants. Organizers executed a branding campaign for the movement based on the slogan "we are the 99%," meant to highlight the growing income gap between the top 1 percent of earners and the remaining 99 percent. Critics of the Occupy Wall Street movement deride the movement for lacking clear focus and actionable objectives. However, the Occupy Wall Street movement, which was still ongoing in early 2012, though at a dramatically reduced scale, emphatically underscores the growing public discontent with the traditional big business.

At the very beginning of the protests, the Occupy Wall Street movement received minimal television and newspaper media coverage, but much coverage through social media such as Twitter.

Indeed, through the years, the traditional news media have played a major role in conveying, filtering, and obstructing messages from corporations as well as government and activist groups (see Chapter 6 for more on the media's influence on business). By the late 1990s, the Internet also began to shape attitudes toward business as activist groups gained access to a broadcast forum for their arguments against business. Today, environmental activists, animal rights groups, and shareholder rights proponents have the ability to get messages out instantaneously to like-minded individuals throughout the United States and the world. In the case of the Occupy Wall Street movement, the videos of protestors posted on YouTube and the flood of tweets coming out of Zuccotti Park quickly became "too viewed" for traditional media to ignore.

---

[11] Stephen Grocer, "Wall Street Compensation–'No Clear Rhyme or Reason,'" *The Wall Street Journal,* July 30, 2009.
[12] Louise Story, "Topics: Executive Pay," *The New York Times,* December 5, 2011.
[13] Alan Taylor, "In Focus: Occupy Wall Street Spreads Worldwide," *The Atlantic,* October 17, 2011.

Although the media and the Internet are powerful channels for views on business to be expressed and debated, nowhere are the attitudes that prevail in the external environment more clearly defined than in television and film.

Hollywood: A Window on Main Street and Wall Street

Throughout history, literature and the arts have both affected and reflected perceptions about institutions. Greek attitudes about government and religion manifested themselves in theater; Shakespeare shaped notions about English history for generations; and in the United States, cinema and television over the past several decades have reflected some of the public's negative attitudes about business.

For many Americans today, what they see in fictional or "factional" accounts in films and on television helps shape their attitudes more than educational institutions. In fact, Americans spend far more time in front of the television set, or watching media content online, than they do in the classroom. According to research undertaken by a number of different organizations, the average American household spends approximately 40–50 hours per week watching television programming. Many have written about what this habit has done to American society in a broader context over the last 30 years, but in this textbook, we will focus on the relationship between popular culture and business.

The Media Institute, a research organization funded by corporations, has been tracking media coverage of business for over 20 years. Each time it issues a report, the results are the same: Businesspeople are portrayed negatively in almost two-thirds of all television programs. Researchers have concluded that half of the time, businesspeople portrayed on television were involved in criminal activities.

In addition, most Americans get their news from television. As a result, the negative portrayals viewers see in fictional programming blend into the negative news they watch on the nightly news. An individual might, for example, watch an episode of *Law and Order* in which a woman is framed for murder after raising questions about her company's back-dating of stock options one night, then see an in-depth story about United Health doing the same thing on *Dateline NBC* the following evening. This information all comes from television, all of it is bad, and the net result is the reinforcement of negative perceptions of business.

Films also contribute to a negative business image. One of the most successful films of the late 1970s was called *The China Syndrome*, a movie about a narrowly averted meltdown at a nuclear reactor. A week after the release of the film, a real nuclear accident occurred at Three Mile Island. While everyone would agree that Metropolitan Edison did a poor job of communicating about this accident, few would say that the company was as bad as the one portrayed in the movie. For many Americans, however, the two events were linked, which made their reaction to the events at Three Mile Island that much stronger.

It is eerie how Hollywood has mirrored events in business at exactly the right time. The movie *Wall Street* is another such example. Oliver Stone's movie came out just ahead of the great scandals that rocked the real Wall Street in the late 1980s. Even within the film itself, reality and fiction were intertwined. Gordon Gekko, the evil financial genius meant to represent someone like the notorious arbitrageur Ivan Boesky, makes a speech in the film about greed. "Greed is good, greed purifies, greed cuts through and captures the essence of the evolutionary spirit," Gekko says in a passionate speech at an annual meeting. Months earlier, the real Ivan Boesky had made a similar speech to a group of graduates at the University of California's Berkeley campus.

Are these examples instances of "life imitating art"? More likely, it is the other way around. As long as business has a negative public image, movies and television will continue to dramatize real-life tales of corporate wrongdoing. As Hollywood exports a large number of American films to countries around the world, these images become part of a global informational tapestry that we explore in more detail in the next section.

The Global Village

Technology has strengthened communication channels around the globe, disintegrating national borders to produce what Canadian philosopher Marshall McLuhan foresaw decades ago—the creation of a world so interwoven by shared knowledge that it becomes a "Global Village."[14] This trend has had a monumental impact on business, particularly over the last two decades.

In 2002, the U.N. Conference on Trade and Development published an article stating that 29 of the world's top 100 economies were multinational businesses rather than countries.[15] Thus, it may not be surprising that individuals have begun to turn to large companies to provide the direction that distinct national cultures, communities, and inspirational narratives offered more strongly in the past. Coupled with this shift is a heightened level of interest in social responsibility on the part of organizations. Later in this book, we will discuss the growing importance of corporate social responsibility and its implications for corporate reputation, but generally, the public is looking for companies to demonstrate care for the communities in which they operate from both an environmental and human perspective.

In his book *The Mind of the CEO,* Jeffrey Garten explains, "As the world gets smaller, CEOs will be unable to escape involvement in some of the most difficult political, economic and social problems of our times. There will be no way to avoid operating in countries with fragile economies, weak democratic structures and mega-cities with severely overburdened infrastructures."[16]

Disintegrating national borders, coupled with the liberalization of trade and finance in today's Global Village, also have fostered an increase in cross-border corporate mergers and the number of multinational corporations. Today, companies tend to specialize in their core competencies and outsource what remains or,

---

[14] Marshall McLuhan and Bruce R. Powers, *The Global Village: Transformations in World Life and Media in the 21st Century* (New York: Oxford University Press, 1989).

[15] Progressive Policy Institute, "The World Has over 60,000 Multinational Companies," April 27, 2005, http://www.ppionline.org/ppi_ci.cfm?knlgAreaID=108&subsecID=900003&contentID=253303.

[16] Jeffrey Garten, *The Mind of the CEO* (New York: Basic Books, 2001), p. 24.

alternatively, merge to integrate the suppliers into their own organizations. As lower-cost professional service companies in developing markets diversify their offerings, a Goldman Sachs study estimated that as many as 6 million jobs could move overseas by 2013.[17] The 2011 PricewaterhouseCoopers Global CEO survey reveals that 59 percent of CEOs were planning to deploy more staff on international assignments, 34 percent planned to complete a cross-border merger or acquisition, and 31 percent planned to outsource a business process or function.[18]

With international mergers and acquisitions diluting once-definitive borders and empowering big business further, many individuals and communities object to the enormous political clout that large corporations wield today. This sentiment gave rise to the "antibrand" and "antiglobalization" movements that flourished in the mid-1990s—a decade in which global companies began to replace government bodies as the primary target of many activists worldwide. This movement continues to percolate today, as supported by Yankelovich Partner Peter Rose's comments during a January 2007 speech made to the Inland Empire United Way: "Ten years ago, 52% of Americans said that 'the brands you buy tell a lot about the person you are.' In 2005, just 41% agreed with that statement." He continued by rationalizing this shift in perspective, referring to the following quote from the Clue Train website (launched in 1999):

> A powerful global conversation has begun. Through the Internet, people are discovering and inventing new ways to share relevant knowledge with blinding speed. As a direct result, markets are getting smarter—and getting smarter faster than most companies. These markets are conversations.

Rose then incorporated the concept of brands into this "conversation," saying, "The Internet hasn't put brands into the conversation. The Internet has simply changed the technology people use to come together with one another. In the process, the Internet has emerged as the new medium of Social Engagement. . . . Looking ahead, the success of brands will be tied to the success in connecting people with each other, not to connecting people with brands."[19]

This analysis circles back to the public's overwhelming distrust in business as it continues to gain momentum in a global context and the subsequent challenges businesses have in delivering their brands to an accepting audience. The 2011 Edelman Trust Barometer revealed that 56 percent of Americans say they trust business to do what is right.[20]

This "global conversation" also accentuates the volume at which these negative feelings can be heard. With it, the antiglobalization movement extends beyond traditional union bodies to include young and old consumers, concerned parents, and vocal student activists alike. An anticorporation sentiment was formalized on paper in October 1997, when Earth First! produced a calendar listing important anticorporate protest dates and announcing the first "End Corporate Dominance

---

[17] Sue Kichhoff and Barbara Hagenbaugh, "Economy Races Ahead, Leaving Jobs in the Dust," *USA Today*, October 1, 2003.
[18] 2011 PricewaterhouseCoopers Global CEO survey.
[19] Peter Rose, Partner, Yankelovich, speech delivered to the Inland Empire United Way, January 31, 2007.
[20] 2011 Edelman Trust Barometer.

Month."[21] Since then, organizations such as Vancouver-based Adbusters Media Foundation, which was founded in 1989, have risen to a dominant position as nonprofits that devote themselves to deriding corporate giants—a practice now officially referred to as *culture jamming*.[22] Plastering the image of Charles Manson's face over a Levi's jeans billboard, hurling pies at Bill Gates, and dumping garbage bags full of shoes outside of Nike Town to protest Pakistani children manufacturing Nike soccer balls for six cents an hour are some of the routine tactics culture-jamming activists have employed to make anticorporate statements to the public.[23]

"The Authentic Enterprise," a report released by the Arthur W. Page Society in the fall of 2007, neatly summarized the reality of a global economy as major driver of the changing business environment:

> Free trade agreements, the Internet and the emergence of highly skilled populations in developing regions have created a 'flat world.' This is reshaping the footprint—and even the idea—of the corporation. It's shifting from a hierarchical, monolithic, multinational model to one that is horizontal, networked and globally integrated. Because the operations and responsibilities of organizations can now be componentized, 'virtualized' and distributed over an ecosystem of business relationships, work can now be located wherever it makes sense, driven by the imperatives of economics, expertise and open business conditions.[24]

The continual technological advances of the Internet—namely, blogs and social networks—also have made it difficult for companies to prevent both positive and negative news about them from reaching individuals in virtually all corners of the world. Media outlets have expanded their reach such that events are no longer confined to local communities; rather, they can create reverberations felt worldwide. In 2011, the United Nation's International Telecommunication Agency estimated that there are 6 billion cellular-mobile phone subscriptions worldwide and that global cellular-mobile phone penetration is 87 percent.[25]

The organization also estimates that one-third of the world's population is online, with 45 percent of users under the age of 25.[26] In 2010, there were an estimated 152 million blogs, 25 billion tweets sent on Twitter, and more than 600 million Facebook users.[27] Data suggest that these numbers will only continue to increase as consumers assume further control of corporate reputations and communicate with one another in real time, 24/7.

Business leaders today therefore must be prepared not only to handle the international media spotlight but also to proactively counter the advocacy groups looking to use today's media environment to compromise their corporate reputation—and bottom line—globally.

---

[21] Naomi Klein, *No Logo: Taking Aim at the Brand Bullies* (New York: Picador USA, 1999), p. 327.
[22] Ibid., p. 280.
[23] Ibid.
[24] "The Authentic Enterprise," Arthur W. Page Society, 2007.
[25] "The World in 2011: ICT Facts and Figures," International Telecommunication Union, November 2011.
[26] Ibid.
[27] Joshua Norman, "Internet in 2010: 107T E-Mails, 255M Websites," *CBS News*, January 13, 2011.

How to Compete in a Changing Environment

Even well-respected companies face attacks in this antibusiness environment. Gillette (now part of Procter & Gamble), for example, was the target of animal rights groups that successfully used teachers and children to create a stir over the company's research methods. One letter to Gillette's former chairman, Alfred Zeien, said: "Let this be a warning to you. If you hurt another animal, if I find out, one month from [the day] this letter arrives to you, I'll bomb your company. P.S. Watch your back." The letter came from a sixth grader at a school in Philadelphia. As homework, his teacher had assigned letters to companies about animal testing.[28] While the children's campaign had no effect on market share, the company worried about potential long-term effects: "Long term, this could be a very bad trend for the business," said CEO Zeien.[29]

When Walmart faced allegations of unfair treatment of employees, including forcing hourly wage earners to work off the clock, favoring men over women in pay and promotion, and locking employees in stores after closing until managers visited every department, the media pounced on the opportunity to deface the corporate behemoth. In 2000, a female Walmart employee named Betty Dukes filed a sexual discrimination suit against the company that would eventually become a class-action suit representing 1.6 million females. The case finally made its way to the highest court in the United States, the Supreme Court, in 2011, and although the court ruled that the plaintiffs had too much variation in their complaints to merit a class-action suit, Walmart endured negative press for 11 years during the proceedings. A journalist who covered the story turned her research into a book called *Selling Women Short: The Landmark Battle for Workers' Rights at Wal-Mart*, and likened Betty Dukes to civil rights activist Rosa Parks.

Beyond the scrutiny it receives in traditional media outlets, Walmart is also the target of vitriolic social commentary online, with an ever-growing list of anti-Walmart blogs and social groups forming to collectively criticize its controversial business practices. This added dimension of communication, coupled with the reputational risk factors it fosters, raises a key question: How can managers adapt to the challenges of a business environment that is constantly in flux but seems to be moving in the direction of greater scrutiny and less favorable impressions of corporations? In the next section, we will look at some of the ways companies can stay on course while navigating these choppy waters.

Recognize the Changing Environment

First, managers need to recognize that the business environment *is* constantly evolving. The short-term orientation of today's managers rarely gives them an opportunity to look at the big picture of how this changing environment affects the company's image with a variety of constituencies. Over the long term, this perspective can have damaging results.

---

[28] Barbara Carton, "Gillette Faces Wrath of Children in Testing of Rats and Rabbits," *The Wall Street Journal*, September 5, 1995, p. A1.
[29] Ibid.

Coca-Cola took note when, in January 2006, the University of Michigan suspended the purchase of its products on campus.[30] This now classic business case had nothing to do with pricing or the products themselves; rather, it was taken based on concerns over environmental concerns in India and labor issues in Colombia. Among the allegations was a contention that products contained unacceptable levels of insecticides (PepsiCo's products were also found to contain unacceptable levels of pesticides).

The business and communication implications of this revelation and the university's subsequent reaction are manifold: First, the University of Michigan's decision was prompted by one man, Amit Srivastava, who ran a small nonprofit out of his home in California. He mobilized students on campus to petition for the ban—an organizational feat that, just a few years before, would have been unthinkable. Second, these visceral reactions on the part of students applied so much pressure that the company agreed to open its overseas facilities to independent, transparent, third-party environmental and labor audits.[31] Third, the event points to a major evolution in business: Sustainable business practices are becoming core brand values that can inspire change. Coca-Cola's sustainability efforts changed dramatically over the course of a year, and the company appeared among the 2007 Global 100 Most Sustainable Corporations in the World. It is still considered a leader in sustainability today, as we will discuss further in Chapter 5.

One of the most important challenges facing senior managers is the profoundly unsettling impact of technological change. Andrew Grove, cofounder and senior advisor to the executive management of Intel Corporation, explained, "We make a cult of how wonderful it is that the rate of [technological] change is so fast. But . . . what happens when the rate of change is so fast that before a technological innovation gets deployed, or halfway through the process of being deployed, [an] innovation sweeps in and creates a destructive interference with the first one?"[32] Although many agree that technology has helped business, it also has led to greater uncertainty for business leaders and consumers alike.

Unlike many shifts in the market that companies can anticipate by keeping their fingers on the pulse of change, such as evolving consumer tastes, technological innovations can happen swiftly and have profound effects. Companies need to quickly determine what, if anything, they need to do to respond to such changes.

Adapt to the Environment without Compromising Principles

Second, companies must adapt to the changing environment without changing what they stand for or compromising their principles. Chemical giant Monsanto faced challenges when its foray into genetically engineered crops met with resistance from protesters who labeled its products "Frankenfoods." Protests were

[30] http://www.umich.edu/news/?BG/procmemo.
[31] Ibid.
[32] Garten, *Mind of the CEO*, p. 32.

not limited to the company's headquarters in St. Louis but spread to some of Monsanto's large, visible customers, forcing McDonald's, for one, to announce that it would no longer use the company's genetically modified (GM) potatoes.[33]

This issue ultimately took its toll on the company's stock price in the late 1990s, even though the company met Wall Street expectations. In response, Monsanto adopted a new approach to handling the "GM backlash" through education and outreach. Historically, the company had been perceived as aggressively marketing products that the public did not understand or trust. Now, Monsanto communicated "The New Monsanto Pledge," which outlined five key elements: dialogue, transparency, respect, sharing, and delivering benefits.[34] Although the company continued to produce GM foods, its collaborative approach to working with consumer groups and farmers to foster greater understanding of biotechnology's role in food production was viewed positively by many who had previously opposed Monsanto.

Arie de Geus of the MIT Sloan School of Management analyzed the strengths of what he defined as "living companies"—a group of 30 companies ranging in age from 100 to 700 years scattered throughout North America, Europe, and Japan.[35] One of the primary reasons these companies—including DuPont, W.R. Grace, Sumitomo, and Siemens—have managed to endure has been their ability to adapt to the rapidly evolving environment in which they live. De Geus explains: "As wars, depressions, technologies, and politics surged and ebbed, they always seemed to excel at keeping their feelers out, staying attuned to whatever was going on. For information, they sometimes relied on packets carried over vast distances by portage and ship, yet they managed to react in a timely fashion to whatever news they received. They were good at learning and adapting."[36]

Don't Assume Problems Will Magically Disappear

Third, assume things will only get worse in today's complex environment, especially with the ever-growing prevalence of consumer-generated media and online communications platforms. For example, twice, Sony executives have let a bad situation turn worse before addressing the fallout. In October 2005, a blogger broke the story that Sony BMG Music Entertainment distributed a copy-protection scheme CD that contained rootkit software, which self-installs on computers and allows hackers to access the systems, posing huge security threats. Within hours, the story was percolating throughout the blogosphere, but Sony executives turned a blind eye, and Sony BMG's president of global digital business, Thomas Hesse, made matters worse with this statement to NPR on November 4: "Most people don't even know what a rootkit is, so why should they care about it?"

Needless to say, the problem didn't disappear over time. Bloggers, traditional media, and consumers grew increasingly incensed by the company's disregard,

---

[33] Jonathan Low and Pam Cohen Kalafut, *Invisible Advantage: How Intangibles Are Driving Business Performance* (Cambridge: Perseus Books, 2002), p. 114.

[34] Ibid., p. 115.

[35] Arie de Geus, "The Living Company," *Harvard Business Review,* March 1, 1997.

[36] Ibid.

and class-action lawsuits soon followed. Six years later, in April 2011, Sony again found itself facing public outrage for its poor handling of the aftermath of a serious hacking incident. The hacking, which affected an estimated 77 million customers and compromised sensitive user personal data, occurred between April 17 and April 19. Sony did not inform customers of the breach until April 26. As a result of the delayed communication, Sony faced consumer wrath in the blogosphere and inquiries from multiple governments, including the U.S. House of Representatives and the city of Taipei, Taiwan. One senator sent an angry public letter to Sony, writing, "Sony's failure to adequately warn its customers about serious security risks is simply unconscionable and unacceptable."[37]

In both cases, had the company's executives anticipated the length of the story's appeal, and had they addressed the issues at their inception in the blogosphere, they no doubt would have changed their communication strategy. Most managers assume that the American public has a short memory about the problems companies face. In fact, consumers have longer memories than you might think, as witnessed by boycotts of companies such as Coors, Walmart, Nike, and Shell.

Some companies seem to be getting it right, but most are still getting it wrong. What's more, all constituent groups—from employees to investors to consumers— are taking advantage of changes in the business environment that empower them to increase their personal gains. For example, in autumn 2007, two separate situations took place on opposite coasts, in New York City and Los Angeles, that illustrate unique communication strategies.

On November 5, 2007, screenwriters took to the streets of Hollywood, initiating the first industry-wide strike in more than 19 years.[38] Under the representation of the Writers Guild of America, approximately 12,000 movie and television writers formed picket lines in response to failed negotiations with Hollywood producers over their stake in new media revenue, including downloaded movies and online promotional showings of movies and television shows.

The strike crippled the industry, as networks such as CBS and ABC had to shut down production of major primetime shows. Clearly, producers could not just hope the problem would disappear, but their communications and negotiation strategies posed interesting nuances. For example, a *BusinessWeek* article entitled "Behind the Hollywood Strike Talks" highlights an underlying factor:

> What makes the often fractious negotiations particularly interesting this time are the underlying business-model challenges confronting both sides. Business models enable companies (and organizations such as the 12,000-member Writers Guild of America or the Alliance of Motion Picture & Television Producers) to create and capture value. . . . The traditional business models of both sides worked well when there were a handful of movie studios and three major TV networks. But now everyone can be a writer or a producer, and every computer is potentially a studio, able to create and publish content. More than 1 billion people on the planet are connected to the Internet, a healthy portion of them via high-speed broadband.[39]

[37] Steven Musil, "Senator Slams Sony's Response to Security Breach,"CNET.com, May 3, 2011.

[38] http://www.nytimes.com/2007/11/06/business/media/06strike.html?_r=1&oref=slogin.

[39] Henry Chesbrough, "Behind the Hollywood Strike Talks," *BusinessWeek*, November 1, 2007, http://www.businessweek.com/innovate/content/nov2007/id2007111_779706.htm?chan=search.

The author of the article, Henry Chesbrough, executive director of the Center for Open Innovation at the Haas School of Business at University of California Berkeley, also highlighted another detail that will continue to play a more prevalent role in management and communication:

> As it happens, consensus between the studios and the writers is not even necessarily the biggest challenge each faces. Much of the new online entertainment content is not coming from professional writers or producers at all. Rather, as others have noted, it is coming from users and user communities that stimulate one another to create content. . . . How this will shake out in the negotiations between the screenwriters and the AMPTP is hard to say. Both sides need to change some strongly held business models to seize new opportunities—a process that has many risks, but potentially lucrative rewards. However, if Hollywood cannot rise to the challenge, the independent, online creative communities stand ready to pounce. The one thing that seems sure is that neither side has a choice.

Coincidentally, as this contention heated up in Hollywood, a similar situation percolated in the Big Apple. On November 10, 2007, stagehands announced a strike of their own, and Broadway went dark. It was the first in the stagehand union's 121-year history, and it darkened 31 theaters.[40] Unlike the writers' strike, which hinged in the proliferation of new media and its role in generating revenue, the stagehand dispute focused on work rules in their contracts that the producers' league claimed to be expensive and inefficient. The league wanted to change these rules, and the consideration was not well received by the stagehands.

The strike lasted 19 days, during which time New York Mayor Michael Bloomberg offered to provide a mediator and a neutral place to negotiate; both offers were declined. What *The New York Times* called "a series of back-channel conversations between league members and union officials" eventually precipitated talks that ended with a resolution.

Again, it is difficult to assume a problem such as one that left Broadway dark would magically disappear, but the communications strategy proved to be much more traditional, and the strike itself was relatively brief compared with the writers' strike. Negotiations focused on work rules and were not clouded by the nebulous laws governing cyberspace. However, with digital communications platforms playing an increasingly integral role in overall management and communications, competition in the changing business environment continues to evolve.

Keep Corporate Communication Connected to Strategy

Fourth, corporate communication must be closely linked to a company's overall vision and strategy. Few managers recognize the importance of the communication function, and they are reluctant to hire the quality staff necessary to succeed in today's environment. As a result, communication people are often kept out of the loop.

Successful companies connect communication with strategy through structure, such as having the head of corporate communication report directly to the CEO.

---

[40] http://www.nytimes.com/2007/11/29/theater/29broadway.html?em&ex=1196485200&en=e23b4406b383964e&ei=5087%0A.

The advantage of this kind of reporting relationship is that the communications professional can get the company's strategy directly from those at the top of the organization. As a result, all of the company's communications will be more strategic and focused (see Chapter 3 for more on structure).

The aforementioned Arthur Page "Authentic Enterprise" report also urged enterprises to define and activate their core values in new ways, which "demands increased delegation and empowerment, while maintaining consistency of brand, customer relationships, public reputation and day-to-day operations. Values are the 'glue' shaping behavior and uniting coals. However, building a management system based on values is a significant challenge. Understanding what the company and its people truly value and turning that into pervasive behavior require new kinds of leadership, tools and skills."[41]

In Chapter 10, we will take a look at how Johnson & Johnson (J&J) handled the Tylenol cyanide crisis of the early 1980s. Part of what helped the company deal so successfully with this dire situation was the existence of the J&J Credo, a companywide code of ethics that spells out J&J's promises to its many constituencies. This credo helped guide the company's actions during an episode that could have irreparably damaged the Tylenol brand and possibly J&J itself. Thirty years later, the company was again under attack for its faulty production practices but still feeling the halo effect from its handling of this situation.

Companies' corporate communications teams play a pivotal role in defining a corporate mission—the cornerstone of a company's overarching strategy—and communicating that mission to internal and external constituents. Given today's rapidly changing environment, a clear-cut corporate mission not only keeps employees aligned with what the company is striving to be but also can act as a source of stability for consumers weary of the constant change surrounding them.

Conclusion   The business environment is constantly changing. Everyone in business today, whether at a large corporation with a national union to deal with or a small business looking to make its mark in the international arena, needs to communicate strategically. The way organizations adapt and modify their behavior, as manifested through their communications, will determine the success of American business in the twenty-first century.

---

[41]"The Authentic Enterprise," Arthur W. Page Society, 2007.

Case 1-1

Google, Inc.

After agreeing to censor Internet search results in China, Google, Inc. found its corporate mantra—breezily summarized by its founders as "Don't Be Evil"—under heavy fire in January 2006. The search engine giant knew bad publicity could be part of any trade-off if it wanted to become a major player in China's burgeoning economy.

Google had faced little besides fawning publicity from the tech press since its founding in 1998, though hints of the public relations headaches on the horizon for the company first surfaced at the close of 2005, when data and privacy concerns intersected with the U.S. Department of Justice. Google had refused to provide user information in a case the government was building against child pornographers, and as it watched its stock price fall, it had already begun wrestling with how to reconcile that decision with its stance on "Evil."

Public appetite for the company's products seemed only to have intensified since Google's successful albeit unorthodox—initial public offering in 2004, but the company still feared that the January 25, 2006, launch of its new portal in China, Google.cn, would direct criticism back on the company. To operate the backend of its search engine, Google agreed that the portal would automatically filter results containing content considered objectionable by the Chinese government.

Knowing full well that it could become the poster child for the controversy surrounding market entry into the still-reforming China, Google's top executives also had to grapple with the reality that the company might truly be at odds with the golden image of its own making. Whereas once Google was able to tout its free-wheeling, creative culture, whispers in the press suggested that Google might be the next Microsoft Corp.—just another soulless, inflexible, corporate behemoth. A December 2005 cover story in *BusinessWeek* magazine blared "Googling for Gold"[1] and suggested that the company's true interests were more pragmatic than pie-in-the-sky ideals, and a February 2006 headline of *Time* magazine asked, "Can We Trust Google with Our Secrets?"[2]

The situation escalated when U.S. Representative Tom Lantos, a Holocaust survivor and human rights advocate, began to speak out publicly against Google's entry into China, comparing its actions to those of U.S. companies that collaborated with Nazis prior to World War II.[3] Already struggling to stay in front of the story, and sending out mixed messages when it came to its stances on privacy and open access to information, Google knew it needed to do something to clear its name—or risk becoming just another dot-com company.

TWO KIDS IN A SANDBOX

Google's well-documented roots began at Stanford University, where cofounders Sergey Brin and Larry Page met as doctoral computer science candidates in the mid-1990s. A shared love of technology made it only a matter of time before the two overcame their initial dislike of one another and began collaborating on projects outside the classroom.

Over the next few years, the duo worked tirelessly on a way to scan and index the information scattered across the Internet. Other search engines had attempted to do the same

[1] Roben Farzad and Ben Elgin, "Googling for Gold," *BusinessWeek*, December 5, 2005, pp. 60–66.

[2] Adi Ignatius, "In Search of the Real Google," *Time*, February 20, 2006, pp. 36–49.

[3] "Q&A: Congressman Tom Lantos," *Red Herring*, February 17, 2006, http://www.redherring.com/PrintArticle.aspx?a=15779&sector=Q&And A.

thing, but none has worked as well as Brin and Page's method, and the pair believed they had hit on a way to revolutionize use of the Internet.

In 1998, the pair founded Google's predecessor, a company called BackRub, named after the technology's use of backward links to find useful websites. Once it received its first investment, a check for $100,000 from an angel investor, BackRub upgraded its dorm-room operating center for space in a friend's garage and traded in its name for Google. ("Googol" is the math term for the figure 1 followed by a hundred zeroes, a nod to the company's vast goal of organizing all of the Internet's data.) The company was founded with a mission to "Organize the world's information and make it universally accessible and usable."[4]

By 1999, Page and Brin had secured more than $25 million in venture capital funding. Google had grown to just over 60 employees, a rapid growth pace that would continue in the coming years, and begun to develop a relaxed culture of its own. Employees were encouraged to spend part of their workweek on projects that interested them, and tales of the recreational amenities routinely offered to Google staffers spread in the close-knit Silicon Valley. That same year, the company relocated to the "Googleplex," a complex in Mountain View, California, that seemed more sprawling college campus than stuffy office space.

Competitors such as Microsoft's MSN relied on traditional advertising, but Google grew solely by word of mouth. The search engine's speed and ability to deliver highly accurate results drove its increasing popularity. The company also developed multiple products meant to complement its search engine, including the Google Toolbar, Google Image Search, and Froogle, an Internet shopping tool.

At the same time, Google successfully developed a business model that brought in

large advertising revenues while maintaining its image as a free, uncluttered, user-friendly search engine. Programs such as AdWords (introduced in 2002) and AdSense (introduced in 2003) allowed Google advertisers to target users according to keywords used in searches—a far cry from the intrusive pop-up ads that were industry standard at the time.

Still, while the company was successful, the leadership styles of Page and Brin led observers to believe that the two executives were little more than kids playing in a sandbox. The long-term financial success of the company was widely doubted, in both the press and on Wall Street.

DON'T BE EVIL

In 2001, Page and Brin decided to bring aboard Eric Schmidt as Google's first chief executive, though they would retain their executive roles to impart their unique vision for the company. Schmidt had 20 years of management experience in tech companies and most recently had been the top executive at software developer Novell Inc. The trio agreed to make decisions by committee, and in an even more unique twist on power sharing, they agreed that in the case of any major decision, all three would have to reach consensus before taking action.

Despite minor Securities and Exchange Commission investigations and a dubious attitude from Wall Street, in August 2004, the men decided to take the company public. The initial public offering (IPO) made many of the company's employees instant millionaires and resulted in a market capitalization of some $23 billion for Google.

Even as industry observers speculated that the company would have to adopt a more buttoned-down image when it came time to answer to the expectations of outside shareholders, Google's founders pledged to look to the long term to vindicate their business strategy, not to quarterly or even annual earnings reports. Shortly before the IPO, the company released a public letter from the founders, reasserting the mission of the

[4] Google, "Corporate Information," http://www.google.com/intl/en/corporate/index.html.

company: "We believe strongly that in the long term, we will be better served . . . by a company that does good things for the world even if we forgo some short-term gains. We aspire to make Google an institution that makes the world a better place."[5] Fittingly, Google's unofficial company motto, "Don't Be Evil," had long embodied that very ideal.

In the year following the IPO, Google grew to become the fifth most popular website in the world, with more than 380 million visitors per month and half of all users coming from outside the United States. The company also continued its tradition of branching out into multiple product lines. By 2006, it had wide-ranging projects in development, including such long-shot ideas as a program that would allow individuals to track their own DNA history online. Google noted that the project was grounded in the practical and might one day provide people with the ability to take ownership of their own health care through the identification of hereditary health risks.

But Google's "Don't Be Evil" policy hit a stumbling block in December 2005, when the Department of Justice requested all major search providers to submit user information in an effort to investigate the prevalence of child pornography on the Internet. The investigation was part of an attempt to enforce the Child Online Protection Act, which required websites to shield minors from harmful materials. The subpoena requested a sampling of 1 million searches initiated through Google over the course of one week. Whereas Google refused to provide any information to the government and elected to fight the subpoena in federal court[6]—citing the importance of user privacy—search engine competitors MSN and Yahoo quickly complied with the government's request.

THE CHINESE MARKET

Meanwhile, with a population of 1.3 billion and a growing economy, China represented an enormously important market where Google felt it needed to gain a stronger foothold. The number of Internet users in the country had grown substantially in recent years, estimated to have reached more than 110 million regular users in 2006, making China the second-largest Internet market in the world.

The Chinese government gave all Internet search providers operating in the country a difficult choice: Either censor results deemed "objectionable" by the government or do not do business in China. Google already had a presence in the Chinese market prior to the Google. cn launch but had been unwilling to censor information on behalf of the Chinese government. A typical search request initiated through the Google.com website would be filtered by the Chinese government to remove objectionable material—a process that slowed Google's response time significantly and made it difficult for the company to compete.

The filtered search results would remove any reference to a number of subjects. Any content mentioning topics such as Tibet, Taiwan, Falun Gong, or the Dalai Lama was banned. For example, a search on Google.cn for the phrase "Tiananmen Square" returned results showing a smiling couple in the square at spring or the large mural of Chairman Mao on permanent display in the area. Absent were any links to the massacre of 1989. The same search on Google. com would include pages showing the all-too-familiar image of a student standing in front of line of tanks in protest.[7] (See Appendix A for images.)

The primary Internet search provider in China at the time was Baidu.com, a Chinese company that owned approximately 48 percent of the market. Baidu had been ranked with the

---

[5] Ibid.

[6] "Google Refusal Raises Online Privacy Issue," *PRWeek*, January 30, 2006, p. 10.

[7] "How Google Censors Its Chinese Portal," *The San Francisco Chronicle*, February 2, 2006.

fastest responsiveness rate by users in China and was accepted as a clear leader in the market in terms of both brand recognition and usage rates.

But a study by Keynote Customer Experience Rankings acknowledged that the competitive advantages maintained by Google in the United States would be easily transferable to the Chinese market. Chinese customers ranked Google first, beating Baidu, Yahoo, and MSN, in categories such as search quality, image search, and reliability. According to the director of research and public services for Keynote, "We see that Chinese consumers really like the overall Google experience better. Eventually, this promises to translate into increased market share, particularly given Google's strong resources and focus on the market."[8]

With the introduction of Google.cn, Chinese users would be able to access the same search engine with a speed similar to that of Google.com in the United States. Although Chinese users would have previously received the same limited results, it would now be Google—and not the Chinese government—routing the inquiry through its own servers to remove banned content.

NGOS, COMPETITORS, AND CONGRESS MAKE NOISE

On the heels of Google's announcement of its official launch in China, a number of nongovernmental organizations (NGOs) soon voiced strong opinions against Google, a leader by any means in the online arena, engaging in any form of censorship.

Reporters Without Borders (RWB), a Paris-based public interest group acting as a media watchdog on an international level, had already established itself as the leading critic of U.S. search engines that agreed to censor material to gain access to international markets. Beginning in 2004, the group wrote to top U.S.

officials, pleading for a code of conduct regarding overseas Internet filtering and condemning attacks on what it considered to be the rights and freedoms of the press.[9] When Google decided to enter the Chinese market two years later, the interest group leapt on the opportunity to bring the issue back into the spotlight.

"Google's statements about respecting online privacy are the height of hypocrisy in view of its strategy in China," said RWB in a January 25, 2006, press release, issued in response to Google's announcement.[10] The group argued that continued censorship would only lead China to become even more isolated from the outside world, a worrisome prospect considering that a 2005 RWB survey of press freedom had ranked China 159 out of 167 countries.[11] "When a search engine collaborates with the government like this, it makes it much easier for the Chinese government to control what is being said on the Internet," said Julien Pain, head of RWB's Internet desk.

Meanwhile, Human Rights Watch (HRW), the largest human rights organization based in the United States, was preparing its testimony for a February 1 hearing before the U.S. Congressional Human Rights Caucus. The group had a history of investigating key human rights abuses, both within the United States and internationally, and then publishing its findings in an effort to draw exposure to the issue.[12]

The centerpiece of its argument was that the Chinese government would be unable to carry out censorship effectively without the cooperation of U.S. search engines. According to the group, the United States' dominance in the search engine market gave providers considerable leverage against any country that hoped

[8] "Google Poses Strong Challenges to Leader Baidu in China, Reports Keynote," *Business Wire*, January 18, 2006.

[9] Reporters Without Borders, "About Us," http://www.rsf.org/rubrique.php3?id_rubrique=280.

[10] "Google Move 'Black Day' for China," BBC News, January 25, 2006, http://news.bbc.co.uk/1/hi/technology/4647398.stm.

[11] Howard French, "Despite Web Crackdown, Prevailing Winds Are Free," *The New York Times*, February 9, 2006.

[12] Human Rights Watch, "About HRW," http://www.hrw.org/about/whoweare.html.

to benefit in the Information Age. The group proposed that if all the search engines acted together in refusing to comply with Chinese censorship rules, they would be in position to push for free access within the country.[13]

But despite the strong stance, the group had yet to act on its threat of organizing an international boycott of search engine providers. "How much choice do you have if all of these companies are doing this?" asked Mickey Spiegel, Senior Researcher at HRW. "We're not going to stop using the Internet."[14]

Censorship issues aside, both Yahoo and Microsoft's MSN were already posing tough competition to Google's aims to advance its market share in China. Whereas Yahoo elected to place its bets on for the evolution of search engines, Microsoft had the resources to challenge Google in search capability and advertising.[15] Both companies were already complying with Chinese censorship regulations before Google joined the fray and had grappled with their own negative publicity.

Yahoo! came under furious fire for giving the Chinese government information that was used to convict the Chinese Internet journalists Shi Tao in 2004 and Li Zhi in 2003.[16] The company defended its actions by saying that it didn't know how the information would be used. "I do not like the outcome of what happens with these things," said Yahoo cofounder Jerry Yang. "But we have to follow the law."[17] While Yahoo publicly encouraged the American government to handle the issue, the company said that it was too early for itself to recommend how.[18]

In December 2004, MSN complied with an order from the Chinese government to close a site belonging to Michael Anti, a Beijing-based employee of *The New York Times* and one of China's most popular bloggers, who had been addressing sensitive political issues.[19] Microsoft chairman Bill Gates responded by stating that "The ability to really withhold information no longer exists" and outlining a policy in which sites blocked by government restrictions would still be available in all other parts of the world.[20]

Although not a direct competitor for Chinese market share, the physical networking provider Cisco Systems was one of the two U.S.-based companies that the Chinese government relied on for a 2004 network upgrade to improve substantially the government's ability to track Internet searches.[21]

Yahoo and Microsoft issued a joint statement on February 1 in support of a collaboration with Google, Cisco, and the U.S. government to create industry guidelines for handling governmental restrictions on their services in the future.[22]

The Congressional Human Rights Caucus also met on February 1 to address "Human Rights and the Internet: The People's Republic of China," billed as an effort to encourage policy discussion among Internet companies. Attendance at the briefing was optional, and Yahoo, Google, MSN, and Cisco all chose not to attend.[23] Google released a statement the day of the briefing, thanking the caucus for the invitation and citing a previously scheduled commitment as its reason for not attending.[24] (See Appendix B for the official Google Blog.)

[13] Tom Malinowski, "U.S.: Put Pressure on Internet Companies to Uphold Freedom of Expression," testimony before the Congressional Human Rights Caucus, February 1, 2006, http://hrw.org/english/docs/2006/02/01/china12592.htm.

[14] Carrie Kirby, "Chinese Internet vs. Free Speech Hard Choices for U.S. Tech Giants," *San Francisco Chronicle,* September 18, 2005.

[15] Ignatius, "In Search of the Real Google."

[16] Kevin Delaney, "Yahoo Outlines Stance on Privacy and Free Speech," *The Wall Street Journal,* February 13, 2006, p. A8.

[17] Kirby, "Chinese Internet vs. Free Speech."

[18] Delaney, "Yahoo Outlines Stance."

[19] French, "Despite Web Crackdown."

[20] Robert McMahon, "Q&A: The Great Firewall of China," The Council on Foreign Relations, February 16, 2006.

[21] Ibid.

[22] "Lawmakers Blast Internet Firms over China," *Associated Press,* February 1, 2006, http://www.msnbc.msn.com/id/11134689/from/ET/.

[23] Ibid.

[24] Andrew McLaughlin, "Human Rights Caucus Briefing," Google Blog, February 1, 2006, http://googleblog.blogspot.com/2006/02/human-rights-caucus-briefing.html.

The statement also outlined Google's strategy for its operations in China, emphasizing the protection features it had put in place to minimize the harmful effects that its filtering system would have on information seekers. First, Chinese users would be notified when their search had been altered by the filtering system. Second, services such as GMail, chat rooms, and blogging—all involving users' personal information—would not yet be offered out of concern that the Chinese government could demand such information, as it had from Yahoo in prior instances. Third, large investments would encourage continued research and development within China.[25]

For Representative Tom Lantos, head of the House International Relations Committee, the statement was not enough. "These massively successful high-tech companies, which couldn't bring themselves to send representatives to this meeting today, should be ashamed," he said. "They caved in to Beijing for the sake of profits."[26]

A follow-up February 15 hearing was demanded by Representative Chris Smith, and before Congress could follow through on its threats to subpoena the four major companies, all indicated their plans to attend.

PRESSURE FROM SHAREHOLDERS AND CHINA

In the weeks leading up to the caucus hearing, Google's stock had already fallen nearly 7.5 percent, from a high of $471.63 on January 11, which the company partly blamed on the Department of Justice request. Although the January 25 announcement to enter the Chinese market had been met favorably by investors, the release of poor final numbers in the fourth quarter of 2005 marked the first time Google had missed its earnings expectations, causing the stock to open with a loss in value of nearly $15 billion the morning of the caucus hearing.

It would hit another low on the day of the hearing, falling to $342.40.

In addition to its problems at home, Google found itself facing further headaches in China. A state-run newspaper would soon report that Google was under investigation by Chinese authorities for operating in China without a proper license[27] and, in an accompanying editorial, criticize Google for entering the Chinese market, only to complain about being required to follow Chinese law.

Chinese authorities had recently begun pressuring Google to eliminate the very protective measures the company had hung its hat on in its original statement to the caucus. The government wanted the notification that appears on the bottom of every filtered page gone, wanted to cut off roundabout access to Google's unfiltered search engine, and demanded that the company offer GMail and blogging services.[28]

Page, Brin, and Schmidt had continued to argue that the benefits outweighed the costs of entering China. In a statement defending Google's original January 25 announcement, Senior Policy Counsel Andrew McLaughlin said, "While removing search results is inconsistent with Google's mission, providing no information—or a heavily degraded user experience that amounts to no information [which is what the Chinese government had been providing]—is more inconsistent with our mission."[29]

Vint Cerf, who holds the title of Chief Internet Evangelist at Google, further justified the move in an interview. "There's a subtext to 'Don't be evil,'" he said, "and that is 'Don't be illegal.'"[30] As McLaughlin explained it on Google's blog, "We ultimately reached our decision by asking ourselves which course would most effectively

[25] "Google Censors Itself for China," BBC News, January 25, 2006, http://news.bbc.co.uk/1/hi/technology/4645596.stm.

[26] "Lawmakers Blast Internet Firms over China."

[27] Philip Pan, "Chinese Media Assail Google," *The Washington Post*, February 22, 2006, p. A09.

[28] Ibid.

[29] John Yang, "Google Defends Censorship of Web Sites," ABC News, January 25, 2006, http://abcnews.go.com/Technology/print?id=1540568.

[30] Ignatius, "In Search of the Real Google."

further Google's mission to organize the world's information and make it universally useful and accessible. Or, put simply: How can we provide the greatest access to information to the greatest number of people?"[31]

Still, Brin, Page, and Schmidt knew they had to develop a winning strategy to convince the market that Google could handle the balancing act between commerce and conscience and, in the process, reestablish their company as the innovative leader with a soul that it had been in the past.

Copyright ©2006. Eugene D. Fanning Center for Business Communication, Mendoza College of Business, University of Notre Dame. This case was prepared by Research Assistants Brynn Harris and Allison Ogilvy under the direction of James S. O'Rourke, Professor of Management. Information was gathered from corporate as well as public sources. Reprinted by permission.

CASE QUESTIONS

1. How does the changing environment for business affect Google's ability to communicate in this situation?
2. Where is the company most vulnerable, from a communications standpoint?
3. What are the key problems Google faces in this situation?
4. What advice would you give Brin, Page, and Schmidt?

Source: This case was originally prepared by Research Assistants Brynn Harris and Allison Ogilvy under the direction of James S. O'Rourke, Concurrent Professor of Management, Mendoza College of Business, University of Notre Dame, as the basis for class discussion rather than to illustrate either effective or ineffective handling of an administrative situation. Information was gathered from corporate as well as public sources. Reprinted with permission from James S. O'Rourke, University of Notre Dame.

---

[31] Andrew McLaughlin, "Google in China," Google Blog, January 27, 2006, http://googleblog.blogspot.com/2006/01/google-in-china.html.

Appendix A

Web   Images   Groups   News   Froogle   Local   Scholar   more »

tiananmen square   [Search]   Advanced Image Search
Preferences

Moderate SafeSearch is on

Images Showing: [All image sizes ▼]   Results 1 - 20 of about 13,400 for tiananmen square [definition]. (0.06 seconds)

The **Tiananmen Square** photo
494 x 449 pixels - 23k - jpg
www.rollins.edu

An icon at **Tiananmen Square** 1989
600 x 380 pixels - 102k - jpg
multigraphic.dk

... on protesters in **Tiananmen Square** .
220 x 168 pixels - 13k - jpg
www.cnn.com

... Demonstrations in **Tiananmen
Square**
400 x 282 pixels - 71k - gif
www.historywiz.com

Remember **Tiananmen Square**
640 x 403 pixels - 58k - jpg
www.loc.gov

Tiananmen Square, 15 Years After
450 x 383 pixels - 24k - jpg
www.lilithgallery.com

Tiananmen Square, 15 Years After
705 x 742 pixels - 133k - jpg
www.lilithgallery.com
[More results from
www.lilithgallery.com]

... massacre since **Tiananmen Square**
440 x 330 pixels - 40k - jpg
crisispictures.org

网页   图片   资讯   更多 »

tiananmen square   [搜索]   高级图片搜索
使用偏好

使用了 SafeSearch 功能。(了解更多)

图片 显示: [所有尺寸图片 ▼]   约有68项符合tiananmen square的查询结果，以下是第1-20项。 （搜索用时 0.05 秒）

... Mrs. Gutierrez at **Tiananmen
Square**
1728 x 1152 像素 - 317k - jpg
www.usembassy-china.org.cn

... Raising Ceremony in **Tiananmen
Square**
320 x 334 像素 - 21k - jpg
blogs.msdn.com

The Revamped **Tiananmen Square** ...
240 x 158 像素 - 18k - jpg
english.people.com.cn

... **Tiananmen Square** to celebrate ...
200 x 302 像素 - 19k - jpg
www2.chinadaily.com.cn

Clean of **Tiananmen Square** kicks off
400 x 609 像素 - 32k - jpg
english.people.com.cn
[english.people.com.cn站内的其它相关

Tiananmen Square protest planners ...
400 x 225 像素 - 10k - jpg
www.chinadaily.com.cn
[www.chinadaily.com.cn站内的其它相关

Full day tour to **Tiananmen Square**, ...
640 x 480 像素 - 160k - jpg
www.etours.cn

... **Tiananmen Square**. China has ...
360 x 252 像素 - 18k - jpg
www2.chinadaily.com.cn
[www2.chinadaily.com.cn站内的其它相

GOOGLER INSIGHTS INTO PRODUCT AND TECHNOLOGY NEWS AND OUR CULTURE.

Human Rights Caucus briefing

2/01/2006 08:26:00 AM

Posted by Andrew McLaughlin, Senior Policy Counsel

For today's Member Briefing of the U.S. Congressional Human Rights Caucus on "Human Rights and the Internet—The People's Republic of China," we've submitted the following statement:

CONGRESSIONAL HUMAN RIGHTS CAUCUS MEMBERS' BRIEFING "HUMAN RIGHTS AND THE INTERNET—THE PEOPLE'S REPUBLIC OF CHINA" SUBMISSION OF ANDREW MCLAUGHLIN, GOOGLE INC. FEBRUARY 1, 2006

On behalf of Google, I would like to thank the Members of the Human Rights Caucus for inviting Google to participate in today's Member Briefing on Human Rights and the Internet in China.

Though previously scheduled commitments prevent me from appearing in person today, I reiterate Google's offer to participate in a Member Briefing on another date, to brief Members individually, and to continue briefing staff on our activities in China.

I. GOOGLE.CN IN CHINA

The rationale for launching a domestic version of Google in China—a website subject to China's local content restrictions—is that our service in China has not been very good, due in large measure to the extensive filtering performed by Chinese Internet service providers (ISPs). Google's users in China struggle with a service that is often unavailable, or painfully slow. According to our measurements, Google. com appears to be unavailable around 10% of the time. Even when users can reach Google. com, the website is slow, and sometimes produces results that, when clicked on, stall out the user 's browser. The Google News service is almost never available; Google Images is available only half the time.

These problems can only be solved by creating a local presence inside China. By launching Google.cn and making a major ongoing investment in people, infrastructure, and innovation within China, we intend to provide the greatest access to the greatest amount of information to the greatest number of Chinese Internet users. At the same time, the launch of Google.cn did not in any way alter the availability of the uncensored Chinese-language version of Google. com, which Google provides globally to all Internet users without restriction.

In deciding how best to approach the Chinese—or any—market, we must balance our commitments to satisfy the interests of users, expand access to information, and respond to local conditions. Our strategy for doing business in China seeks to achieve that balance through improved disclosure, targeting of services, and local investment.

A. IMPROVED DISCLOSURE TO USERS OF GOOGLE.CN.

In order to operate Google.cn as a website in China, Google is required to remove some sensitive information from our search results. These restrictions are imposed by Chinese laws, regulations, and policies. However, when we remove content from Google.cn, we disclose that fact to our users. This approach is similar in principle to the disclosures we provide when we have altered our search results to comply with local laws in France, Germany, and the United States. When a Chinese user gets search results from which one or more results has been filtered, the Google webpage includes an explicit notification—an indication that the search results are missing something that might otherwise be relevant. This is not, to be sure, a tremendous advance in transparency to users, but it is at least a meaningful step in the right direction.

B. TARGETING OF SERVICES ON GOOGLE.CN.

Google.cn today includes three basic Google services (web search, image search, and Google News), together with a local business information and map service. Other products—such as Gmail and Blogger—that involve personal and confidential information will be introduced only when we are comfortable that we can provide them in a way that protects users' expectations about that information. We are conscious of the reality that data is subject to the laws and regulations of the country in which it is stored, and we make decisions about where to locate our services with that reality squarely in mind.

C. LOCAL INVESTMENT AND INNOVATION.

Looking beyond the Google.cn launch, we will continue to make significant investments in research and development in China. We believe these investments—and the innovations that will result—will help us to better tailor our products to user demands and better demonstrate how the Internet can help advance key objectives supported by the Chinese government, such as building stronger, more efficient, and more equitable markets, promoting the rule of law, and bolstering the fight against corruption.

While China has made great strides in the past decades, it remains in many ways closed. We are not happy about governmental restrictions on access to information, and we hope that over time everyone in the world will come to enjoy full access to information. Information and communication technology—including the Internet, email, instant messaging, weblogs, peer-to-peer applications, streaming audio and video, mobile telephony, SMS text messages, and so forth—has brought Chinese citizens a greater ability to read, discuss, publish and communicate about a wider range of topics, events, and issues than ever before. We believe that our continued engagement with China is the best (and perhaps only) way for Google to help bring the tremendous benefits of universal information access to all our users there.

II. NEXT STEPS

1. Expanded Dialogue and Outreach. For more than a year, Google has been actively engaged in discussion and debate about China with a wide range of individuals and organizations both inside and outside of China, including technologists, businesspeople, government officials, academic experts, writers, analysts, journalists, activists, and bloggers. We aim to expand these dialogues as our activities in China evolve, in order to improve our understanding, refine our approach, and operate with openness.

2. Voluntary Industry Action. Google supports the idea of Internet industry action to define common principles to guide technology firms' practices in countries that restrict access to information. Together with colleagues at other leading Internet companies, we are actively exploring the potential for Internet industry guidelines, not only for China but for all countries in which Internet content is subjected to governmental restrictions. Such guidelines might encompass, for example, disclosure to users, and reporting about governmental restrictions and the measures taken in response to them.

3. Government-to-Government Dialogue. In addition to common action by Internet companies, there is an important role for the United States government to address, in the context of its bilateral government-to-government relationships, the larger issues of free expression and open communication. For example, as a U.S.-based company that deals primarily in information, we have urged the United States government to treat censorship as a barrier to trade.

On behalf of Google, I would like to thank the members of the Human Rights Caucus for their attention to these important and pressing issues.

CHAPTER TWO

Communicating Strategically

In the first chapter, we examined the changing environment for business over the last half century. In this chapter, we explore how these changes have affected corporate communication and why it has become imperative for modern companies to communicate strategically.

Strategic communication can be defined as "communication aligned with the company's overall strategy, [intended] to enhance its strategic positioning."[1] An effective strategy should encourage a company to send messages that are "clear and understandable, true and, communicated with passion, strategically repetitive and repeated, [and] consistent (across constituencies)."

We begin this chapter with a summary of the basic theory behind all communication, whether individual or organizational in nature. We will also briefly discuss influential models in modern communication theory. Although many communication experts have adapted these theories to help leaders communicate in writing and speaking, few have looked at how these same basic theories apply in the corporate communication context—that is, the way organizations communicate with various groups of people.

Communication, more than any other subject in business, has implications for everyone within an organization—from the newest administrative assistant to the CEO. Thanks in part to important strategy work by academics such as Michael Porter, Gary Hamel, and C. K. Prahalad, most managers have learned to think strategically about their business overall, but few think strategically about what they spend most of their time doing—communicating.

This chapter discusses what it means to develop a cohesive communication plan within an organization, emphasizing the critical link between corporate communication and the firm's overall corporate strategy.

Communication Theory

Most modern theories associated with communication can be traced back thousands of years to a single common ancestor, the Greek philosopher Aristotle.

[1] Paul A. Argenti, Robert A. Howell, and Karen A. Beck, "The Strategic Communication Imperative," *MIT Sloane Management Review*, Spring 2005.

Aristotle, who studied under Plato and taught in Athens from 367–347 BCE, is most often associated with the development of rhetoric, the ancient antecedent to modern persuasive communication. In his book *The Art of Rhetoric*, Aristotle defined the three basic components of every speech, which have been adapted to meet the needs of the modern corporation as follows:

This strategy depends on thinking carefully about the same three parts that Aristotle used to describe the components of speech: (1) a "speaker," or in our case, a corporation, with something to say; (2) a "subject," or message that needs to be conveyed; and (3) a "person" or group to whom the message will be delivered.

Aristotle's observations on message communication laid the foundation for modern communication theory, which developed in the United States along with several other social sciences following World War II. In 1948, law professor and political scientist at Yale University Harold Lasswell proposed a communications model that he believed applied especially well to mass communications.[2] His linear model can be summarized as "who (Aristotle's speaker) says what (Aristotle's subject or message) in which channel (medium) to whom (Aristotle's recipient) with what effect (effect)." Several years later, professor of communication skills Richard Braddock proposed an expansion of the Lasswell model to include more reflection on the intent of the message, as well as more analysis of the circumstances under which the message was being delivered.[3]

Further in 1948, mathematician and engineer Claude Shannon published his *"A Mathematical Theory of Communication"* in the in-house scientific journal at Bell Labs. The following year, Warren Weaver helped Shannon to publish the article as a book, and as a result this communications model is called both the Shannon-Weaver model and the Shannon model. The model, used today in social sciences, mathematics, and engineering, is linear and focuses on the physical transmission of information. It follows the creation of a signal by an information source (using a transmitter) to the reception of the signal by the recipient. The model also includes a "noise source," which can be anything that interferes with the integrity of the signal.[4]

In 1956, professor of communications George Gerbner proposed a communication model that built on both the Lasswell and Shannon-Weaver models and emphasized the important role that perception plays in communication as well as the transactional nature of communications.[5]

The Corporate Communication Strategy Framework presented in Figure 2.1 incorporates these and other communication models to provide a valuable framework for effectively analyzing corporate communications.

Looking at the framework, one can easily visualize the connections between each component. As communication theorist Annette Shelby states: "The unique

[2] Harold D. Lasswell, "The Structure and Function of Communication in Society," in Lyman Bryson, ed., *The Communication of Ideas: A Series of Addresses* (New York: Institute for Religious and Social Studies), pp. 203–243.

[3] Richard Braddock, "An Extension of the 'Lasswell Formula,'" *Journal of Communication*, 8, no. 2 (June 1948), pp. 88–93.

[4] Claude Elwood Shannon and Warren Weaver, *The Mathematical Theory of Communication* (University of Illinois Press), 1964.

[5] George Gerbner, "Toward a General Model of Communication," *Audio-Visual Communication Review*, 4 (1956), pp. 171–199.

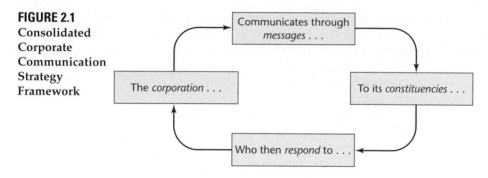

FIGURE 2.1
Consolidated
Corporate
Communication
Strategy
Framework

interrelationships of these variables determine which messages will be effective and which will not." These interrelationships will also determine the most effective tools for communicating the message. In addition, this framework is circular rather than linear, which reflects the reality that communication of any kind is an ongoing process rather than one with a beginning and an end.

Whether an organization is trying to enhance its reputation through social media, to communicate a new health care plan to employees, or to convince shareholders that the company is still worth investing in, it is critical to use a coherent communication strategy. An effective strategy should take into account the impact that the message will likely have on its audience.

Developing Corporate Communication Strategies

Let's further develop each of these variables and apply them to real situations and see how they operate in practice.

Setting an Effective Organization Strategy

The first part of an effective corporate communication strategy relates to the organization itself. The three subsets of an organization strategy include (1) determining the *objectives* for a particular communication, (2) deciding what *resources* are available for achieving those objectives, and (3) diagnosing the organization's *reputation*.

Determining Objectives

An organization, like an individual, has many different reasons for deciding to communicate. For example, a company might want to announce to employees a change in its benefits package for the upcoming year. Let's suppose the organization has decided to eliminate retiree health benefits as a result of increasing health care costs. In this case, its objective is more than just announcing the change; it also must convince employees it has a good reason for taking something away from them. Thus, the objective here is to get employees to accept the change with a minimal amount of protest.

In contrast, let's suppose that a Japanese candy manufacturer has decided to enter the U.S. market. To stimulate interest in its confections, the company decides to produce a brochure that will show and explain what the product is and how it is an extension of Japanese culture. The company's objective, then, is to create a demand among American consumers for something that they have not previously known about or wanted.

Notice that in both of these cases, the audience's *response* to the message is what is most important. That is the basis for defining an objective: *what does the organization want each message recipient to do as a result of the communication?* Management communication expert Mary Munter writes in her *Guide to Managerial Communication* that managerial communication is only successful if you get the desired response from your audience.[6] To get that response, you must think strategically about your communication, including setting measurable objectives for it.

Deciding What Resources Are Available

Determining how to communicate about something like an employee benefits plan or introducing a new product into a market depends heavily on what resources are available within the organization, including money, human resources, and time.

Money In our earlier example involving cutbacks in health benefits for employees, the company must decide whether it is better to simply announce the program as clearly as possible to its employees—for example, through the company newsletter, via e-mail, or on the company's intranet—or to hire a benefits consultant with experience in helping other companies sell employees on benefits reductions. The first option looks less expensive than the second in the short term, but if the employees revolt because they feel they are losing something for no good reason, the company might end up spending far more than it would have if it had hired the more experienced consultant in the first place.

Most companies, unfortunately, often err on the side of short-term, inexpensive solutions to communication problems because they are not looking at the problem from the perspective of the constituency in question. This issue is similar to a problem individuals often have in communicating: they look at their own needs rather than the needs of their audience and end up having difficulty reaching their communication objective.

Human Resources Human resources are also an important factor in determining the success or failure of a company in achieving its objectives. Typically, too few are assigned to deal with communication tasks, and those involved are often inexperienced or unqualified.

Imagine a company that has just gone public and has decided to create an investor relations function to deal with shareholder relations and communication with financial analysts. It could assign one person to do all of these things, or it could decide that it really needs three. The best approach depends on the size of the company and its shareholder base. Let's look at the case of a well-known, multibillion-dollar company that turned this function over to one person with

---

[6] Munter, *Guide to Managerial Communication.*

weak communication skills rather than devote two or three experts to deal with the different constituencies involved. In this company's case, it wasn't a question of whether it could afford to pay more people to do the job correctly; it was the lack of understanding about how important corporate communication really is and the limitations put on the human resources needed to accomplish a specific task.

This *Fortune* 500 company changed its approach after analysts started to downgrade its stock despite healthy prospects for the company's future. The CEO discovered that the analysts felt that the investor relations person at the company was not interested in giving them sufficient information to rate the company's stock. This perception led them to believe that something was wrong at the company. The investor relations person, on the other hand, was actually trying to do two or three tasks at the same time and simply could not keep up with the demands of the job. After this incident, the company hired two more professionals to handle the job properly, creating a more effective and efficient investor relations function, and its stock price shot back to where it should have been all along.

Time   Time, like human resources and money, is also a critical factor in determining an organization's corporate communication strategy. Let's look at two approaches for dealing with the same problem involving the allocation of time.

In the case of the Japanese confectioner mentioned earlier, the company decided to produce a brochure (with the help of a communications consulting firm) describing its product more than two years before it was actually necessary. So much time was involved, however, in getting everyone in the company to buy into both the proposed text and the design for the brochure that it took almost the full two years to produce an eight-page pamphlet. Cultural differences between Japanese and American business styles contributed to the tremendous amount of time needed to develop the brochure.

For an American firm, it is unheard of to devote so much time to what would be viewed as such a simple project. American firms produce brochures like this from start to finish in a matter of weeks. But is this really a better approach?

The allocation of time, like the allocation of all resources, should be determined by what it will really take to achieve the company's objective rather than to seek a short-term solution. In some cases, this might mean allocating more resources than the organization would like to achieve the desired result, but almost always, the organization is better off allocating the resources up front. Correcting mistakes in corporate communication after the fact can be a costly proposition. Too often, qualified communicators are brought in only after a crisis has erupted or to combat rumors that have materialized to fill a "communications void." This scenario is often the case when a company is in the midst of a merger or acquisition and employees hear details about the company's merger plans through media outlets before they hear it from the company itself. When rampant rumor mills and third-party information inspire fear and uncertainty among employees, productivity and customer service typically suffer—in some cases enough to reduce shareholder value.[7] The company then suddenly has a much larger—and potentially more costly—problem to deal with.

[7] Michael Kempner, "When RUMORS Thrive Your Deal's in Trouble: Damage Control Techniques to Seize the Communications High Ground, *"Mergers & Acquisitions,* May 1, 2005, pp. 42–47.

Diagnosing the Organization's Reputation

In addition to setting objectives for a communication and deciding what resources are available to accomplish that objective, organizations must determine what kind of image credibility they have with the constituencies in question. An organization's overall reputation with constituencies is based on several factors. We will get into this in greater detail in Chapter 4 when we talk about image, identity, and reputation, but it is also a critical factor in the development of all communication strategies, whether specifically related to image or not.

Image credibility is based on the constituency's perception of the organization rather than the reality of the organization itself. As an example, think about a university that is trying to generate positive publicity in the national press. If the university is not well known outside its region, this effort might prove very difficult. Its image credibility in this situation would be low because national press representatives would have limited experience with the institution compared with an institution that already has a national reputation. Thus, no matter what kind of resources the university puts behind this effort, it will be an uphill battle.

Worse than limited image credibility is credibility that is lacking or damaged. In the fall of 2010, top toy company Mattel had to recall 7 million of its Fisher-Price-brand tricycles when 10 young children reportedly injured themselves on the sharp, protruding plastic ignition key. During this same period, 3 million Fisher-Price toys were recalled due to concerns that small parts could cause choking. The 2010 recalls unfortunately followed a series of other reputation-damaging recalls for Mattel, including a recall of magnetic toys with faulty designs in November 2006, a recall of Fisher-Price-brand toys with high levels of lead paint in August 2007, and a recall of lead paint-laden Barbie accessories in September 2007.

Once the most credible of toy makers, Mattel had damaged its credibility with investors and customers. During the height of the high-profile recalls, the stock value fell as much as 25 percent, in the fall of 2007. However, Mattel executives took aggressive action to help upend the credibility crisis, opting for complete transparency and leveraging digital communications channels to deliver messages to constituents. Mattel's communications team also launched an advertising campaign with the headline "Because your children are our children, too," and spokespeople constantly reiterated the company's investigation of the safety breaches and communicated openly with the media. Mattel's response to the recalls of 2006 and 2007 likely reassured investors during the 2010 recalls because the company's stock price actually increased slightly immediately following the September 30, 2010, tricycle recall announcement.

Sometimes, damaged image credibility can result from circumstances beyond an organization's control, rather than from any specific actions or missteps by the company itself. Mattel fits this description to some degree, as some of its recalls were caused by issues with overseas manufacturing partners. Although Mattel's executives should have ensured more stringent safety requirements and monitoring standards, there are really two credibility crises at play: the handling of the product recall by Mattel, and the reputation crisis at Fisher-Price, which was responsible for regulating the overseas production of its toys.

Also victims of circumstances beyond their control, global energy companies faced a collective image credibility challenge in the wake of the Enron collapse. Many began having problems with bondholders, regulators, and investors following the scandal, as they were presumed guilty of engaging in similar practices as the former energy giant. One possible strategy to combat this "guilt by association" would have been for a company to craft a communication program that would actively seek to distinguish it from Enron in a highly visible way.[8]

We can see that an organization's reputation is an important factor in setting a coherent communication strategy. For simple tasks, this is not a problem, but in other cases, the image credibility an organization has built with a specific constituency can make a huge difference in determining the success or failure the organization has in achieving its objectives. Companies increasingly are recognizing this fact and, accordingly, are dedicating resources to assessing their corporate reputation. One such company is FedEx. Once a year, the company's senior executives gather at its Memphis headquarters to assess the different risks the company faces. In addition to considering the possible financial impact and implications for the business continuity of each scenario, they examine what would happen to the company's reputation. "We believe that a strong reputation can act as a life preserver in a crisis and as a tailwind when the company is on the offensive," explained Bill Margaritis, FedEx's senior vice president of worldwide communications and investor relations. In addition to this hypothetical scenario analysis, FedEx conducts a survey to find out how the company is perceived by external stakeholders and performs a similar exercise with its employees annually.[9]

The three considerations for creating an effective organization strategy—setting objectives, deciding on the proper allocation of resources, and diagnosing the organization's reputation—are the building blocks upon which all other steps in communication strategy depend. A second set of issues the organization can turn to is an assessment of the constituents involved.

Analyzing Constituencies

Analyzing constituencies is similar to analyzing your audience when you want to plan a speech or write a memo. This analysis determines (1) who your organization's constituencies are, (2) what each thinks about the organization, and (3) what each knows about the communication in question. We will look at each of these in turn.

Who Are Your Organization's Constituencies?

Sometimes the answer to this question is obvious, but most of the time, it will take careful consideration to analyze who the relevant constituencies are for a particular corporate message. Do not be fooled into thinking that it is always obvious who the main constituency is. Usually, constituencies come from a group that is primary to the organization, but a secondary group also can be the focus for a particular communication (see Table 2.1).

[8] Duncan Wood, "Not Cleaning Up Your Act Can Be Costly," *Treasury & Risk Management*, September 2004.
[9] Ibid.

TABLE 2.1
Constituents of
Organizations

Primary	Secondary
• Employees	• Traditional media
• Customers	• Suppliers
• Shareholders	• Creditors
• Communities	• Government
	• Local
	• Regional
	• National
	• Individual bloggers and activists

Companies have different sets of constituencies depending on the nature, size, and reach (i.e., global or domestic, local versus regional or national) of their businesses. Although a company may list its constituencies on a piece of paper, as in Table 2.1, it should resist thinking of them as too fixed or too separate. An organization's primary constituency or constituencies can change over time. In a time of crisis, for example, it may be wise for a company to focus more intently on its relations with the media—which it may normally consider a secondary constituency—to manage its reputation and attempt to minimize negative press. Additionally, constituencies should not be thought about in "silos," as the lines between them can blur. When employees are also shareholders in a company, for instance, they belong simultaneously to two constituency groups. For example, Starbucks formally blends employees and investors by offering all employees "bean stock" based on the number of hours they work, a practice that Starbucks began in 1991 and considers to be core to its mission.[10]

It is also important to recognize that constituencies interact with one another, and that an organization must sometimes work through one constituency to reach another. For instance, if a department store is focused on revitalizing a customer service focus to drive more loyalty (and sales) from its customer constituency, it must reinforce this mission with employees before customers will see results. A classic example can be seen in the employee-customer-profit chain model created by Sears, which tracked success from management behavior through employee attitudes to customer satisfaction and, ultimately, financial performance.[11]

Companies should acknowledge the role of their own employees as "brand ambassadors"—given that they interact with a large number of external constituencies, the potential for "word-of-mouth" goodwill and image building is significant when employees fully understand what the corporation aims to be in the mind of its customers and other constituencies. Software Company SAS has established itself as a leader in this area, ranking #1 in *Fortune* magazine's

---

[10] Howard Schultz and Dori Jones Yang, *Pour Your Heart into It: How Starbucks Built a Company One Cup at a Time* (New York: Hyperion, 1997).
[11] Anthony J. Rucci, Steven P. Kim, and Richard T. Quinn, "The Employee Customer Profit Chain at Sears," *Harvard Business Review*, January–February 1998, pp. 83–97.

2011 "100 Best Companies to Work For" list. Its executives offer employees some compelling (and unusual) reasons to advocate the brand: car cleaning, a health care facility on-site, a kids' summer camp, child care, a beauty salon, and a 66,000-square-foot gym are just some of the perks employees at SAS enjoy. Zappos, a $1.2B online retailer specializing in shoe sales, and #6 in *Fortune*'s 2011 "100 Best Companies to Work For" list, is another company where employees are valued as significant corporate communication assets. As Zappos CEO Tony Hsieh wrote in his 2010 book *Delivering Happiness,* "we trust our employees to use their best judgment when dealing with each and every customer . . . we want our reps to let their true personalities shine during each phone call so that they can develop a personal emotional connection (internally referred to as PEC) with the customer."[12]

However, keep in mind also that constituencies can have competing interests and different perceptions of a company. For example, cutting employee benefits may be welcomed by shareholders but will likely be highly unpopular with employees. Finally, keep in mind that communications intended for one constituency often reach others.

The individual communication experience of one marketing vice president (VP) brings this last point to life. The executive VP to whom he reported had decided to cut the group's administrative support staff due to the increased use of technology to handle communications while professionals were away from their desks. This vice president detailed his plan for cutting the support staff by almost two-thirds in a memo to the vice president in charge of human resources. The plan involved laying off five assistants in the department over a period of six months. Many of them had been with the firm for several years.

As usual, the marketing VP typed up his thoughts in rough form and e-mailed it to his assistant, asking her to format the letter and print the final draft on his letterhead. Although his assistant was not one of the five affected by the layoffs, she couldn't help but empathize with her colleagues of many years, and within an hour, the marketing VP had a revolt on his hands. Now, with a constant news cycle that is aided and abetted by online communications, a scenario like this one could be prompted by information that gets into the hands of, for example, a blogger, as we'll see later on in this chapter.

Now obviously, the aforementioned VP didn't intend for his assistant to be a part of his constituency, nor did he stop to think about her reaction to the change when he asked her to print the letter to the human resources VP. Nonetheless, she became a conduit to a more important constituency—the employees who would actually be affected by the plan.

This simple example is instructive to organizations seeking to communicate at a more macro level as well. Just as we cannot always control the flow of information to one constituency alone on an individual level, on the corporate level, the same set of problems arises.

---

[12] Tony Hsieh, *Delivering Happiness: A Path to Profits, Passion, and Purpose* (New York: Business Plus, 2010), p. 145.

What Is the Constituency's Attitude toward the Organization?

In addition to analyzing who the constituencies for a particular communication really are, organizations also need to assess what each constituency thinks about the organization itself.

We know from personal experience that it is easier to communicate with people who know and like us than it is with those who do not. The same is true for organizations. If a company has built goodwill with the constituency in question, it will be much easier to reach its objective.

The classic example of good corporate communication is Johnson & Johnson's redemption of the Tylenol brand in 1982, when poisoned capsules killed seven people in Chicago. (See Chapter 10 for more on the Tylenol crisis.) That the company was able to succeed against all odds—when people like advertising executive Jerry Della Femina and several other experts in communication declared Tylenol impossible to save at the time—was a tribute to the hard work the organization had done before the tragedy actually happened. The company was known in the industry, by doctors, by consumers, and by the press as rock solid—willing to stand by its products and do the right thing, no matter what the cost. In this case, the cost ran into the hundreds of millions of dollars, when the company decided to recall over 31 million bottles of Tylenol capsules.

Convincing people to buy a product that had been laced with cyanide was not an easy proposition, but because the company had the trust of many different constituencies, it was able to achieve its objective, which was to revive the brand. If people hadn't trusted the company, or if they had questioned its behavior in any way, this revival would not have been possible.

When goodwill or trust is lacking, communication can be a struggle. And companies cannot expect to be trusted until they prove themselves trustworthy through concerted actions that demonstrate care, concern, and understanding for their constituencies. As stated in "Authentic Enterprise," mentioned in Chapter 1 as a document produced by the Arthur W. Page Society:

> In addition to the familiar intermediaries and constituencies with whom corporations have interacted in the past, there is now a diverse array of communities, interests, nongovernmental organizations and individuals. Many of these new players represent important interests, while others are not legitimate stakeholders, but rather simply adversarial or malicious. Regardless of motive, all are far more able to collaborate among themselves around shared interests and to reach large audiences. At the same time, companies and institutions themselves are seeking similar kinds of engagement with multiple constituencies . . . Constituent relationships have always been important for businesses and institutions, but the proliferation and empowerment of new kinds of stakeholders have profoundly altered the landscape. First, in a radically more transparent world, organizations can no longer be different things to different constituencies; an enterprise must be one thing across its entire ecosystem.
>
> Source: Reprinted with permission from the Arthur W. Page Society.

Building trust often must start from within the organization—by communicating up and down with employees, hearing them out on the topics that concern them, and making constructive changes based on their input. Companies with high levels of trust with employees are also those that take the time to clearly communicate the company's business goals to employees and help them understand the vital roles they play in achieving those goals.[13]

What Does the Constituency Know about the Topic?

In addition to the constituents' attitudes toward the company, we also must consider their attitudes toward the communication itself. If they are predisposed to do what the organization wants, then they are more likely to help the organization reach its objective. If they are not, however, the organization will have great difficulty in trying to achieve its goals.

Consumers are often wary of new or unknown products. The Japanese confectioner mentioned earlier was a victim of such bias as it tried to convince Americans to buy a product that was well known and liked in Japan but completely foreign to Americans. In Japan, the company is seen as the highest-quality manufacturer of *wagashi*, or candy. The company, Toraya, is one of the oldest companies on earth. It can trace its roots back to the ninth century, and the same family has been in control of the firm for 17 generations. It has been serving the imperial family since its inception.

Given its long history and aristocratic roots, the president of the company assumed that the product would speak for itself in the U.S. market. Because no one else was around to compete with the firm, middle managers in charge of the U.S. operation assumed that its introduction of *wagashi* would be a huge success.

Unfortunately, they didn't think about how American palates would react to the taste of a candy made out of red beans and seaweed. Most of the people who heard about the product couldn't even pronounce its name, and when they tasted the gelatinous form of the product, known as *yokan*, they didn't like it.

To get consumers in the United States interested in the product, Toraya had to educate people about the role of *wagashi* in Japanese history and its exclusivity, as demonstrated by its aristocratic roots. Those who tasted the product in focus groups early in the process of its introduction to the United States likened the experience to the first time they had tasted caviar or espresso.

Japanese candy isn't the only example of misjudged consumer feelings. Take Walmart, for example. The retail behemoth tried to break into the German market for nine years before retreating with its proverbial tail between its legs in 2006. Walmart had 85 stores in the country but eventually lost the battle to local rivals such as Aldi and Lidl because it failed to adapt to the German consumer and business culture. Among the many missteps: German Walmarts imported the U.S. practices of bagging groceries for customers at check-out counters and requiring employees to smile and greet every customer. The service-with-a-smile approach was seen as distasteful and unnecessary by shoppers. Executives also imparted the company's American policy of forbidding romances between employees. This restriction was seen as inappropriately intrusive by German standards. In misjudging its target

---

[13] Shari Caudron, "Rebuilding Employee Trust," *Workforce Management*, October 2002, pp. 28–34.

consumer and subsequently abandoning its German business, Walmart took a $1 billion hit.

Companies that try to sell an idea to the public are always in danger of failing as a result of the lack of information or the negative feelings consumers may have about it. The U.S. automaker General Motors (GM) realized, after several failed attempts to penetrate the U.K. market with Cadillacs, that rather than spending money on a U.K. advertising campaign, it was better served to hire an automotive public relations specialist to help the company educate people about Cadillac's new approach to the market, including an increased range of right-hand-drive models.[14]

When companies are communicating to their employees about something like a change in benefits—from a defined benefit pension plan to a cash balance plan, for instance—understanding what employees know about the topic, as well as how they feel about it, is critical. Without this insight, valuable time and resources can be spent on a communications campaign that ends up completely missing the mark. For example, a company may assume that employees' greatest concern is the competitiveness of their new benefit plan relative to other companies, when, in fact, they are most concerned about understanding how the new plan differs from the existing one. Absent this knowledge, the company's communication strategy may focus too heavily on the benchmarking issue and fail to address the issue of most concern to this constituency.[15]

Clearly, then, after a firm has set objectives for its corporate communication, it must thoroughly analyze all the constituencies involved. This requirement means understanding who each constituency is, finding out what each thinks about the organization, and determining what each already knows and feels about the communication in question. Companies should consider allocating a portion of their marketing budget to this kind of research. Armed with this intelligence, the organization is ready to move to the final phase in setting a communication strategy: determining how to deliver the message.

Delivering Messages Effectively

Delivering messages effectively involves a two-step analysis for companies. A company must decide how it wants to deliver the message (choose a communication channel) and *what approach* to take in structuring the message itself.

Choose an Appropriate Communication Channel

Determining the proper communication channel, or channels, is more difficult for organizations than it is for people. An individual's channel choices are usually limited to writing or speaking, with some variation in terms of group or individual interaction. For organizations, however, the channels available for delivering the message are several.

As you can see from Table 2.2, there are now more communication channels than ever before for an organization's internal and external communications. For

[14] Richard Cann, "Cadillac Media Push Aims to Crack the UK," *PRWeek*, July 9, 2004.

[15] "Communicating Cash Balance Plans," *Watson Wyatt Insider*, April 2000, http://www.watsonwyatt.com.

TABLE 2.2
Communication
Channels

Old Channels	New Channels
• Spoken word	• E-mail
• Letter	• Blogs
• Print media	• Digital newsrooms
	• Television
	• Podcasts
	• Text messaging
	• Internet
	• Voicemail
	• Electronic meetings
	• Video teleconferencing
	• RSS feeds
	• Facebook
	• Twitter

example, a company looking to reveal a change in top management may decide to announce the change through a press release, which gets the message out to a broad set of constituencies. In addition, it may announce the change in a memo and an e-mail to employees, as well as posting it on the company's intranet.

Even this simple example has multiple channel possibilities. Should the press release go to local media or national media? If the company is global, should it get the message out on an international newswire, such as Reuters or PR Newswire? Should it display the message on its website? Should the message go to employees as part of a video teleconference or over the company's intranet? Should the company post an update on its Twitter account? Then there is the whole question of *timing*. Should the employees hear about it first? Should the story be given to one reporter before all others, on an exclusive basis?

In December 2010, Yahoo announced the layoff of roughly 4 percent of its workforce, or around 650 employees. Yahoo notified affected employees on December 14. The same day, CEO Carol Bartz issued an explanatory memo beginning with "Yahoos, I want to share some tough news with you" to put the layoffs into context.[16] However, she wasn't the first to deliver the bad news to the affected employees; rumors of impending layoffs percolated throughout both respected media outlets and the blogosphere in the days leading up to the official announcement. With the real-time, 24/7 news cycle that is increasingly dominated by social media platforms, the company didn't address prevalent online rumors, announcing more than a week later that layoffs would indeed be necessary. Thus, despite Bartz's "personal" appeal to employees, online commentary from sources such as the UK's *The Guardian* and *The New York Times* suggests that Yahoo staffers caught wind of the impending layoffs well in advance of the official announcement. It's just one example of the necessary actions companies must take to protect sensitive information and to deliver it to the intended audience before it leaks in cyberspace.

[16] http://www.guardian.co.uk/technology/blog/2010/dec/17/yahoo-closing-problems.

After GM announced 25,000 planned job cuts by 2008 at the company's annual meeting in 2005, the company had to work fast to calm worker uncertainty about what lay ahead. Sue Melino, staff director of GM's global internal communications department, explained that the company's goal was to ensure that employees were informed about the layoffs as soon as they were released to the public. The company delivered the news to employees through multiple channels: a webcast of CEO Rick Wagner's speech at the annual meeting, newsletters in company plants, and a segment on GM's daily employee television show. "What we are trying to do," Melino explained, "is provide as much context as we can internally."[17]

Each time a corporate communication strategy is developed, the question of which channels to use and when to use them should be explored carefully. Before this step, the company needs to think about the best way to structure the message and what to include in the message itself.

Structure Messages Carefully

According to most experts in communication, the two most effective message structures are commonly referred to as *direct* and *indirect*. Direct structure means revealing your main point first, and then explaining it in more detail; indirect structure means giving context first, and then revealing your main point.

When should a company choose to be direct and when should it decide to be indirect? Normally, organizations should be as direct as possible with as many constituencies as possible, because indirect communication is confusing and harder to understand.

Take the example of Nissan when it introduced the Infiniti series in the United States. Instead of just coming out with photographs of the new cars (as it does now), the company took a more indirect (and typically Japanese) approach by showing impressions of landscapes and creating a mood without actually showing the car. This effort was a creative success compared with the approach its direct competitor, Toyota, took by showing traditional pictures of its comparable Lexus model. Unfortunately, Nissan's campaign didn't sell many cars. The company wanted to create a strong identity in the American market through this type of advertising, but this mixture of product and image advertising was completely lost on American consumers.

A third option in terms of message structure is to simply have *no* message. Today, this approach simply doesn't work with a public hungry for the next sound bite and the media looking for an "angle" on the story. Usually, saying that the company cannot talk about the situation until "all the facts are in" is better than just saying "no comment" or nothing at all, but managers (especially in the United States) are often influenced by lawyers who are thinking about the legal ramifications of saying anything. Deciding to be direct often means taking the court of public opinion into consideration as well, which, to some companies, is often far more important than a court of law.

---

[17] John N. Frank, "GM Pushes Growth Message in Light of Announced Job Cuts," *PRWeek*, June 9, 2005.

Constituency Responses

After communicating with a constituency, you must assess the results of your communication and determine whether the communication had the desired result. In some instances, this feedback can be gathered nearly immediately after the delivery of an important message or set of messages. For example, employees can be provided a short questionnaire to confirm an understanding of the main points of the communication and uncover areas where they would have wanted more information or clarification. In other cases, it may take some time to measure the success of the communication, such as determining whether sales rose in response to an advertising campaign. After the results are in, you must determine how you will react. Has your reputation changed? Do you need to change your communication channel? Hence the circular nature of the corporate communication framework.

Creating a coherent corporate communication strategy, then, involves the three variables we have discussed in detail: defining the *organization's* overall strategy for the communication, identifying and analyzing the relevant *constituencies*, and delivering *messages* effectively. In addition, the organization needs to analyze constituency *responses* to determine whether the communication was successful. Figure 2.2 summarizes this more complete version of the corporate communication strategy model introduced earlier.

FIGURE 2.2
Expanded
Corporate
Communication
Strategy
Framework

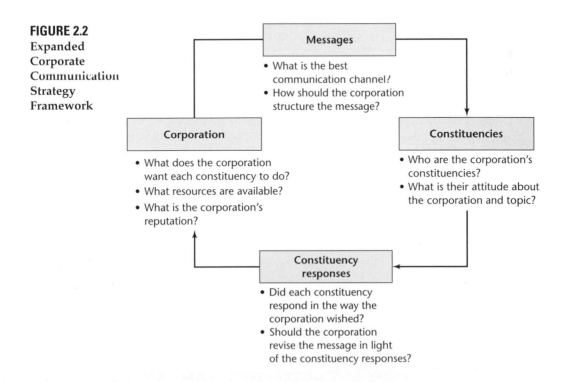

Messages
- What is the best communication channel?
- How should the corporation structure the message?

Corporation
- What does the corporation want each constituency to do?
- What resources are available?
- What is the corporation's reputation?

Constituencies
- Who are the corporation's constituencies?
- What is their attitude about the corporation and topic?

Constituency responses
- Did each constituency respond in the way the corporation wished?
- Should the corporation revise the message in light of the constituency responses?

Conclusion: The Corporate Communication Connection to Vision

By creating a coherent communication strategy based on the time-tested theories presented in this chapter, an organization is well on its way to reinventing how it handles communications. Just as important for the firm, however, is its ability to link its overall strategy to its communication efforts.

As discussed in Chapter 1, firms are facing increased scrutiny from external groups (for example, the NGO Environmental Working Group, which launched the Skin Deep Cosmetic Database to "create online safety profiles for cosmetics and personal care products") as well as their key constituencies (such as activist shareholder groups that protest executive compensation increases, and recently have become more empowered to do so by the 2010 Dodd-Frank Wall Street Reform and Consumer Protection Act.)[18] By linking corporate strategy to corporate communication, managers can mitigate the potential loss in reputation (see Chapter 4) that can result from a weak or negative response from an organization to challenges from external groups and its own constituencies.

The extent to which an organization is affected by external forces also is determined by what industry the firm is in, where it does business, and how public its operations are. In addition to staying competitive, then, the question of how the firm is perceived externally must be considered. Just as the company's awareness about competitive forces protects it from competitors, its awareness of external forces also protects it from attacks.

Despite a recent series of very public recalls, consumer products company Johnson & Johnson has benefitted from a largely undamaged reputation, simply because the company reinforced its commitment to the J&J Credo, the company's written values system, in its communications with its key stakeholders during moments of crisis (see Chapter 10).

When developing an overall strategy, firms need to consider their corporate communication effort as manifested in the company's vision and mission statement. By doing so at the inception of an overall strategy, the firm avoids repercussions later. Because all organizations operate at the behest of the public will, this egalitarian approach to communications will be appreciated by a society that has come to depend on its organizations more than ever before.

[18] "Brief Summary of the Dodd-Frank Wall Street Reform and Consumer Protection Act," 2010, http://banking.senate.gov/public/_files/070110_Dodd_Frank_Wall_Street_Reform_comprehensive_summary_Final.pdf.

Case 2-1

Galen Healthcare System

The Galen Healthcare System (GHS), a leading and innovative healthcare organization founded years earlier, provided a full spectrum of healthcare services in upstate New York. As the largest health system in the area, Galen consisted of a diverse network of healthcare facilities, including a large academic medical center, six community hospitals, a physician organization that included a network of physician practices throughout the area, a rehabilitation center, nursing homes, and hospice and palliative care services. Prior to joining Galen, member organizations operated as independent entities. Although now under the same organizational umbrella, member organizations continued to operate quite independently, and functions such as procurement were not coordinated. Each site had its own Purchasing Department that issued contracts and purchase orders with local suppliers. The individual purchasing managers determined needs, identified suppliers, and managed inventory logistics without oversight from corporate headquarters. This decentralized approach was indicative of the system's practice over the last decade of loose management of its affiliated sites.

Leaders within the Galen System, however, began to recognize a trend of rising health care costs and intensified competitive pressures in the region. New specialty practices and outpatient surgical centers sprouted in communities where Galen previously commanded the market. In addition, the system continued to see lower reimbursement rates from Medicare and Medicaid, making it increasingly challenging for the system to meet its annual budget.

Health systems around the country were also experiencing similar challenges, but unlike Galen, many health systems had already joined group purchasing organizations (GPOs). In the 1970s and 1980s, GPOs arose as a means to combine health organizations into larger unified groups to leverage aggregated purchasing power. GPOs worked with members—which could include academic medical centers, community hospitals, and large physician practices—to help them save money on medical supplies. The purchasing power of a GPO enabled it to command higher discounts when negotiating contracts with manufacturers, distributors, and other vendors, and allowed it to pass on these cost savings to its individual members.

Recognizing the problems associated with a fragmented supplier base, Galen's senior leadership team made the coordination of system purchasing a priority for the coming year. Dr. Jeffrey Dahmer, the President and CEO of the Galen Healthcare System, created the first corporate Purchasing Department at Galen, and hired an experienced procurement manager, Richard Gunerson, as Vice President of General Services/Supply Chain Management to oversee the procurement of all equipment, supplies, and services. Preparing Galen to join a GPO was one of Gunerson's primary tasks.

Dahmer gave Gunerson great autonomy and flexibility, and placed Stacie Friberg as Gunerson's executive assistant. Friberg had worked for 12 years within the Galen system in several different positions, and thus was familiar with many of the local hospital and practice leaders as well as their historic purchasing practices. Gunerson's appointment was announced in the employee newsletter, published by senior leadership at Galen's headquarters, and on the GHS intranet.

The end of the hospital budget cycle was nearing its end, and Gunerson was eager to notify each of the local Purchasing Department heads that the entire system would begin to purchase its required supplies in bulk through a GPO contract. To estimate the large purchase orders its member hospitals were making,

Gunerson intended to request that local purchasing heads report contracts with suppliers over $100,000. Gunerson discussed this idea with Jeffrey Dahmer, who discussed it with the Board of Directors.

Gunerson then drafted the following memo to the local Purchasing Departments:

Dear Purchasing Department lead,

The Board of Directors has approved a new procurement process. Henceforth, all heads of Purchasing at local sites of care will alert the Vice President of General Services/Supply Chain Management about contracts above $100,000, which they plan to negotiate with suppliers at least a week before the day they will be signed.

I know you must understand that this change is critical to coordinate procurement requirements and to eliminate inefficiencies within the system. Today, our member organizations' purchasing habits result in unnecessary redundancies and high prices. This step will provide the head office with the information it needs to ensure that each site of care procures the optimal volume of supplies at the best prices. As a result, the goals for each site of care and for Galen as a whole will more likely be achieved.

Sincerely,
Richard Unger Gunerson II, FACHE
Vice President of General Services/Supply
Chain Management

Gunerson gave Stacie Friberg the memo and asked for her input. She told him she thought the memo was great. She also suggested, however, since he had met only two of the local Purchasing Department managers that he might like to meet all of them and discuss this major change in procurement with each individually. Gunerson declined, stating that he was unable to travel to each local site because he had so many things to do at home and at the office and it would be too expensive.

Over the next few days, responses came in from all but a few local Purchasing Departments. Some managers wrote longer responses, but the following e-mail message was typical:

Dear Dick,

Welcome to Galen! We are pleased to hear that you are settling into your position as the new procurement coordinator. We received your recent communication about notifying headquarters a week in advance of our intention to sign contracts with suppliers. This suggestion seems very practical. We would like to let you know that you can be sure of this hospital's cooperation in your new job. ☺

Best regards,

Over the next several weeks, Gunerson heard nothing from the local Purchasing Departments about contracts being negotiated with suppliers. When he asked Friberg to follow up with several of the local managers, they frequently reported that they were quite busy. As a result, the usual purchasing procedures continued and Gunerson struggled to build momentum towards GPO membership.

CASE QUESTIONS

1. What problems does Galen Healthcare have that will affect its communications?
2. What specific problems does Mr. Gunerson have as a result of his communications to materials managers?
3. What advice would you give to Gunerson to help solve his and Galen's problems?

Source: This is a fictional case based on real events as well as ideas presented in both the "Dashman Company" case (9-462-001) published by HBS Case Services, Harvard Business School, Boston, 1947, and the "Marathon Plastics" case published in W. H. Newmann, E. K. Warren, and J. E. Schnee's "The Process of Management," 5th ed., Prentice Hall, Englewood Cliffs, NJ, 1982. This revision was developed by Catherine Augustyn T'13 and Abigail Isaacson T'13 under the direction of Professor Paul A. Argenti. © 2011 Trustees of Dartmouth College. All rights reserved. Reprinted by permission.

CHAPTER THREE

An Overview of the Corporate Communication Function

The past two chapters painted a broad picture of the business environment and provided a framework for communicating strategically. Against this backdrop, we will now discuss the corporate communication function itself. More and more companies recognize the value of corporate communication and are adapting their budgets and internal structures accordingly. The sixth Public Relations Generally Accepted Practices (GAP) Study, released by the Strategic Public Relations Center at the University of Southern California, revealed that despite the economic downturn in 2008, 57.8 percent of businesses saw an increase or no change in their public relations (PR) and/or communications budgets between 2008 and 2009.[1] The Communications Executive Board of the Corporate Executive Board Company reported that 96 percent of communication teams expected their staff to grow in size or remain the same between 2010 and 2011.[2] Dan Bartlett of leading public relations and public affairs agency Hill & Knowlton observed in 2011 that PR and communications "will fare better in budget pressures because the CEOs are seeing the strategic value"[3] even while other departments face significant budget reductions.

This chapter traces the evolution of corporate communication and the developments in recent years that have led to heightened recognition for the field. After examining corporate communication's roots, we discuss the most appropriate structure for the function within an organization, including reporting relationships. We also briefly showcase each corporate communication subfunction, all of which are developed in greater detail later in this book.

---

[1] GAP Study, http://annenberg.usc.edu/CentersandPrograms/ResearchCenters/SPRC/PrevGAP.aspx.

[2] Communications Executive Board, Corporate Executive Board Company, "2010 Resource Allocation Benchmark."

[3] *PR Week*/Hill & Knowlton, "Corporate Survey 2010," http://www.hillandknowlton.com/content/pr-weekhill-knowlton-corporate-survey-2010-document.

From "PR" to "CorpComm"

Public relations (PR), the predecessor to the corporate communication (CorpComm) function, grew out of necessity. Although corporations had no specific strategy for communications, they often had to respond to external constituencies whether they wanted to or not. As new laws forced companies to communicate in many situations they hadn't previously confronted, the constant need for a response meant that dedicated resources were required to manage the flow of communications.

This function, which was tactical in most companies, was almost always called either *public relations* (PR) or *public affairs*. Typically, the effort was focused on preventing the press from getting too close to management. Like a Patriot missile, designed to stop incoming missiles during war, the first PR professionals were asked to protect the company from bad publicity, often by "spinning" damaging news in a positive light. Thus, the term "flak" came to be used to describe what PR people were actually doing: shielding top managers from "missiles" fired at them from the outside.

The "flak" era of public relations lasted for many decades, and when companies needed other communications activities, public relations personnel were the obvious choice to take them on. In the 1960s, for instance, it was fairly typical to find public relations officials handling speechwriting, annual reports, and the company newsletter. Given that the majority of work in this area involved handling print media (television wasn't truly a factor until the early 1970s), many companies hired former journalists to manage this job. The former-journalist-turned-flak brought the organization the first dedicated expert in the area of communication.

Until recently, the top managers in large companies came from backgrounds such as engineering, accounting, finance, production, or, at best (in terms of understanding the company's communication needs), sales or marketing. Their understanding of how to communicate depended on abilities they might have gained by chance or through undergraduate or secondary school training rather than years of experience. Given their more quantitative rather than verbal orientation, these old-style managers were delighted to have an expert communicator on board who could take the heat for them and offer guidance in times of trouble.

PR professionals often were seen as capable of turning bad situations into good ones, creating excellent relations with their former colleagues in journalism, and helping the chief executive officer become a superb communicator. In some cases, this reputation was true, but for the most part, the journalists were not the answer to all of the company's communications problems. When situations turned from bad to worse, they were the obvious ones to blame—easy scapegoats for irresponsible managers.

The First Spin Doctors

In addition to the internal PR staff, outside agencies often helped companies that either couldn't afford a full-time person or needed an extra pair of hands in a crisis. The legends of the public relations field—such as Ivy Lee and Edward Bernays and, later, Howard Rubenstein and Daniel Edelman—helped the public relations function develop from its journalistic roots into a more refined and respected profession.

For many years, PR agencies dominated the communications field, billing companies hefty fees for services they could not handle in-house. Few large companies were willing to operate without such a firm for fear that they might be missing an opportunity to solve their communications problems painlessly by using these outside "spin doctors."

Some of the top public relations firms today—such as Fleishman-Hillard and Edelman in the United States, Weber Shandwick in the United Kingdom and the United States, and Ogilvy and Cosmo PR in Japan—still provide some of the best advice available on a number of communications-related issues. But outside agencies cannot handle all the day-to-day activities required for the smooth flow of communications from organization to constituents. Therefore, they often work alongside in-house communication professionals on strategic or project-based communications activities.

A New Function Emerges

By the 1970s, the business environment required more than the simple internal PR function supplemented by the outside consultant. The rise in importance and power of special-interest groups, such as Ralph Nader 's Public Interest Research Group (PIRG), and environmentally oriented nongovernmental organizations (NGOs), such as Greenpeace, forced companies to increase their communications activities. During the Arab oil boycott and embargo in the 1970s, the entire oil industry came under fire as consumers had to wait hours for a tank of gasoline while big oil companies reported what many consumer groups felt were "obscene" profits running into the hundreds of millions of dollars.

This situation led Mobil Oil to develop one of the most sophisticated public relations departments of its time. Mobil's Herb Schmertz revolutionized the field by solving communications problems with strategies that no one had thought of before. His series of advertisements, called "issue ads" (see Chapter 4 for more on this subject), which ran on *The New York Times* and *The Wall Street Journal* op-ed pages once or twice a week, directly attacked the allegations of both "obscene" profits and hoarding of oil to inflate prices. Instead of merely reacting to these allegations, the Mobil issue ads put the blame on the government, explained why the oil companies needed hefty profits for exploration, and refocused discussion on other issues the company's CEO thought were important to shareholders.

With a budget in the tens of millions of dollars, Schmertz created a new communications function that changed the nature of Mobil's communications effort from old-style public relations to the first significant corporate communication department. A senior vice president of the corporation, Schmertz was also one of the very few communications executives with a seat on the board of directors—further proof of Mobil's commitment to enhanced communications.

Thus, as individual corporations and entire industries were increasingly scrutinized and had to answer to a much more sophisticated set of journalists, the old-style public relations function was no longer capable of handling the flak. As a result, what at first had been deemed a waste of resources at Mobil in the early

1970s became the norm in corporate America. The focus now shifted to structuring these new corporate communication departments effectively to fit the function into the existing corporate infrastructure.

Corporate Communications Today

In more recent years, the corporate communication function has continued to evolve to meet the demands of the ever-changing business and regulatory environments. At the outset of the millennium, a string of financial scandals at corporations including WorldCom and Enron resulted in the Sarbanes-Oxley Act of 2002, which although only legally affecting public companies, increased the public's expectations for transparency, responsiveness and corporate responsibility for all companies large and small. The need to maintain this level of transparency has elevated the corporate communication function within companies to a new strategic level. Messages, activities, and products—from investor conferences and annual reports to philanthropic activities and corporate advertising—are now analyzed by regulators, investors, and the public at large with unprecedented scrutiny. And the proliferation of online communication vehicles, including web portals, instant messaging, and blogs, has accelerated the flow of information and the public's access to it to record speeds. (Chapter 6 covers media relations in more detail and Chapter 8 discusses investor relations at length.)

In its report, "The Authentic Enterprise," the Arthur W. Page Society, an association of chief communications officers of large corporations, characterizes the changes in corporate communications into three groups: (1) new audiences, (2) new channels and new kinds of content, and (3) new measurements. It also asserts that corporate communications professionals today must not only "position" their companies but also must "help to define" them.[4]

Under today's higher-resolution microscope, the clarity, alignment, and integration of communications to all constituents can make or break a corporate reputation. As a result, 77 percent of in-house communicators cite spending a "moderate amount" or "great deal" of time developing integrated communications.[5]

Specific Responsibilities of Corporate Communications

While organizations may have unique combinations of needs at any given time, in its 2010 Resource Allocation Benchmarks report, the Communications Executive Board of consultancy Corporate Executive Board, identified almost 30 distinct responsibilities that a modern communications team is likely to have.

Responsibilities categorized as *external communication activities* include:

- Investor Relations
- Financial Communications
- Annual Report
- Corporate Web Site

---

[4] Arthur W. Page Society. "The Authentic Enterprise: Relationships, Values and the Evolution of Corporate Communications." May 17, 2007.

[5] Weber Shandwick, "Corporate Survey 2005," *PRWeek*, June 27, 2005.

- Corporate Advertising
- Marketing Communications
- Executive Communications
- Community Relations
- Government Relations
- External Social Media
- External Social Media Monitoring
- Reputation Monitoring

Responsibilities categorized as *internal communication activities* include:

- Employee Communications
- Corporate Intranet
- Internal Social Media
- Employee Surveys
- Leader/Manager Communications Training

Responsibilities categorized as *miscellaneous communication activities* include:

- Graphics or Creative Services
- Communications Measurement
- Event Management
- Corporate Social Responsibility—Reporting
- Corporate Social Responsibility—Programs
- Charitable Activity and Donations
- Corporate Sponsorship
- Communications Staff Development[6]

To Centralize or Decentralize Communications?

One of the earliest problems organizations confronted in structuring their communication efforts was whether to keep all communications focused by *centralizing* the activity under one senior officer at headquarters or *decentralizing* the activities and allowing individual business units to handle communications. The more centralized model provided an easier way for companies to achieve consistency in and control over all communication activities. The decentralized model, however, gave individual business units more flexibility in adapting the function to their own needs.

The same structural challenges persist today, and the answer to the centralization/decentralization debate often depends on a company's size, the geographic dispersion of its offices, and the diversity of its products and services. For organizations as large and diversified as General Electric, for example, the question is moot: There is no way such a sprawling organization involved in activities as diverse as aerospace and health care could remain completely centralized in all of its communication activities.

[6] ibid.

The same is true for Johnson & Johnson (J&J): With more than 110,000 employees in more than 200 operating companies in 57 different countries, complete centralization of communications would be difficult, if not impossible. Instead, Bill Nielsen, the legendary former corporate vice president of corporate communication at J&J, described the function as "a partnership of professionals in communication."[7] J&J even avoids centralizing its external communications counsel with a single public relations firm. Instead, the company uses both small firms on a project basis and large, global agencies with resources around the world, amounting to a total of over 20 different agencies worldwide to support various elements of its business.

Global events and economic trends also affect decisions about the structure of an organization's communication function. Not only did the shock of the September 11, 2001, attacks teach companies the importance of expecting the unexpected in terms of crises, but it also gave decentralized communication structures a new appeal for many companies. As Jim Wiggins, first vice president of corporate communication for Merrill Lynch (now part of Bank of America), explained at the time, "Companies will have to look at less centralization of key activities if we now live in a world where terrorism is a key possibility."[8]

Increased security threats are not the only catalyst for the decentralization of communications; economic downturns can have a similar effect. Consider a major international airline that imposed significant staff reductions on its corporate communication department due to across-the-board cost cuts. As a result, the director of communications explained that the department became more selective about what they committed to, saying: "We don't do everything for everybody anymore." Instead, other departments throughout the company established communication positions, doing some of the activities formerly handled by the centralized corporate communication department.[9]

In instances of scaled-back budgets, delegating tasks is doubly important because economic uncertainty also can force the communications department to handle activities it would generally outsource to a full-time PR agency. This appears to be happening particularly at government agencies and at NGOs, who report decreasing use of external PR and communication agencies.

Although decentralization allows for more flexibility in tough economic times, these advantages are not without accompanying risks. Dispersing corporate communications across individual operating units without some central oversight significantly raises the potential for inconsistent messages. In decentralized structures, a company's communication professionals must be diligent about assuring quality, consistency, and integration of messages across the board.[10] Companies often require formal mechanisms to ensure that this integration takes place.

---

[7] Interview with Bill Nielsen, February 2002.

[8] Shane McLaughlin, "Sept. 11: Four Views of Crisis Management," *Public Relations Strategist*, January 1, 2002, pp. 22–28.

[9] Jack LeMenager, "When Corporate Communication Budgets Are Cut," *Communication World 3* (February 3, 1999), p. 32.

[10] "Sixth Annual Public Relations Generally Accepted Practices Study," Strategic Public Relations Center, University of Southern California, April 30, 2010, http://annenberg.usc.edu/CentersandPrograms/ResearchCenters/SPRC/PrevGAP.aspx.

Perhaps, then, finding a middle ground between a completely centralized and a wholly decentralized structure is preferable for large companies. For example, a strong, centralized, functional area can be supplemented by a network of decentralized "operatives" who adapt the function to the special needs of the independent business units. Dell Computer Corporation organizes its corporate communication staff using an approach that follows how its businesses are organized: a "matrix" based on customers, products, and geography. Although the more than 80 team members are physically located within the businesses they support, the person who heads up the team sits at headquarters among Dell's corporate staff, interacting constantly with senior management. This combination of centralized communication management with "operatives" dispersed throughout the various business units has proved successful for Dell, a company with over 50,000 employees who live and work in more than 30 countries.

Between 2008 and 2010, corporate communication departments were "being asked to not only drive reputation, awareness and PR, but also to contribute to business results."[11] As an example, 45 percent of companies now actively involve the marketing team when planning their communications calendar.

The ways in which the marketing and PR functions work together may also depend on how much of the marketing function is communication focused versus marketing focused. In their book *Marketing Communication*, James G. Hutton and Francis J. Mulhern outlined five core relationship situations that companies may fall into:

1. Communication and marketing issues are both relatively small = a "separate-but-equal" model may work best.

2. If there are moderate communication and marketing issues = an "overlapping" model might work best.

3. For companies that have multiple brands, such as consumer products companies = a "marketing-dominant" model may deliver more success.

4. Professional services, hospitals, universities, and NGOs may benefit from = a "PR-dominant" model.

5. In situations where marketing issues are communication issues, primarily for smaller companies where there is very little difference = a "marketing-equals-public-relations" model might be best.[12]

Where Should the Function Report?

Surveys conducted over the last decade have consistently shown that a high percentage of the average CEO's time is spent communicating. Research conducted at the Tuck School of Business suggests that, on average, *Fortune 500* company CEOs spend between 50 and 80 percent of their time on communication activities.

---

[11] LeMenager, "When Corporate Communication Budgets Are Cut."

[12] *PRWeek*/Hill & Knowlton, "Corporate Survey 2010," http://www.hillandknowlton.com/content/pr-weekhill-knowlton-corporate-survey-2010-document.

As an example, former Johnson & Johnson CEO James Burke estimated that he spent over 40 percent of his time as CEO communicating the J&J Credo alone (see Chapter 10 for more on the Credo).[14]

CEOs generally devote their time to communicating their company's strategic plan, mission, operating initiatives, and community involvement both internally and externally. Michael Useem, a management professor at the Wharton School of the University of Pennsylvania, estimates that due to investors' increasing demands for companies to deliver short-term results, about one-third of a CEO's time is devoted to capital markets and to communications with up to four dozen analysts and investors.[15]

In many respects, CEOs themselves are an embodiment of the corporate brand. As such, their behavior and commentary can easily and markedly affect a company's financial performance. Recall Martha Stewart, founder and CEO of Martha Stewart Living Omnimedia, sentenced to five months in prison in 2004 after being found guilty on four counts of obstructing justice and lying to investigators regarding a stock sale. Expectations of a not-guilty verdict caused the company's stock price to rally prior to the announcement, only to nose-dive 22 percent on the New York Stock Exchange following the guilty verdict.[16]

All of this evidence implies that the CEO should be the person most involved with both developing the overall strategy for communications and delivering consistent messages to constituencies. Ideally, the corporate communication function will have a direct line to the CEO. (See Figure 3.1 for a sample corporate

FIGURE 3.1 Ideal Structure for CorpComm Function for Larger Companies.

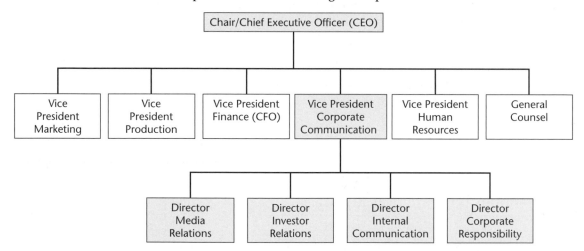

[13] Hutton, James G. and Francis J Mulhem. "Marketing Communications: Integrated Theory, Strategy & Tactics". Pentagram. Jan 2002.

[14] James C. Collins and Jerry I. Porras, *Built to Last* (New York: Harper Business, 1994, 1997), p. 80.

[15] Nanette Byrnes, "The 21st Century Corporation: The New Leadership: Chief Executive Officer: The Boss in the Web Age," *BusinessWeek* 3696 (August 28, 2000), p. 102.

[16] Jim Robinson, "Leader of the Brand—Keeping the Best CEOs in Step," *Management*, June 1, 2005, p. 26.

TABLE 3.1
Where Corporate
Communications
Reports

Source: Hill & Knowlton,
"Corporate Survey 2010,"
PRWeek.

Communications Head Reports to:	Total (Percentage of Companies)
Chairman/CEO/president	51
Head of marketing	19
Other	12
Chief operating officer	6
Chief of human resources	5
Chief financial officer	2
General counsel	5

TABLE 3.2
Title of Senior
Communication
Executive

Title	Total (Percentage of Companies)
Senior vice president	17.1
Vice president	31.1
Director	26.3
Manager	17.1
Other	8.3

communication reporting structure.) Nearly half (51 percent) of respondents to a *PRWeek* Corporate Survey said their company's head of communications reported directly to the CEO, president, or chairman.[17] (See Table 3.1 for details of the study.) Even if reporting lines do not, on paper, go directly to the CEO, it is vital that the head of corporate communication have access to the highest levels of senior management and that those executives believe in the value and necessity of corporate communication as a means to achieve corporate goals. Without this connection, the communications function will be less effective and far less powerful. The 2010 *PRWeek*/Hill & Knowlton Corporate Survey demonstrates that communications teams, now more than ever, should have a seat at the table. This is as a result of increased use of social media and a greater emphasis on reputation management.[18]

To keep the number of direct reports to the CEO down to a handful of senior executives (often the biggest stumbling block to getting the corporate communication function "plugged in" at the top), some companies are now integrating the corporate communication function into the strategic planning function. Given the importance of tying communications to the overall strategy of the firm, this approach may benefit the growing corporate communication function.

---

[17] Weber Shandwick, "Corporate Survey 2005," *PRWeek*, June 27, 2005.
[18] ibid.

In some cases, however, the function still reports to the catch-all executive vice president (EVP) in charge of administration. This person also has responsibility for areas such as HR, security, and buildings and grounds. This structure can present tremendous problems for the communication function—especially if the EVP has little knowledge of or lacks an interest in communications.

In this classic example, when Union Carbide Corporation was dealing with the aftermath of its Bhopal plant accident in India in 1984, the company transferred its communication responsibilities to the vice president of strategic planning. In a letter to executives, the chairman and CEO of the company at the time, Robert Kennedy, said:

> The Corporation's strategic direction is a key element of our communication to shareholders, employees and the public at large. . . . It is therefore more important than ever to be open and consistent in our communications to all of these groups, to keep them informed of our progress as we implement strategy, and to make sure that we address the special concerns and interests of all the groups and constituencies with a stake in Union Carbide's future. . . . To ensure the closest possible alignment of our communications with management directed at strategic planning developments, the management of those functions is being consolidated under . . . [the] Vice President of Strategic Planning and Public Affairs.[19]

Gerald Swerling, head of the graduate public relations program at the University of Southern California's Annenberg School of Communication, observed that the increased recognition of communications and public relations by senior management over the last decade has caused an increasing number of CEOs to "demand that there be PR professionals at the strategic planning table for new products and initiatives."[20] According to a research report by Spencer Stuart and Weber Shandwick, entitled "The Rising CCO II" released in early 2008, 54 percent of today's corporate communication officers (CCOs) report directly to the CEO. A more recent study from the CCO Council confirmed "that the most effective CCOs are those who report directly to the CEO." Similarly, CCOs in *Fortune's* "Most Admired Companies" list are more likely than other contenders to have longer tenures (six years, one month versus five years), have prior PR agency experience (41 percent versus 26 percent), and have no interdepartmental rivals (32 percent versus 9 percent).[21]

Often such studies focus on larger corporations, but many of the lessons that we learn from large companies can be translated and applied to smaller companies. Small and mid-size businesses may not have the luxury of an experienced communications team, but these companies still have important communications needs just as larger companies do. All companies need to communicate with a consistent voice to all of their constituencies.

One difference between larger and smaller companies is that in a small or even mid-size company, it can be easier to involve all employees in internal and external communications. For example, Sweet Leaf Tea, a mid-size beverage company

---

[19] Letter from Union Carbide's CEO, Robert B. Kennedy, to Executive List, dated March 5, 1992.

[20] Richard Nemec, "PR or Advertising—Who's on Top?" *Communication World 4* (February 3, 1999), p. 25.

[21] Craig McGuire, "Market Focus: Corporate Boards—Board Games," *PRWeek*, June 27, 2005, p. 25.

in Austin, Texas, eliminated the director of marketing role and replaced it with a director of branding. Today the communications manager reports to the director of branding, who reports directly to the CEO. Charla Adams, the current director of branding, helps to develop strategy, to develop core messages, and to ensure consistent implementation across the all communication channels (internal and external). Reporting to Charla are a communications manager and an assistant marketing manager. Sweet Leaf Tea spends less than 1 percent of its revenues on PR, compared to the industry average of 1.3 percent.

Working Strategically with External PR and Communication Agencies

No matter how they structure their marketing and PR functions, all companies may benefit from the strategic use of external PR and communication agencies. For the best results, external agencies need to be actively managed and provided with clear expectations for deliverables. In some cases, companies and organizations engage external PR and communication agencies due to a lack of internal manpower or expertise. This is particularly common in smaller organizations where the external agency may report to an internal employee who has responsibilities in several functions. Table 3.3 highlights the main reasons that companies choose to work with external PR and communications agencies. Despite the economic downturn, corporations have continued to value external PR agencies, with 82.1 percent reporting use of external PR agencies in 2010.[22]

Companies should thoughtfully consider how and when to integrate the work of external agencies. The 2010 GAP study cited earlier reports the interesting finding that PR/communications functions that work with outside agencies are more likely to have responsibilities in corporate intranet, CSR, executive communication, issues management, and philanthropy, whereas PR/communications functions that do *not* work with outside agencies are more likely to have responsibilities in corporate image advertising and product, the corporate external website, the corporate image (including graphics and logo), and customer relations.

TABLE 3.3
Why Companies Choose to Work with External PR and Communications Agencies

Source: Sixth "Public Relations Generally Accepted Practices Study," Strategic Public Relations Center, University of Southern California, April 30, 2010.

They provide additional "arms and legs"	85.4
They complement our internal capabilities	71.2
They provide strategic and/or market insight and experience	69.2
They offer unique expertise	61.1
They provide an objective point of view	51.0
They have resources in geographies or markets where I need them	48.5
We have a limit on internal "head count"	37.9
They are cheaper than adding staff	34.3
They provide expertise in digital/social media that we lack internally	26.8
They provide an ability to quantify results	20.7

[22] Spencer Stuart and Weber Shandwick, "The Rising CCO," 2008.

No matter how it staffs and organizes the communications function, when senior management shows commitment to communications by allocating significant resources, all employees will begin to rightfully appreciate communications as a critical management tool.[23] Now let's take a look at what that function should include.

The Subfunctions within the Function

According to recent surveys, over half of the heads of corporate communication departments oversee communications functions that include corporate communications/reputation, crisis management, executive communications, employee/internal communication, and marketing PR/product. Nearly 70 percent of surveyed CEOs rely on their PR/communications counsel to manage corporate reputation. With today's increased use of social media, almost 30 percent of PR/communications counsel also provide expertise in digital and social media where most CEOs lack experience. Although not every company can include all the subfunctions and responsibilities listed here under one umbrella, to operate most effectively, a majority of these functions must be included in the overall communications function.

The best approach to building a corporate communication function is to begin with the most global and strategic issues and then move into the narrower aspects of the function. We begin this section with a discussion of identity, image, and reputation, and then move on to the various subfunctions of corporate communication.

Identity, Image, and Reputation

Difficult to classify as a separate subfunction, an organization's identity, image, and reputation strategy is the most critical part of any corporate communication function. (In the next chapter, we explore these constructs in greater detail.) What is the difference between image, identity, and reputation, and how do they shape the operations of a corporate communication department?

Image is the corporation as seen through the eyes of its constituencies. An organization can have different images with different constituencies. For example, cigarette companies might be reprehensible to many American consumers looking for a healthier lifestyle but a delight to shareholders reaping the profits from international sales of the same product. On the other hand, customers might have been perfectly happy with what Macy's had to offer in its many stores throughout the United States, but securities analysts were reluctant to recommend the parent company's stock, knowing that inevitably it would enter bankruptcy.

Determining the organization's image with different constituencies is usually less obvious than in these examples—particularly given the increasingly blurred lines separating one constituency from another, as discussed in Chapter 2. For

---

[23] LeMenager, "When Corporate Communication Budgets Are Cut."

example, many employees today are encouraged to own stock in their own company and can be the most visible ambassadors of their company's brand by also being consumers of its products or services. The corporate communication department should conduct research to understand and monitor each constituency's evolving needs and attitudes. Obviously, the organization cannot please everyone, but by monitoring what constituencies are thinking about, it can make a conscious effort to avoid hostility with a particular group. A similar monitoring system can also help to gauge the impact and success of the company's communication activities.[24]

Unlike its image, however, the organization's *identity* should not vary from one constituency to another. Identity consists of a company's defining attributes, such as its vision and values, its people, products, and services. An organization has an identity whether it wants one or not, based in part on the reality it presents to the world. People all over the world know Coca-Cola's red can and white script lettering and McDonald's golden arches in front of a store, whether they are in Singapore, or California.

Because identity building and maintenance require a variety of skills, including developing strategy and the ability to conduct research, to design attractive brochures, and to enforce identity standards and cohesion, it should be spread around several different functions in the absence of a single, centralized corporate communication function. For example, the research needed to determine a firm's image with various constituencies might be a minor by-product of the over-all marketing research effort currently under way at a company, to determine customer attitudes toward particular products and services rather than the firm as a whole.

Determining how a firm wants to be perceived with different constituencies and how it chooses to identify itself is the cornerstone function of corporate communication. If the firm is making serious changes in its identity, this subfunction can easily be a full-time job for a team of corporate communicators for a period of time.

At nearly all companies, outside agencies specializing in identity and image, such as Lippincott, Siegel and Gale, or Landor, would definitely be involved in the makeover as well, if the company alters significant components of its identity. These changes can range from the merely cosmetic—to keep the "look" of the company up to date—to the more momentous—such as a name change or a new logo.

Whereas identity represents the reality of an organization and image its reflection by key constituents, *reputation* is the sum of how all constituents view the organization. As a result, the idea that an organization can manage its reputation is unrealistic. Instead, corporations should focus on developing and implementing strategies in an integrated fashion across constituencies.

---

[24] Heyman Consulting, "State of U.S. Corporate Communications," prepared for Janis Forman, The Anderson School at UCLA, May 31, 2001, p. 32.

Corporate Advertising and Advocacy

A company's reputation also can be enhanced or altered through *corporate advertising*. This subfunction of corporate communication is different from its product advertising or marketing communication function in two ways. (See Chapter 4 for more on corporate advertising.)

First, unlike product advertising, corporate advertising does not necessarily try to sell a company's particular product or service. Instead, it tries to sell the company itself—often to a completely different constituency from customers. For example, in 2007 General Electric committed a substantial portion of its $90 million corporate advertising budget to "Ecomagination"—a marketing campaign not only promoting its environmentally friendly products but also positing GE as an eco-friendly company and leader in corporate responsibility.[25] Underpinning the campaign was GE's promise to improve its energy efficiency 30 percent while also cutting greenhouse-gas emissions by 1 percent by 2012.[26]

Adding a new layer to traditional television and print campaigns, corporations also are turning to the ever-growing Internet "blogosphere" and social media to create viral marketing campaigns that can influence consumers' opinions. As early as 2005, Microsoft employees were writing approximately 1,500 blogs, some devoted to corporate recruiting by focusing on the employee experience at Microsoft, as well as available positions and hiring trends.[27] Whether online or off, employees are important word-of-mouth advertising vehicles for a company's advocacy efforts. GE kept this in mind in its "Ecomagination" launch, conducting a simultaneous internal communications program in 2007 that featured a children's magazine conveying Ecomagination's core messages to employees' children and local communities.[28]

When the upscale discount retailer Target began running an extensive corporate advertising campaign back in the late 1990s featuring products ranging from satin lingerie to earplugs, accompanied only by the product name and Target's bull's-eye logo, the goal was not to sell more of these products but rather to showcase the company's diverse merchandise and potential to be the discount retailer that "looks like Barneys, priced like Kmart."[29] In much the same way, the aerospace and defense firms that advertised extensively in publications such as *The New Republic* in the 1980s were not trying to sell F-15s to liberals but rather to influence public opinion and facilitate approval for increases or allocations in the defense budget.

Even though product advertising is the purview of the marketing department in many large companies, corporate advertising is usually run from the CEO's office or through corporate communication departments instead. During the 1980s

---

[25] Matthew Creamer, "GE Sets Aside Big Bucks to Show off Some Green," *Advertising Age 76*, no. 19 (May 9, 2005), p. 7.

[26] Daren Fonda and Perry Bacon Jr., "GE's Green Awakening," *Time*, July 11, 2005, p. A10.

[27] Sarah E. Needleman, "Blogging Becomes a Corporate Job; Digital 'Handshake'?" *The Wall Street Journal*, May 31, 2005, p. B1.

[28] Creamer, "GE Sets Aside Big Bucks."

[29] Shelly Branch, "How Target Got Hot," *Fortune*, May 24, 1999, p. 169.

and 1990s, this area was the fastest-growing segment of the advertising industry, as senior officers tried to present a coherent company identity for opinion leaders in the financial community.

An important subset of corporate advertising is *issue advertising*. Business and policy groups spent over $29 million on issue ads in 2010 in New York State alone, with some companies using the approach as a key supplement to or in lieu of government lobbying.[30] This type of advertising attempts to do even more than influence opinions about the company; it tries to influence the attitudes of a company's constituencies about specific issues that affect the company. Recall Mobil's extensive issue advertising campaign during the oil crisis, described earlier in this chapter. Over the years, Philip Morris spent millions on issue ads covering topics ranging from domestic violence to youth smoking prevention in an effort to put a more caring face on a company many held in contempt for its role in producing addictive, carcinogenic tobacco products. Another example is when U.S. home mortgage lender Fannie Mae spent $87 million on an advertising campaign in 2003 to help curtail Congress's efforts to create a more stringent regulator to oversee its operations and have the authority to alter its capital standards. It is interesting to consider the impact of such a campaign in light of the credit crisis that started in 2008.[31]

As we will see in Chapter 4, however, issue advertising is risky. By taking a stand on a particular issue, the company is automatically creating a negative image with one or several constituencies. Many companies take this risk nonetheless, facing the consequences of adding their opinions to debates that they consider important.

Corporate Responsibility

Many companies have a separate subfunction in the human resources area to deal with community relations and a foundation close to the chairman that deals with philanthropy, but the two should be tied closely together as companies take on more responsibilities in communities in which they operate.

Taking on these social responsibilities has a number of positive outcomes for corporate leaders. (See Chapter 5 for more on corporate responsibility.) According to the Edelman Trust Barometer, the highest percentage of respondents— 39 percent—said that if they considered a company to be socially responsible, they would be most inclined to purchase their products or services; 17 percent said the would "recommend them to others."[32]

There are also serious internal implications of a strong corporate citizenship record: A survey by Net Impact revealed that, assuming all compensation and benefits are the same, 60.3 percent of respondents said they would be very likely to leave their current employer for one that they believed to be more socially responsible. What's more, when asked to rank factors based on their importance when considering working for a company, the number one concern for the survey

---

[30] Jimmy Vielkind, "Issues Ads Skyrocket," *Times Union*, June 13, 2011.

[31] Bloomberg News, "Fannie Spent $87 Million on Ad Campaign," *Los Angeles Times*, April 13, 2005, p. C-3.

[32] Edelman 2007 Trust Barometer.

respondents was the "belief that your job will make a positive difference in society," which ranked above "opportunity for career advancement" and "reputation of the organization" (2 and 3, respectively).[33]

Corporate philanthropy also has become increasingly important as companies are expected to do more than just give back to the community. Firms now feel a greater obligation to donate funds to organizations that could benefit the firm's employees, customers, or shareholders. Examples include donations to universities that might be conducting research in the industry and organizations representing minority interests.

And with increased globalization and international corporate expansion, constituents' expectations for corporate citizenship also have grown more global in scope. In December 2004, the devastating tsunami that struck 11 countries in Southeast Asia, killing 180,000 people, demonstrated this broadened focus; the U.S. Chamber of Commerce's Center for Corporate Citizenship reported that more than 400 U.S. companies donated $528 million to the tsunami relief efforts, many of these representing a company's first-time disaster relief donation.[34]

In turn, many companies are publishing environmental and social performance information in the same manner as they would traditionally report financials.[35]

Media Relations

Although the old-style public relations function, focused almost exclusively on dealing with *media relations*, may be a thing of the past, the subfunction we now refer to as media relations is still central to the corporate communication effort. Most of the average company's corporate communication staff typically reside within this subfunction, and the person in charge of the communications department as a whole must be capable of dealing with the media as a spokesperson for the firm. Although the media relations subfunction started off as a "flakking" service for managers in response to requests from news organizations, today the best corporate communication departments actively set the discussion agenda of the firm in the media. (See Chapter 6.) There is little debate about whether media relations, unlike other subfunctions, should come under the purview of corporate communication versus other corporate functions.

Technology has helped companies communicate through the hundreds of media services available from virtually anywhere in the world. Satellite uplinks are available at most corporate headquarters, and companies can put their press releases out to wire services electronically or through the Internet without making a single phone call. In addition, social media has changed the nature of how we think about media in general. Despite these advances, the relationship between business and

---

[33] Edelman, Boston College Center for Corporate Citizenship, Net Impact, and the World Business Council for Sustainable Development, "Corporate Responsibility & Sustainability Communications: Who's Listening? Who's Leading? What Matters Most?" January 2008.

[34] Michael Casey, "Tsunami Prompts Companies to Play Greater Role in Humanitarian Relief Efforts," Associated Press, June 28, 2005.

[35] "Corporate Sustainability: A Progress Report," KPMG and The Economist Intelligence Unit, 2011, http://www.kpmg.com/Global/en/IssuesAndInsights/ArticlesPublications/Documents/corporate-sustainability-v2.pdf.

media remains largely adversarial, though positive relationships between sources and reporters are much more common today than in the past. Because the media and business rely on one another to a certain extent, most companies try to make the best of these relationships.

Marketing Communications

The marketing communications department coordinates and manages publicity relating to new or existing products and also deals with activities relating to customers. It also may manage corporate advertising. Product publicity almost always includes sponsorship of events for major corporations, such as product introductions, golf tournaments, car races, and marathons. In addition, celebrities often are involved in these activities, which requires coordination within the company. Given how important such events and sponsorship agreements can be in shaping a company's image, corporate communication experts are often involved in setting the events' agenda.

With corporate sponsorships reaching the hundreds of millions of dollars for the 2012 Olympics, such sponsorships required board level approval and were closely watched by senior managers.

Customer relations activities have increasingly become a part of corporate communication as a result of pressure groups among consumers that try to exert their influence on an organization. Rather than simply making sure the customer is happy with the product or service, as in the past, companies today must get involved in quasi-political activities with constituencies claiming to represent a firm's customers.

For example, the conservative Reverend Donald Wildmon pursued a family-oriented agenda for decades against a number of companies that sold products he deemed unfit for families. Waldenbooks, for example, was vilified in the mid-1990s for selling sexually explicit literature in its stores. By organizing conservative church groups, Wildmon was able to apply pressure on Waldenbooks to stop selling literature ranging from what most people would consider simply erotic to literature with bad language in it. Disney was the focus of a boycott by Southern Baptists in 1997 for its liberal policies toward homosexuals. A year later, the Concerned Women for America (CWA) joined the protests for a Christian boycott of the company, asserting that Disney consciously laced their movies with messages of witchcraft, exemplified by such classics as *Fantasia, Peter Pan*, and *Escape from Witch Mountain*.[36] In 2005, the American Family Association, Focus on the Family, and the American Decency Association successfully lobbied for companies including Kellogg, Lowe's, Tyson Foods, and S.C. Johnson to stop buying additional advertising space on U.S. television shows with gratuitous violence and adult content, such as ABC's *Desperate Housewives*.[37]

---

[36] Sylvia Weedman, "Bothered and Bewildered," *American Prospect*, May 1, 1998, p. 10.
[37] Jay Greene and Mike France, with David Kiley, "Culture Wars Hit Corporate America," *BusinessWeek*, May 23, 2005, p. 90.

More informed consumers—able to examine the messages and advertising presented to them with a discerning eye—and the proliferation of social media mean that marketing communications teams must ensure that product and brand promotions are sending the right messages.

Internal Communications

As companies focus on retaining a contented workforce given changing values and demographics, they have to think strategically about how they communicate with employees through *internal communications*. (See Chapter 7 for more on this subfunction, also referred to as *employee communication*.) Although strong internal communications have always generated a more engaged, productive, and loyal workforce, the bursting of the dot-com bubble, the collapse of several of America's most respected firms, the proliferation of outsourcing jobs to foreign countries in recent years, and cuts in workforces as a result of the financial crisis have further necessitated strong communication channels between management and employees to win back employee trust and loyalty.

Often, internal communications is a collaborative effort between the corporate communication and human resources departments, as it covers topics from employee benefit packages to the company's strategic objectives. More and more, companies are making sure their employees understand the new marketing initiatives they are communicating externally and are uniting the workforce behind common goals and corporate strategies. This type of communication requires the expertise of strong corporate communicators who are also well-connected to senior management and the corporate strategy process. The CementBloc, a small health care advertising agency, has all employees meet regularly with senior partners and the founders of the company to ensure that everyone understands the strategic goals of the company. The founders and the partners make sure that everyone who enters the doors to their company receives the same communication regardless of whether they are employees or clients.

Additionally, difficult economic times, layoffs, and uncertainty require open, honest communication from senior management to all employees. The sensitive nature of some of these messages further speaks for the involvement of seasoned communications professionals alongside their human resources counterparts and, most important, of the CEO or of senior executives who are the individuals communicating messages to internal and external audiences most frequently. When communicating bad news to The CementBloc employees, the two founders, Rico Viray and Sue Miller, ensure that the news comes directly from them. This reduces gossip and poor morale within the company.

Finally, as mentioned previously, due to the blurring of constituency lines, companies must recognize that employees now also may represent investors and members of community advocacy groups—making thoughtful communications even more critical.

Investor Relations

Investor relations (IR) has emerged as the fastest-growing subset of the corporate communication function and an area of intense interest at all companies. (See Chapter 8 for more on investor relations.) Traditionally, investor relations was handled by the finance function, often reporting to the company's chief financial officer (CFO), but the focus in recent years has moved away from "just the numbers" to the way the numbers are actually communicated to various constituencies.

IR professionals deal primarily with shareholders and securities analysts, who are often a direct source for the financial press, which this subfunction cultivates in conjunction with experts from the media relations area. IR professionals interact heavily with both individual and institutional investors. They also are highly involved with the financial statements and annual reports that every public firm must produce.

Given the quantitative messages that are the cornerstone of the IR subfunction, as well as the need for IR professionals to choose their words carefully to avoid any semblance of transferring inside information, this subfunction must be a coordinated effort between communications professionals and the chief financial officer, comptroller, or vice president for finance. The need for this coordination has only increased in recent years with more stringent regulatory demands in the age of Sarbanes-Oxley and Reg. FD. (Regulation Fair Disclosure was an SEC ruling implemented in October 2000. It mandated that all publicly traded companies must disclose material information to all investors at the same time.)[38]

Government Relations

The *government relations* function, also referred to as *public affairs*, is more important in some industries than others, but virtually every company can benefit by having ties to legislators on both a local and a national level. (See Chapter 9 for more on government relations.) Many companies have also established offices in Washington to keep a finger on the pulse of regulations and bills that might affect the company. Because of their critical importance in heavily regulated industries such as public utilities, government relations efforts in such companies are often both staffed internally and supplemented by outside government relations specialists in Washington.

Either firms can "go it alone" in their lobbying and government affairs efforts, or they can join industry associations to deal with important issues as a group. For example, the Edison Institute acts as a lobbying group for electric companies. Either way, staying connected to what is happening in Washington through a well-staffed and savvy government relations team is important to virtually all businesses given the far reach of government regulations within industries from pharmaceuticals to computer software. As companies expand internationally, building or outsourcing government relations efforts in key major foreign hubs—for example in Brussels to concentrate on European Union legislation—will become equally important.

Crisis Management

Although not really a separate function requiring a dedicated department, crisis communications should be coordinated by the corporate communication function,

---

[38]Weber Shandwick, "Corporate Survey 2005."

and communications professionals should be involved in crisis planning and crisis management. Ideally, a wider group of managers from throughout the organization—including the senior management spokesperson who will be facing the public—is included in all planning for such eventualities. (See Chapter 10 for more on crises.)

Although company lawyers typically need to be involved in crises, this need presents problems for both the organization and the corporate communication function, because lawyers often operate with a different agenda than that of their communications counterparts and do not always consider how actions might be perceived by specific constituencies or the public at large. A research study on the subject of communication versus legal strategies stated: "legal dominance is shortsighted and potentially costly . . . organizations [must] reconcile the often contradictory counsel of public relations and legal professionals and take a more collaborative approach to crisis communication."[39]

Working collaboratively with in-house counsel and, importantly, senior management, corporate communications professionals can make the difference between good and poor crisis management. We will see examples of both in Chapter 10.

Conclusion

The success of a company's communication strategy is largely contingent on how closely the communication strategy is linked to the strategy of the business as a whole.[40] In addition to thoughtful design and careful planning of firm strategy, a company must have a strong corporate communication function to support its mission and vision.

Although the investor relations function could be in the finance function of a company, the internal communications function within the human resources department, and the customer relations function within the marketing department, all of these activities require communication strategies that are connected to the central mission of the firm.

Corporate communications professionals must be willing to perform a wide variety of subfunctions within the function, and their roles will continue to broaden and diversify as globalization and information flows from a variety of sources demand that communications be strategic and purposeful. The greater number of global firms and the increasing demand for senior management to travel and speak in international venues place additional pressure on the communication function to communicate successfully with even more diverse, foreign audiences.[41]

Many corporations have made strides in building strong corporate communication functions that are closely aligned with overall strategy; however, there is

---

[39] Kathy R. Fitzpatrick and Mareen Shubaw Rubin, "Public Relations vs. Legal Studies in Organizational Crises Decisions," *Public Relations Review 21* (1995), p. 21.

[40] David Clutterbuck, "Linking Communication to Business Success: A Challenge for Communicators," *Communication World*, April 1, 2001, p. 30.

[41] Norm Leaper, "How Communicators Lead at the Best Global Companies," *Communication World 4* (April 5, 1999), p. 33.

still much work to be done. A 2011 poll released by Gallup revealed that public confidence in "big business" was the second-lowest rated of all institutions in the United States, tied with health maintenance organizations (HMOs) with only 19 percent votes of confidence.[42] In this light, managing reputation and building trust are more important than ever, and a strong corporate communication program is a means to achieve those goals.

[42] Gallup, "Confidence in Institutions," June 23, 2011, http://www.gallup.com/poll/148163/Americans-Confident-Military-Least-Congress.aspx.

Case 3-1

Sweet Leaf Tea*

In March 2010, Clayton Christopher announced to his employees and all of the fans of Sweet Leaf Tea that he was leaving the company and handing leadership to an outsider, Dan Costello. Clayton's final act was to send an email to his employees praising their performance and integrity, which helped grow Sweet Leaf Tea to the number one ready-to-drink tea at Whole Foods. In his email he stated, that after his departure he hoped:

"we will not sacrifice the magic of this brand and thus jeopardize the love affair we have created with our consumers in order to save a few dollars on our path up the mountain."[1]

TEA DRINKERS HEAVEN

The ready-to-drink tea market was consistently ranked as one of the fastest growing new product entries in the early 21st century. In 2007 total sales of tea equaled $6.85 billion, almost a third of which were ready-to-drink (RTD) or bottled tea.[2]

Tea is high in antioxidants, has health-boosting properties and is either all natural or organic.[3] Which might be the reason more people were moving from traditional carbonated drinks to healthy options like Sweet Leaf Tea. Between 2003 and 2008 the RTD category grew by 65%.[4]

In the Southern United States, tea has long been a popular beverage choice. It is usually served cold and sweetened. It can be found in any restaurant, mom and pop store, or at a road-side stop.

FROM MOONSHINE TO BIG TIME

Clayton Christopher and David Smith were always fans of sweet tea and loved their Grandma's recipe. But, they couldn't find any good bottled tea that tasted as good as what Grandma made. Their Grandma made sweet tea by brewing teabags for 3–4 minutes. Then she would pour the freshly brewed tea over ice and add natural sugarcane sugar to make it sweet. In 1997 they founded Sweet Leaf Tea (SLT) to fill this void, and hoped that others would enjoy their Grandma's recipe as much as they did.[5]

Initially their production and marketing more closely resembled that of a moonshine producer than that of the multi million-dollar brand it is today. They used giant crawfish pans to boil the water and pillowcases as giant tea bags to brew the tea. To distribute the bottled beverage they had an old run down van.[5]

Clayton and David moved Sweet Leaf Tea from Beaumont to Austin, TX after a couple of years and started using an automated system to make the tea. But, they always remained true to their Grandma's recipe.[5]

Sweet Leaf Tea's only competitive advantage was its superior flavor compared to other ready-to-drink teas like Arizona, Snapple, Lipton, and Nestea.[6] In 2008 SLT had $12 million in revenue,[6] and was available in 30% of the US Market.[7] In March 2009, Nestle Waters purchased a third of Sweet Leaf Tea for $16.5 Million.[7]

With the large investment from Nestle Waters also came a new president, Dan Costello, a former executive at Nestle Waters North America.

GROWING THEIR BRAND

Beverages tend to be low-involvement products, but also a unique business in that everyone has a favorite drink, which makes it very personal.[9] Sweet Leaf Tea had to find ways to develop a following of customers without a big advertising budget. They focused on sampling at music festivals, product placement on shows like MTV's Real World and CBS's Big Brother as well as making sure they had a clear brand personality.[8]

Clayton and David worked with Lyon Advertising to create a brand personality that would represent who they were, laid back and fun, but didn't forget Clayton's Grandma Mimi.[8] SLT could be described as a cool product for "good times and happy moments."[8]

Before SLT could build a large fan base they needed people to try their product. In an interview conducted by *Inc. Magazine* Clayton stated: "Sampling is the best form of marketing. You've got to get the product past people's lips."[6]

In 2002 they started partnering with music festivals like Austin City Limits (ACL), Lollapalooza, South-by-South West (SXSW), and Country Thunder. The folks attending the festivals were thirsty and willing to try new drinks. This also allowed them to target their core target audience—young (25–45), laid back, hip, and health conscious beverage drinkers.[8]

Initially, ninety percent of their advertising budget went to sampling (Inc.). Focusing their sampling program during music festivals allowed them to target masses of people at a time when they were thirsty such as the dead of summer in Texas, Chicago, and Arizona.[6]

SLT realized very early, however, that they needed to be on store shelves if they wanted their customers to find and buy the product. Selling their product at a few music festivals a year wouldn't be enough to keep them in business.[12]

Their first major store partnership was with Whole Foods (WF) in 2002 which put them on store shelves in the greater Texas market.[12] Adi Wilk, the former marketing manager at SLT, stated the Whole Foods partnership "lifted the brand." In 2006 Whole Foods expanded the SLT market to all of their stores in the US.[12]

Along with the Whole Foods partnership they also found distribution through partnerships with 7-11 stores, placement in Texas school vending machines, and at army bases.[12] This allowed fans that may have tried SLT at Lollapalooza to find it in Chicago, or people that traveled to Austin for ACL or SXSW to find it in New York at their local Whole Foods stores.[12]

SLT's communication strategy had traditionally focused on connecting to their customers.

Their partnership with WF, however, definitely helped them grow the brand by being on the shelves of a national chain store and also helped them connect to other distributors. But, more importantly was that SLT had built a strong connection with their customers, with most of them willing to search far and wide for a bottle for SLT. This truly helped them become a successful company.[8]

In 2008, with an infusion of money from Catterton Partners, a Connecticut-based private equity firm, and Nestle Waters North America Inc., SLT had expanded their marketing beyond sampling and store partnerships. In 2009 they had three major advertising updates— 1) Launched a new website 2) Dedicated twitter writer 3) Team of Facebook managers for the fan page.[8, 11] SLT empowered every employee to be a spokesperson for the brand—with even their receptionist taking a core part in their twitter and facebook posts.

COMMUNICATIONS ROLE

Initially the core communication for Sweet Leaf Tea was through direct-to-customer marketing at music festivals. The owners, Clayton and David, were at the music festivals handing out their product. This allowed customers to meet the people behind the beverage and link friendly faces to a good beverage. Also the association with music festivals may have helped the brand develop the "cool and fun" image they were pushing.[8, 12]

Once SLT had expanded beyond the central Texas market they needed to find a way to stay connected to their customers and keep that direct to customer communication active. The infusion of money from both Catterton Partners and Nestle Waters allowed them to expand their communication strategy beyond music festivals and their core website.

Their facebook fans matched the same target audience they first had at music festivals. Their blog, Facebook, and Twitter pages allowed them to continue that face-to-face communication Clayton and David started at the music festivals, just in the digital world.[8, 11] Their Facebook and Twitter communication reflected the brand and the culture of the company.[8]

At the time, these two sites were used for announcements about the brand or to communicate special events taking place during music festivals. As an example, during 2009's SXSW they also used Twitter to announce a free concert and used Facebook to get people to RSVP to the event. In one week they had 4,500 people registered on Facebook for the free concert.[8]

Another advertising advance SLT made in 2009 was using mobile technology to get consumers to search for their product and receive free samples. SLT was one of the first companies to give away real samples using Gowalla.[8] Gowalla was a location game that encouraged people to find a virtual item in a real-world place.[12] Gowalla allowed them to use virtual sampling of a product and connect it to real world sampling.[8]

CASE QUESTIONS

1. What are the strengths and weaknesses of SLT's corporate culture in terms of communications, as described in the case?

2. Considering the relationship the brand had with its audience, should Clayton have reached out to customers to announce he was leaving?

3. What challenges do you see for SLT's new management?

4. What role should corporate communication play at SLT to help the company advance its strategic goals?

5. As the Clayton's replacement would you change the way SLT communicated with its customers, or who was allowed to? Why or Why not?

REFERENCES

1. Egan, J., Austin Market Examiner website. Retrieved July 2011. http://www.examiner.com/market-in-austin/clayton-christopher-leaving-as-ceo-of-sweet-leaf-tea

2. Simrany, J.R. (2010). Tea Association of the United States, Industry report. Retrieved March 2010, http://www.teausa.com/general/002ga.cfm

3. (2007). Beverage World, State of the Industry. Retrieved March 2010, www.beverageworld.com/special_reports/State-Industry-2007.pdf

4. Costa, Neil C., Anastasious, Theodore J., Adedeji, Jide. (2010) ACS Symposium Series, Overview of Flavors in Noncarbonated Beverages. Retrieved April 2010, http://pubs.acs.org/doi/full/10.1021/bk-2010-1036.ch001

5. (June 2009). Sweet Leaf Tea website. Retrieved March 2010. SweetLeafTea.com

6. Buchanan, Leigh. (2009, July–August). What It Takes. *Inc.*, *31* (3), p70, Retrieved March 2010.

7. 2010, January. 2008 Deals of the Year. *Nutrition Business Journal, 15* (1), Retrieved March 2010 from Columbia Library Database.

8. In person interview with Charla Adams—Communications Manager Sweet Leaf Tea, Austin Texas.

9. Kaplan, Andrew. (2009, May). Break out Brands. Beverage World, 128 (5), p22. Retrieved March 2010 http://www.nxtbook.com/nxtbooks/idealmedia/bw0509/#/24

10. Sweet Leaf Tea Facebook Fan page.

11. Phone Interview with Adi Wilk – Marketing and Branding Consultant. Former Director of Marketing Sweet Leaf Tea, Austin Texas.

12. http://gowalla.com/

13. Gaar, B. (2010, March). Sweet Leaf Founder says leadership transition came with growth. Austin American Statesman. Retrieved March 23, 2010 http://m.statesman.com/statesman/db_/contentdetail.htm;jsessionid=5062D255F421699A1D8E62C537A89F36?contentguid=eDDpBYIc&full=true#display

Identity, Image, Reputation, and Corporate Advertising

Chapter 3 covered the various components of the corporate communication function. This chapter will examine the first and most critical part of that function: managing a corporation's identity and image. The chapter will then discuss how a close alignment between a company's identity and image generates a strong reputation and discuss what the benefits are for the organization. Finally, it will address paid corporate advertising, one of the easiest and fastest ways for organizations to communicate their identities.

Looking at an example of image at the personal level might be a good place to start. People choose certain kinds of clothing, drive particular cars, or style their hair a certain way to express their individuality. The cities and towns in which we live, the music we prefer, and the restaurants we frequent all add up to an impression, or identity, that others can easily distinguish.

The same is true for corporations. Walk into a firm's office, and it takes just a few moments to capture those all-important first impressions and learn a great deal about the company. The effort is relatively easy to understand at the personal level but significantly more difficult at the organizational level. One reason for this complexity is that many more potential opportunities to interact with a company's identity exist. Take, for instance, the following example from the hotel industry:

> An executive and her husband decide to treat themselves to one of life's great pleasures: a weekend in a suite at the Oriental Hotel in Bangkok. During their stay, their daily copies of the Asian *Wall Street Journal* and *Herald Tribune* are ironed for them to eliminate creases; the hotel staff, omnipresent, run down the hallway to open their door lest they should actually have to use their room keys; laundry arrives beautifully gift-wrapped with an orchid attached to each package; every night, the pillows are adorned with a poem on the theme of sleep; and, outside the lobby, Mercedes limos are lined up, ready to take the couple anywhere at any time of the day or night.
>
> A few weeks later, they return to the United States, and she is giving a presentation to a group of fellow executives at a midwestern resort. A *USA Today* appears on the outside doorknob squeezed into a plastic bag; the staff, invisible if not for their cleaning carts left unattended in the hallway, are unable to bring room service in

FIGURE 4.1 Reputation Framework.

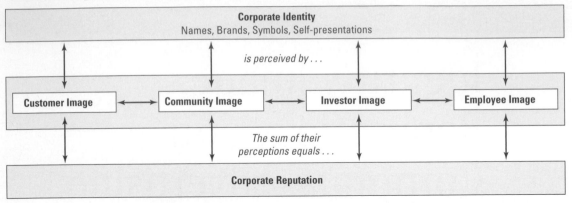

under 45 minutes; her pillow is "adorned" with a room-service menu for the following morning and a piece of hard candy; the vehicle waiting to whisk guests to various destinations is a Chrysler minivan; and for flowers, the resort provides silk varietals in a glass-enclosed case that plays the song "Feelings" when the top is lifted.

Both hotels have strong identities, and the choices each has made about its business are at the heart of what identity and image are all about. These choices contribute to and shape the identities and images of these hotels and, more generally, convey the identity and image of any institution. As we address the various components of identity and image in this chapter, it is important to keep in mind the relationship among identity, image, and reputation. Figure 4.1 presents a visual representation of these relationships.

Just what are identity, image, and reputation? How do organizations distinguish themselves in the minds of customers, shareholders, employees, communities, and other relevant constituencies? How do they use their corporate communications function and corporate advertising to enhance their image? Above all, how does an organization manage something so seemingly ephemeral?

What Are Identity and Image?

A company's *identity* is the actual manifestation of the company's reality as conveyed through the organization's name, logo, motto, brands, products, services, buildings, stationery, uniforms, and all other tangible pieces of evidence *created by the organization* and communicated to all of its various constituencies. Constituencies then form perceptions based on the messages that companies send in tangible form. If these images accurately reflect an organization's reality, the identity program is a success. If the perceptions differ dramatically from the reality (as often happens when companies do not take the time to analyze whether a match actually exists), then either the strategy is ineffective or the corporation's self-understanding needs modification. In a worst-case scenario, when the perceptions differ dramatically from the reality, a company's reputation can be irreparably damaged and the company may even have to cease operations.

As we discussed in Chapter 3, *image* is a reflection of an organization's identity. Put another way, it is the organization as seen *from the viewpoint of its constituencies*. Depending on which constituency is involved, an organization can have many different images. Thus, to understand identity and image is to know what the organization is really about and where it is headed. This understanding is often hard for anyone but the company's top management team, which constantly focuses on the big picture, to grasp. What, for example, is the reality of an organization as large as Walmart, as diversified as Siemens, or as monolithic as Tata?

Certainly the products and services, the people, the buildings, and the names and symbols are a part of this reality. Although there are inevitably differences in how the elements are perceived by different constituencies, it is this cluster of facts, this collection of tangible and intangible things, that provides the organization with a starting point for creating and then communicating about an identity.

Organizations can get a better sense of their image (as conveyed through identity) by conducting research with constituents. This research should be both qualitative and quantitative in nature and should try to determine how consistent the identity is across constituencies. The Susan G. Komen Breast Cancer Foundation, a 30-year-old nonprofit organization dedicated to supporting women affected by breast cancer, realized the importance of good research as its 25th anniversary approached. An in-depth audit of the nonprofit's identity revealed that the organization represented "too many voices," which was confusing to activists. To rectify this identity crisis, Komen executives set out to relaunch the brand via a clarification of its mission and goals. They focused on a precise identity to be embodied by the nonprofit, and then renamed the organization accordingly. Based on the original inspiration for Komen's founding—one woman's promise to her sister, Susan G. Komen, who died of breast cancer at the age of 36—the nonprofit changed its name to Susan G. Komen for the Cure. The latter part of the name resonated because of Komen's most recognizable brand asset, the Race for the Cure fundraising series.[1] Using Komen's 25th anniversary as a springboard, executives launched the newly honed brand identity to the public, and to great result.

Differentiating Organizations through Identity and Image

Today, with increased globalization and product commoditization, an organization's identity and image might be the only way that constituencies can distinguish that company from the next. Is there really any difference between buying a tank of ExxonMobil gasoline and a tank of Shell gasoline, given that both probably came from Saudi Aramco in the first place? Given that the same distributor often sells the same gasoline to dealers in the United States, the answer would seem to be no. Yet consumers make distinctions about such homogeneous products all the time based on what the *company's* image is all about rather than the product itself.

---

[1] "NonProfit PR Awards," PR News, December 3, 2007.

If an ExxonMobil and a Shell station sit two blocks apart and your gas needle is approaching "E", where will you go? You might have strong negative feelings about ExxonMobil, for example, because of the devastating oil spill in Alaska from the *Valdez* tanker. Or, conversely, you may be delighted by the consistent returns to shareholders that this behemoth provides. Perhaps you have a cousin who works at ExxonMobil and loves her job. Maybe you spent one summer collecting plastic cups featuring the ExxonMobil tiger mascot from ExxonMobil mini-marts.

You see the Shell logo and you might recall some of the award-winning environmental responsibility advertising that the company sponsored as part of its "Energy for People Now and in the Future," the cleaner energy/liquefied natural gas (LNG) campaign, "People Like Natural Gas." One of the recent print ads addressed the benefits of LNG by relaying the story of a small Japanese noodle manufacturer's business flourishing when using the energy alternative. Or you may see Shell and think back to the company's attempt in the 1990s to dump the Brent Spar oil platform into the Atlantic Ocean. Rationally, you know that both companies provide gasoline that meets industry standards. Both tanks of gas will keep the car going, both tanks of gas have approximately the same octane rating and per-gallon price, and both service stations should offer comparable service quality.

So, when you finally decide where to buy gasoline, aside from the location of the gas station, these factors of identity, image, and reputation represent the differences between the two companies and are what drive purchasing decisions.

As products become much the same all over the world, consumers are increasingly making distinctions based on their brand impressions, rather than just the products themselves, thereby making image and identity even more powerful differentiators. We will now turn to a more in-depth discussion of identity and image. We then move on to a discussion of how these come together to create an organization's reputation, and what this means for an organization.

Shaping Identity

Because identity building is the only part of reputation management an organization can control completely, we will first discuss some of the things that contribute positively to corporate identity: an inspirational corporate vision, careful corporate branding (with a focus on names and logos), and, very important, *consistent* and integrated self-presentation.

A Vision That Inspires

Most central to corporate identity is the vision that encompasses the company's core values, philosophies, standards, and goals. Corporate vision is a common thread that all employees, and ideally all other constituencies as well, can relate to and repeat to others. Thinking about this vision in terms of a narrative or story of sorts can help ensure the overall coherence and continuity of a company's vision and the collective messages it sends constituencies.[2]

[2] Cees B. M. van Riel, "Corporate Communication Orchestrated by a Sustainable Corporate Story," in *The Expressive Organization*, ed. Majken Schultz, Mary Jo Hatch, and Mogens Holten Larsen (Oxford: Oxford University Press, 2000), p. 163.

Cees B. M. van Riel, a professor at Erasmus University in the Netherlands, links the importance of narratives to successful corporate reputations. He explains that "communication will be more effective if organizations rely on a . . . sustainable corporate story as a source of inspiration for all internal and external communication programs. Stories are hard to imitate, and they promote consistency in all corporate messages."[3] External constituencies rely on articles in publications, television ads, discussions about the company with other people (e.g., family, friends, colleagues), and direct interaction with company employees for information about a company and the story it is telling.

The most appealing of stories, literary and corporate, often involve an underdog—an unsung hero that audiences can admire and rally behind. Going against the grain can instill a sense of noble purpose in the actions of a hero— or an entrepreneur—who hopes to do things differently. Consider Steve Jobs, the founder of Apple Computer. His unwillingness to succumb to IBM and Microsoft had "hero appeal" that did wonders for Apple's brand initially. The narrative of the underdog is particularly popular among startups that wish to leverage the support of their employees' friends and contacts to build the buzz and goodwill that they need to compete with more established companies.

Corporate Brands

Just as our society demands top-10 lists and rejects the full story in favor of sound bites, it also prizes *brands* as identification tags that can allow us to gauge everything around us quickly and effortlessly. Given this phenomenon, a company's value can be significantly influenced by the success of its corporate branding strategy. Coca-Cola, for example, has a valuation that far exceeds its total tangible assets because of its strong brand name.

Branding and strategic brand management are critical components of identity management programs. Although it is beyond the scope of this book to fully explore corporate branding, this chapter will focus on a subset of corporate branding— names and logos—to help illustrate the conscious actions that organizations can take to shape their identity and further differentiate themselves in the marketplace.

Companies often institute name changes to signal identity changes, to make their identities better reflect their realities, or to account for organizational changes due to an acquisition or merger. Andersen Consulting's name change to Accenture is an example of the first reasoning. In late 2000, Andersen Consulting, the global technology and consulting company that had separated from its founding parent Arthur Andersen earlier that year, announced a name change that would take effect January 1, 2001. The new company would be called *Accenture*, a play on the words "accent" and "future" that, according to James E. Murphy, the company's global managing director for marketing and communications, was meant to be "a youthful and dynamic expression of the firm's new positioning as a bridge builder between the traditional and new economies."[4] The name also clearly distinguished

[3] Ibid.

[4] Howard Wolinsky, "Consulting Firm to Change Name; Andersen Consulting to Be Accenture," *Chicago Sun-Times*, October 27, 2000, p. 64.

the company's identity from that of its former and now defunct parent, Arthur Andersen, which had its own, competing consulting division at the time. Today, Accenture is a strong player in the global consulting industry.

Philip Morris provides an example of the second name-change scenario. Recognizing that it was known as a tobacco company despite its reality of being a diversified company with a number of lines of business (the company was also America's largest food company at the time through its Kraft division), Philip Morris proposed a name change for itself in late 2001. The company chose the name *Altria*, derived from the Latin word *altus*, meaning "high."[5]

Reactions to the name-change proposal were not positive. Some saw the move as an attempt by the company to distance itself from tobacco litigation. This possible motive aside, although it was understandable that the company wanted its identity to reflect more accurately its reality as a diversified company, the proposed name change would not achieve that goal—for a name change alone will never single-handedly fix a perception problem. Such a change must be part of a broader identity program that is clearly explained to the company's constituencies. To many people who only read of the name change in the press, it was not clear why a Latin word meaning "high" would better reflect what Philip Morris was all about.

PricewaterhouseCoopers is one of many examples of an organization that underwent a name change for the last reason: a merger or acquisition. Formerly Price Waterhouse and Coopers & Lybrand, the two entities merged in 1998 and rebranded, thus assuming the current moniker. As in any name-change scenario, communication with key constituents is essential; as experienced by Philip Morris executives, motives can be misconstrued, prompting backlash and even litigation. As stated by Mike Davies, director of global communications for PricewaterhouseCoopers, "Communications is paramount when you are trying to bring together two organizations, or when you are trying to communicate worldwide. Communications has to be very high up on the agenda."[6]

Another example that illustrates the importance of properly communicating about name changes (and the risks inherent in not doing so) is AT&T. Long associated with landline telephones in a world increasingly connected by web-based networking, mobile phones, and high-speed Internet connections, the brand sought to readjust its position in the marketplace with an $86 billion merger with BellSouth, a former parent company of Cingular. The union, which was finalized in late 2006, presented an opportunity for the brand to transform itself. All things Cingular, including the moniker, were absorbed by A&T, but the reshuffling of the corporate identity posed major problems that landed front and center in national news coverage.

As covered by *BusinessWeek* in September 2007, the attempt to seamlessly blend the identities of the merged companies led to haphazard communications strategies. For starters, AT&T adopted the Cingular brand's signature orange color in July 2007, when it began appearing in monthly billing statements. Marketing materials and the website followed but over the course of months. Then, on September 11,

---

[5] David Lazarus, "Name Change Is an Exercise in Futility; So What's in a Name? Lots of Spin," *San Francisco Chronicle*, December 5, 2001, p. B1.

[6] "When Multiple Brands Combine Their Identities, PR Mediates," PR News, April 16, 2007.

A young woman in Hanoi, Vietnam, sports a counterfeit version of the Nike swoosh on her hat.

2007, AT&T spokespeople announced plans to further incorporate the orange color into branding. At the same time, they announced plans to launch new TV and Internet marketing campaigns with creative input from the likes of film director Wes Anderson.

While consumers and investors alike struggled to follow the brand identity's rapid evolution, AT&T had its hand in another pot: a partnership with Apple to launch the wildly anticipated iPhone. With this partnership came more TV spots and advertisements, including one that identified three locations where a professional might find himself or herself during the course of business and life: China, London, and Moscow or New York, San Francisco, and South Dakota. The ads merged these locations, concluding, "AT&T works in more places, like Chilondoscow" or "New Sanfrakota."

The motley branding, from a slow infusion of Cingular's orange color to the ads with hard-to-pronounce amalgamations of cities, did little to establish an authoritative corporate identity. According to one branding expert, Bob Giampietro of Giampietro+Smith, "It suggests some lack of brand leadership. What you could end up with is a 'bizzaro' version of what they think their customers' vision of the brand is."[7]

As these examples illustrate, though organizations can differentiate themselves based on identity through names and logos, they also can risk losing whatever identity they have built up very quickly through changes in the use of names and logos that are not communicated properly.

Logos are another important component of corporate identity—perhaps even more important than names because of their visual nature (which can allow them to communicate even more about a company than its name) and their increasing prevalence across many types of media. When upscale discount retailer Target placed an ad in *The New York Times* in 1999 depicting only its bull's-eye logo and inviting readers to call a toll-free number if they knew what the symbol meant, its phone lines were tied up immediately. The company was soon forced to shut down the toll-free number due to the staggering response.[8]

One of the most recognizable logos in the world today (perhaps second only to Coca-Cola's) is Nike's "swoosh," which was designed for Nike founder Phil Knight by Portland State graduate Carolyn Davidson in 1972 for $35. Some experts believe the swoosh is better known today than McDonald's golden arches. Reputationally challenged golfing sensation Tiger Woods wears the swoosh on his hat and clothes. Lance Armstrong cycled through seven consecutive Tour de France triumphs with the swoosh on his yellow jersey. Teams in hockey's Canada Cup and national soccer teams also have worn the swoosh in competition. With Nike as their sponsor, Team USA athletes will wear the swoosh at both Summer and Winter Olympics through at least 2016.[9]

[7] "AT&T Rebrands. Again." *BusinessWeek*, September 11, 2007, http://www.businessweek.com.

[8] Shelly Branch, "How Target Got Hot," *Fortune*, May 24, 1999, pp. 169–74.

[9] http://www.teamusa.org/news/article/44697.

Logos can simply be symbols, like the Nike swoosh, or they can be symbols that represent names, like the Target "bull's eye" or Arm & Hammer's arm and hammer. Logos can be stylized depictions of names or parts of names (like the "golden arches" that form the "M" in "McDonald's"), or stylized names with added mottos or symbols. Accenture's logo, for example, is the company name with a "greater than" symbol above the "t" that is meant to connote the firm's goal of pointing the way forward and exceeding clients' expectations.[10]

In fact, sometimes stylized names can be the most resonant and have the greatest endurance over time. For example, the Helvetica font, first created in 1957, is the typeface of choice for countless corporations, including 3M, Microsoft, American Airlines, and Staples. The font's simple lines and proportionate letters are credited with making these companies' names so iconic.

"We don't have a long name—just a numeral and an alphabetical character. So typography becomes very important to our logo," Karyn Roszak, a manager in the corporate identity and design department of 3M, has said. "Helvetica is straightforward and no-nonsense. Not to mention bold and strong visually."[11]

Firms that specialize in identity management and design should be involved with the process of logo creation for a company. Later in this chapter, we will take a closer look at the processes behind creating new names and logos as part of an overall identity program.

Putting It All Together: Consistency Is Key

An organization's vision should manifest itself consistently across all its identity elements, from logos and mottos to employee behavior. Companies should also be aware of how their constituencies describe the company's vision and activities; opportunities to improve the brand identity can be inspired by popular culture. Overnight package-delivery pioneer FedEx is a good example. In the 1990s, the company had noticed that customers routinely referred to it as "FedEx," rather than using its official name, the multisyllable "Federal Express." Additionally, office workers were beginning to use "FedEx" as a verb; few people said they would "UPS a package" or "Airborne Express a letter." Instead, it was "Let's FedEx this." The company thus decided to use the abbreviation already used by thousands of customers (and competitors' customers) as its official name. On June 23, 1994, Federal Express changed its name to *FedEx* and paired it with a distinctive new motto: "The World on Time." As a launch advertisement read in 1994: "We're changing our look to FedEx. Isn't that what you call us anyway?" (See Figure 4.2 on page 79.)

By officially making the company name synonymous with punctual overnight delivery ("The World on Time"), FedEx demonstrated that it was in touch with what its customers wanted from the company and made an open commitment to reinforce the same message throughout its organization. With the new motto and logo, FedEx's clean and pressed uniforms, immaculate transport vehicles and

[10] Sandra Guy, "Consultant to Launch Big Effort to Advertise Its New Identity," *Chicago Sun-Times*, November 16, 2000, p. 66.
[11] "For Logo Power, Try Helvetica," *BusinessWeek*, May 14, 2007, http://www.businessweek.com.

FIGURE 4.2

service centers, and an employee mantra of "service without excuse" all echoed a consistent commitment.[12]

Michael Glenn—executive vice president, market development and corporate communications for FedEx—explained that by embracing its one-word association, "FedEx and its name have changed their environment from morally neutral to morally charged."[13] By putting its promise to deliver "The World on Time" on every package, truck, and plane, FedEx ensured that every pick up, delivery, and customer interaction would reinforce that promise. The new name and logo showed that the company was in touch with its customers, and FedEx's advertising of this new identity reinforced the message that its customers mattered.

And being in touch with customers is of paramount importance. Many companies have followed in FedEx's footsteps and adapted their names according to how their key constituents already viewed them. Binney & Smith Easton is a prime example. Although this staid corporate name likely won't ring a bell among too many consumers, its widely known product certainly will: Crayola Crayons. That's why, in January 2007, Binney & Smith CEO Mark Schwab unveiled the corporation's new sign, which simply reads "Crayola." Schwab didn't mince words when explaining the decision to rename the century-old company: "The reason is simple. When you think about how Binney & Smith is known, it's for making Crayola crayons."[14]

Likewise was the reasoning behind Citigroup's decision to truncate its name to Citi or for Dell Computer to drop the latter word in favor of just Dell. With modern-day society characterized by short attention spans, easy-to-consume media, and text-message-friendly acronyms, short is proving to be sweet when it comes to corporate branding.

Identity Management in Action

The dual nature of identity and image—embodied in reality such as physical objects yet inextricably tied to constituency perceptions—creates a special dilemma for decision makers. In a world where attention is focused on quantifiable results, the emphasis here is on qualitative issues. Devising a program that addresses these elusive but significant concerns requires balancing thoughtful analysis with action. The end goal is always to have a positive impact on the company's reputation. Here is a method that has been successfully used by many organizations to manage the identity process.

Step 1: Conduct an Identity Audit

To begin, an organization needs to assess the current reality. How does the general public currently view the organization? What do its various symbols represent to

---

[12] "Chapter 11: The Image Is the Reality (If You Work at It)," *The World on Time*, July 1, 1996, p. 115.
[13] Ibid.
[14] "Crayola Brightens a Brand," *BusinessWeek*, January 26, 2007, http://www.businessweek.com.

different constituencies? Does its identity accurately reflect what is currently happening, or is it simply a leftover from the past?

To avoid superficial and politically influenced input, often companies bring in external consultants to conduct in-depth identity audit interviews with top managers and those working in areas most affected by any planned changes. These consultants should review company literature, advertising, stationery, products and services, websites and facilities. They also research perceptions among the most important constituencies, including employees, analysts, and customers. The idea is to be thorough, to uncover relationships and inconsistencies, and then to use the audit as a basis for potential identity changes. The goal is to get a deep understanding of the organization, which means getting as close to the reality as possible from the perspective of managers within and matching it with perceptions from key constituents.

In this process, executives should look for red flags. We saw that FedEx took action after learning that its customer constituency was no longer using its official name. Typical problems include symbols or names that conjure up images of earlier days at the company or just generally incorrect impressions. Once decision makers have the facts, they can move to create a new identity or institute a communication program to share the correct and most up-to-date profile of the company.

Although the identity audit may seem a fairly straightforward and simple process, it usually is not. Often the symbols that exist and the impressions that result are not how the organization sees itself in the present at all. Companies trying to change their image and reputation are particularly difficult to audit because the vision that top executives have of what the company *will be* is so different from what the reality currently *is*. Often executives disregard research that tells them how constituents' perceptions about the organization differ from their own. Such cognitive dissonance is the first challenge in managing identity for executives. The reality of the organization must be far enough along in the change process so that the new reality the company is trying to adopt will actually make sense, at some time at least, to those who will encounter this company in the years ahead.

Step 2: Set Identity Objectives

Having clear goals is essential to the identity process. These goals should be set by senior management and must explain how each constituency should react to specific identity proposals. For instance: "As a result of this change process, analysts will recognize our organization as more than just a one-product company" or "Putting a new logo on the outside of our stores will make customers more aware of dramatic transformations that are going on inside." It is extremely important, however, that emphasis be placed on *constituency response* rather than company action.

That's where problems often start. Most managers—particularly senior managers—are internally focused and thus have great difficulty in getting the kind of perspective necessary to see things from the viewpoint of constituents. Consultants can certainly help, but the organization as a whole must be motivated to change and willing to accept the truth about itself, even if it hurts.

In addition, change for the sake of change, or change to meet some kind of standardization worldwide, is not the kind of objective that is likely to meet with success. Usually, such arbitrary changes are the result of a CEO's desire to leave his or her mark on the organization rather than a necessary step in the evolution of the company's image.

A positive example of clear objectives leading to necessary change is Kentucky Fried Chicken's plan to change its image and menu in the mid-1990s as a result of changes in American dietary habits. The strong corporate identity of this company worldwide (it has one of its biggest restaurants on Tiananmen Square in Beijing and can be found in remote corners of Japan) conjures up images of Colonel Sanders's white beard, buckets of fried chicken, salty biscuits, and gravy.

To an earlier generation, these were all positive images closely connected with home and hearth. Today, however, health-conscious Americans are more likely to think of the intense cholesterol, the explosion of sodium, and gobs of fat in every bucket of the Colonel's chicken. Thus, the company tried to reposition itself with health-minded Americans by offering broiled chicken and chicken salad sandwiches. The company's goal was to change the old image and adopt a more health-conscious positioning.

To do so, executives decided to change the name of the 5,000 restaurants gradually to just "KFC." The obvious point was to eliminate the word "fried." While most identity experts would agree that it is very difficult to create an identity for a restaurant out of initials alone, this one has the well-known Colonel to go along with the change. The communication objective for this particular change made a great deal of sense and put KFC in a better position to sell to a more nutrition-minded set of customers. It also had the side benefit of making the company name more globally translatable.

Step 3: Develop Designs and Names

Once the identity audit is complete and clear objectives have been established, the next phase in the identity process is the actual design. If a name change is necessary, consultants must search for alternatives. This step simply cannot happen without the help of consultants because so many names are already in use that companies need to avoid any possibility of trademark and name infringement. Even so, options for change can still number in the hundreds. Usually, certain ones stand out as more appropriate. The criteria for selection depend on several variables.

For example, if the company is undergoing a global expansion, the addition of the word "international" might be the best alternative. If a firm has a lot of equity built into one product, as Binney & Smith's Easton did with Crayola, changing the name of the corporation to that of the product might be the answer, as happened when Consolidated Foods changed its name to "Sara Lee." We have already seen that Federal Express changed its name to reflect what its constituencies were already calling it and that Andersen Consulting chose a new name, Accenture, that would give it a distinct identity from its former parent, Arthur Andersen.

Companies also should ensure that logos continue to reflect accurately the company's reality and should consider modifications if they do not. Dunkin' Donuts is a good example. The popular chain, founded in Quincy, Massachusetts, and now present in 30 countries, is known as much for the 4 million cups of coffee it sells daily as it is for its doughnuts.[15] As the company expanded into new markets where its brand was unfamiliar, it recognized the importance of emphasizing the "coffee connection," particularly given the proliferation of bagel chains and upscale coffee chains in many of the markets it was entering. Accordingly, it added the image of a steaming cup of coffee to its existing logo, which is simply the Dunkin' Donuts name in balloon-like pink and orange letters.

The process of designing a new look or logo is an artistic one, but despite contracting professionals to develop designs, many company executives get very involved in the process, often relying on their own instincts rather than the work of someone who spent his or her entire career thinking about design solutions. One CEO of a multibillion-dollar company designed what he thought would be the perfect logo for his company on a napkin. After several weeks of design exploration by a reputable design firm, he kept coming back to that same napkin design. Until the designer finally presented an exploration that resembled the napkin design, each of the suggestions was rejected. When the CEO saw his own idea come back at him, he was happy. Everyone else agreed that it was not the best design, but it was adopted and is in use today.

Obviously, there has to be a balance between the professional opinion of a designer and a manager's own instincts. Both need to be a part of the final decision, whether a name change or just a new logo is involved. In some cases, designers and identity consultants are perfectionists or idealistic, presenting ideas that are unrealistic or too avant-garde for typically conservative large corporations. In the end, strong leadership must be exerted to effect the change, no matter what it is, for it to succeed.

Step 4: Develop Prototypes

Once the final design is selected and approved by everyone involved, the next step is to develop models and prototypes using the new symbols or name. For products, prototype packaging shows how the brand image may be used in advertising. If a retail operation is involved, a model of the store might be built. In other situations, the identity is applied to everything, including ties, T-shirts, business cards, and stationery, to see how it works in practice.

During this process, it is common for managers to get cold feet. As the reality of the change sinks in, criticism mounts from those employees who have not been involved in the process and from others because they do not have a good sense of the evolution and meaning of the design. At times, negative reactions from constituents can be so strong that proposals have to be abandoned and work started all over again.

To prevent this failure, a diversity of people and viewpoints should be involved in the entire identity process. The one caveat is to avoid accommodating different ideas by diluting concepts. A company should not accept an identity that is simply the lowest common denominator. Two ways to deal with the task are to let a strong

[15] "News," dunkindonuts.com, January 2011.

leader champion the new design or to set up a strong committee to work on the program. In either approach, everyone has to be informed about the project and involved in it from the beginning: The more people involved in the process from its inception, the less work necessary to sell the idea after much hard work has already taken place.

Step 5: Launch and Communicate

Given the time involved and the number of people included in the process, news about future changes can easily be leaked to the public. Sometimes such publicity is a positive event, as it can create excitement and a sense of anticipation. Still, such chance occurrences are no substitute for a formal introduction of the company's new identity. To build drama into the announcement, public relations staff should be creative in inviting the media without giving away the purpose. One company sent six-foot pencils and a huge calendar with the date of the press conference marked on it to announce their change.

At the press conference itself, the design should be clearly displayed in a variety of contexts, and senior executives must carefully explain the strategy behind the program. As additional communication tools, corporations might want to use advertising (see Figure 4.1), webcasts, or video news releases and satellite links (see Chapter 6). Especially because of the increasingly significant role web platforms play in communications, the latter tools, as well as blogs and social media networks, should be leveraged to reach target audiences in the places that they personally consume media. In October 2010, Gap learned the hard way that an identity change needs to be thoughtfully presented and defended. In the face of weakening sales and increased competition from fast fashion companies such as H&M and Uniqlo, Gap decided to modernize its logo from its iconic blue box to a design featuring "Gap" written in Helvetica with a small blue cube to the upper right of the text. Brandchannel, a branding division of Interbrand, ridiculed the change on its blog: "Yes, after dominating the late 1990s and early 2000s, Gap has dropped its iconic logo in favor of something that looks like it cost $17 from an old Microsoft Word clipart gallery." Gap was criticized for both the logo design and for its decision to "launch" the new logo by quietly updating the logo on its website in lieu of announcing the change to the media. After significant media and customer backlash, Gap quickly returned to its original logo.[16] Whatever the launch strategy, remember that presenting an identity, particularly for the first time, is a complex process, as it is easy for constituencies to interpret the change as merely cosmetic rather than strategic.

Step 6: Implement the Program

The final stage is implementation, which can take years in large companies and a minimum of several months for small firms. Resistance is inevitable, but what is frequently shocking is the extent of ownership constituents have in the old identity.

---

[16] Abe Sauer, "Gap Rebrands Itself into Oblivion," brandchannel blog, October 6, 2010, http://www.brandchannel.com/home/post/2010/10/06/Gap-Rebrands-Itself-Into-Oblivion.aspx.

Usually, the best approach to ensure consistency across all uses for a new identity program is to develop identity standards. A standards manual (usually based online for the most part today) shows staff and managers how to use the new identity consistently and correctly. Beyond this, someone in the organization needs to monitor the program and make judgments about when flexibility is allowed and when it is not. Over time, changes will need to be made in some standards—for instance, when a modern typeface chosen by a designer is not available for use everywhere.

Implementing an identity program is a communication process involving lots of interpersonal savvy and a coordinated approach to dealing with many constituencies. In addition to communicating its new identity program *within* the organization, Accenture, for example, had to train more than 100 other firms, including ad agencies, printers, and Web designers, on how to use its new logo.[17]

Image: In the Eye of the Beholder

We just explored some of the means by which a company can manage its identity. An organization's *image* is a function of how constituencies perceive the organization based upon all the messages it sends out through names and logos and through self-presentations, including expressions of its corporate vision.

Constituencies often have certain perceptions about an organization *before they even begin to interact with it*. The perceptions are based on the industry, what they have read about the organization previously, what experiences their friends have had with the organization, and what visual symbols they recognize. Even if you have never eaten a hamburger at McDonald's, you have certain perceptions about the company and its products. Preconceived notions are especially prevalent today in the age of Internet review sites such as TripAdvisor, Yelp, and CitySearch.

After interacting with an organization or exploring peer views through social media outlets, the constituencies may have a different image of it than they did before. If this happens, the goal is to have that image better, not worse. One bad experience with a company representative, or a bad review from a reputable blogger can destroy a relationship for a lifetime with a customer, as was the case when one individual, displeased with the treatment he received on a customer service call with an AOL representative, posted a recording of the conversation online. The viral nature of digital platforms makes organizations vulnerable to the impressions of consumers, many of whom are quick to judge—and publicly, virtually criticize—based on one negative encounter with a brand. That's why organizations today are so concerned with the quality of each and every interaction. The credibility that a company acquires through the repeated application of consistently excellent behavior will determine its image in the minds of constituents in a much more profound way than a one-shot corporate advertising campaign.

Organizations should seek to understand their image not only with customers, but also with other key constituencies such as investors, employees, and the

[17] Guy, "Consultant to Launch Big Effort," p. 66.

community (keeping in mind, as discussed in Chapter 2, that some of these may overlap). Often, a company's image with a given constituency is driven not only by its own unique corporate identity but also by the image of the industry or group it belongs to. Internet companies rode this phenomenon in both directions from the late 1990s into the new millennium. Before the bursting of the dot-com bubble in 2000, virtually all e-based companies rose together on a tide of investor optimism with a collectively vibrant, cutting-edge image. Similarly, when that tide turned and investors wanted tangible products, real business plans, and seasoned management again, these companies all suffered, and so did their collective image.

Turning to the employee constituency, a company's image with its employees is particularly important because of the vital role employees play with the company's other constituencies. Starbucks Coffee has built one of the strongest brands and reputations in America by creating an equally powerful story and unified culture that begins inside and works its way out. Chairman Howard Schultz explains the philosophy: "We built the Starbucks brand first with our people, not with consumers, the opposite approach from that of the crackers-and-cereal companies. . . . [b]ecause we believed this was the best way to meet and extend the expectations of employees who were zealous about good coffee."[18] The enthusiasm of Starbucks' *baristas* is meant to be contagious, personally connecting them with their customers. Every barista is meant to play such a key role in generating customer loyalty that Starbucks refers to each one as a "partner," the official name for a Starbucks employee.[19]

Disgruntled employees can have a highly damaging effect on customer loyalty and a company's image, as has been the case time and time again for Walmart. When customers swipe their credit cards at many Walmart registers, two questions pop up: "Did the cashier greet you?" and "Was the store clean?" This procedure was implemented by former CEO H. Lee Scott to improve lackluster customer service. However, the plan backfired when, after being asked by a customer who read the survey why she wasn't greeted, the cashier replied, "If Walmart doesn't care for me, why should I care?"[20] With that kind of critique from an employee, how can a customer feel good about shopping at a store?

As former CEO of Procter & Gamble Ed Artz observed, "Consumers now want to know about the company, not just the products."[21] The day-to-day behavior of employees, from Starbucks' baristas to its executives, can rank just as high as product or service quality as the source of a strong corporate image that is aligned with the company's identity.

---

[18] "No Ordinary Joe," *Reputation Management* 4, no. 3 (May–June 1998), p. 54.

[19] Ibid.

[20] "Wal-Mart: A Snap Inspection," *BusinessWeek*, October 2, 2007, http://www.businessweek.com (accessed December 19, 2007).

[21] Kevin L. Keller, "Building and Managing Corporate Brand Equity," in *The Expressive Organization*, ed. Majken Schultz, Mary Jo Hatch, and Mogens Holten Larsen (Oxford: Oxford University Press, 2000), p. 118.

Building a Solid Reputation

The foundation of a solid reputation exists when an organization's identity and its image are aligned. Charles Fombrun, New York University professor emeritus and author of the book *Reputation*, says that "in companies where reputation is valued, managers take great pains to build, sustain, and defend that reputation by following practices that (1) shape a unique identity and (2) project a coherent and consistent set of images to the public."[22]

Reputation differs from *image* because it is built up over time and is not simply a perception at a given point in time. It differs from *identity* because it is a product of both internal and external constituencies, whereas identity is constructed by internal constituencies (the company itself).[23] We first saw the Reputation Framework (Figure 4.1) at the beginning of this chapter, and it is beneficial to examine it again now that we have discussed identity and image in more depth. As the Reputation Framework illustrates, reputation is based on the perceptions of *all* of an organization's constituencies. Thus, reputation is an outcome, and as a result, cannot really be "managed" in any way.

Why Reputation Matters

The importance of reputation is evidenced by several prominent surveys and rankings that seek to identify the best and the worst among them: *Fortune*'s "Most Admired" list; *BusinessWeek* and Interbrand's "Best Global Brands" ranking; and Harris Interactive and the Reputation Institute's Reputation Quotient (RQ) Gold study, featured in *The Wall Street Journal*. Such highly publicized rankings have gained so much attention that some corporate PR executives' bonuses have actually been based on *Fortune*'s list of America's Most Admired Companies.[24] And, according to the *PR Week*/Burson-Marsteller CEO Survey, these media scorecards are extremely influential. Surveyed CEOs ranked the influence of the rankings, with *Fortune*'s "100 Best Companies to Work For" topping the list, followed by *Fortune*'s "Most Admired Companies," *The Wall Street Journal*'s "Shareholder Scoreboard," and the *Financial Times*' "Best Places to Work." The variety of influence represented here, from employees ("Best Places to Work") to shareholders ("Shareholders Scoreboard") suggests the power that constituents have over corporate reputation.[25]

According to the Hill & Knowlton Corporate Reputation Watch survey, almost all analysts agree that if a company fails to look after reputational aspects of its performance, then it will ultimately suffer financially.[26] In response to this demand, many public relations firms and consultancies now offer reputation measurement and management services to their corporate clients.

---

[22] Charles J. Fombrun, *Reputation: Realizing Value from the Corporate Image* (Boston: Harvard Business School Press, 1996), pp. 5–6.

[23] Pamela Klein, "Measure What Matters," *Communication World* 16, no. 9 (October–November 1999), pp. 32–33.

[24] Matthew Boyle, "The Right Stuff," *Fortune*, March 4, 2002, pp. 85–86.

[25] *PR Week*/Burson-Marsteller 2007 CEO Survey, *PR Week*, November 12, 2007.

[26] 2006 Hill & Knowlton Corporate Reputation Watch.

In 2011, a Harris Interactive survey of the U.S. public found that the following companies had the top "Reputational Quotient" scores: Google, Johnson & Johnson, 3M, Berkshire Hathaway, and Apple.[27] A strong reputation has important strategic implications for these and other firms, because, as Fombrun notes, "it calls attention to a company's attractive features and widens the options available to its managers, for instance, whether to charge higher or lower prices for products and services or to implement innovative programs."[28] As a result, the intangible entity of reputation is undoubtedly a source of competitive advantage. Companies with strong, positive reputations can attract and retain the best talent, as well as loyal customers and business partners, all of which contribute positively to growth and commercial success. Reputation management consulting firm Thackway McCord calculated that in 2009, a full 16 percent of the value of the S&P 500 could be attributed to reputation.[29]

Reputation also can help companies weather crises more effectively. For example, strong reputations helped Johnson & Johnson (J&J) survive the Tylenol cyanide tampering crisis in the early 1980s and its recent quality control problems (see Chapter 10 for more on J&J's handling of the Tylenol crisis) and allowed Coca-Cola's contamination cases in India in 2004 to come and go without measurable long-term damage to the firm.

The changing environment for business, as discussed in Chapter 1, has implications for reputation. The proliferation of media and information, the demand for increased transparency, and the increasing attention paid to social responsibility (see Chapter 5) all speak to a greater focus by organizations on building and maintaining strong reputations. According to a recent Burston-Mueller study, the vast majority of business leaders feel that it has become much more difficult to manage reputation as a result of digital and social media. The resulting rapid dissemination of information requires quick responses and internal alignment from companies.[30] Public confidence in business is low, and public scrutiny of business is high. The collapse of the energy giant Enron in 2001 dragged its auditor, Arthur Andersen, down with it in an accounting scandal that not only irreparably damaged both firms' reputations (and indeed their chances for survival) but also heightened public mistrust of large corporations in general—particularly those with complex accounting—and of the entire accounting profession. This negative public sentiment toward business was then aggravated by the subprime mortgage credit crisis and the controversial U.S. government bailouts of banks, insurance companies, and U.S. automobile manufacturers.

Against this backdrop, organizations are increasingly appreciating the importance of a strong reputation. How does an organization know where it stands? Because reputation is formed by the perceptions of constituencies, organizations must first uncover what those perceptions are and then examine whether they coincide with the company's identity and values. Only when perceptions and identity are in alignment will a strong reputation result.

---

[27] 2011 Annual RQ Summary Report, Harris Interactive, http//:www.harrisinteractive.com.

[28] David A. Aaker, *Building Strong Brands* (New York: The Free Press, 1996), p. 51.

[29] http://www.thackwaymccord.com/repvalue1.html.

[30] "Managing Corporate Reputation in the Digital Age," Burston-Mueller, November 2011.

Measuring and Managing Reputation

In assessing its reputation, an organization must examine the perceptions of *all* its constituencies. As mentioned earlier, many PR firms have developed diagnostics for helping companies conduct this research. Although one size does not fit all when it comes to measurement programs, all of them require constituency research.

Employees can be a good starting point, as they need to understand the company's vision and values and conduct themselves in every customer interaction with those in mind. An organization runs into trouble when it does not practice the values it promotes. As an example, IBM long espoused the value of lifetime employment. In the early 1990s, however, the company went through severe downsizing, and a joke that circulated throughout the company was that "IBM means 'I've Been Misled.'" Clearly, employees did not feel that IBM was true to its own values, and this disillusionment caused IBM's reputation to suffer.[31] However, the company took this to heart and made a subsequent turnaround after internal initiatives targeted employees and implored them for help in reversing this value misalignment. One such initiative, dubbed the "ValueJam," drew more than 57,000 employees online to post ideas about how IBM's values could be applied to improve its operations, workforce policies, and relationships.[32]

Customer perceptions of an organization also must align with the organization's identity, vision, and values. In the late 1990s, Burberry learned what can happen to corporate reputation when this is *not* happening, and how the reputation can be saved by taking aggressive steps to restore these connections.

When Rose Marie Bravo became CEO of Burberry in 1997, the company was facing a number of challenges. Profits were plummeting, and although some of it could be explained by the Asian economic crisis of the mid-1990s (by 1996, Asian consumers—at home and abroad—generated two-thirds of the company's revenues, causing the downturn to dramatically affect Burberry's sales),[33] internal factors were also at work. For one, prior to Bravo's arrival, instead of maintaining a cohesive Burberry brand across the globe, the company allowed each country's management team to develop the brand as it desired in the local market and allowed extensive licensing. As a result, when customers thought of Burberry, what came to mind depended on their geographic location. In the United States, it meant $900 raincoats and $200 scarves; in Korea it meant whiskey; and in Switzerland it meant watches. Bravo explained that, before her arrival, "[Burberry] had a disparate network of licensees marketing Burberry around the globe. It wasn't a coherent business. Each country was representing its own version of Burberry. Demand slowed. The business needed a clean up. The brand was over-exposed and over-distributed."[34]

---

[31] Mary Jo Hatch and Majken Schultz, "Are the Strategic Stars Aligned for Your Corporate Brand?" *Harvard Business Review*, February 2001, pp. 129–134.

[32] "Gone to (Google) Hell: Resurrecting a Reputation When the Devil's in the Digital," PR News, June 11, 2007.

[33] Lauren Goldstein, "Dressing Up an Old Brand," *Fortune*, November 9, 1998, pp. 154–56.

[34] Quoted in Nigel Cope, "Stars and Stripes," *Independent*, June 6, 2001, Online Lexis-Nexis Academic, August 2001.

Not only was the company having trouble deciding what it was selling, but it also was struggling with how it was positioning its products. Burberry's inability to decide whether it was targeting upper- or lower-end consumers in Asia, for example, led to its products being sold in bulk to discount retailers. This decision undermined the image the exclusive, high-end Burberry boutiques were trying to generate in that same market. Bravo realized that Burberry had to sharpen its focus and concentrate exclusively on high-end retailing to send a consistent message to consumers. Additionally, she recognized that by speaking primarily to older males as a high-end men's raincoat retailer, the company was not catering to a key consumer constituency—women—as effectively as it could.

Recognizing that the Burberry store portfolio needed to reflect the high-end focus of the brand, Bravo upgraded the flagship store in London and doubled the size of the New York store. Even more important, Burberry began to rein in its detached network of franchises to allow the company greater control over consistency of product and identity. The most visible turning point was a print advertising campaign featuring supermodel Kate Moss in a Burberry plaid bikini. These ads pushed Burberry's sales up dramatically and the average age of its customer down considerably by putting a fresh, playful face on a venerable fashion brand that, though esteemed for its nearly 150-year heritage, was looked upon by younger constituencies as stodgy and by many women as "not for me."

These initiatives, from store renovations to a more unified product focus across all franchises to the elimination of discount retailing, created a cohesive image and firmly established Burberry as a luxury brand, greatly enhancing its reputation around the world.

Corporate Philanthropy

Every organization today needs to consider corporate philanthropy and social responsibility when thinking about its own reputation. We will discuss corporate responsibility in much more detail in Chapter 5. The 2010 Cone Cause Evolution Study results revealed that 85 percent of consumers have a more positive image of a company that supports a cause they care about; and 83 percent wish that more of the companies that they do business with would support causes.[35] Many customers factor philanthropy and social responsibility in when deciding where to purchase goods and services. The Shell and ExxonMobil example earlier in this chapter provides an example.

Despite these findings, corporate philanthropy is not without its perils. As we saw in Chapter 1, trust in business is low, and efforts to publicly "do good" can be perceived as self-serving, particularly in the case of "strategic giving," in which the charitable activity relates directly to the business the company is in. Alternatively, when companies are too silent about what they are doing for the community or the environment, they face criticism for being apathetic.

Philip Morris provides a good example of the former. The company's advertising campaign touting its charitable activities met with skepticism from the public,

---

[35] 2010 Cone Cause Evolution Study, http://www.coneinc.com/files/2010-Cone-Cause-Evolution-Study.pdf.

many of whom viewed these ads as an attempt by Philip Morris to "undo" its negative image as a big tobacco company rather than as a manifestation of true concern for the community. Despite continued spending on promoting its philanthropic activities, the company fell out of the top 60 in the 2011 Harris Interactive RQ Survey, ranking below BP. Philip Morris had been ranked 56 out of 60 in 2006.

The catastrophic events of September 11, 2001, provided another proving ground for companies' social responsibility communications programs. Procter & Gamble provided more than $2.5 million in cash and products to relief efforts, but because it did not publicize these activities, the company was accused in a Harris Interactive–Reputation Institute survey of doing "absolutely nothing to help."[36] P&G had consciously taken a low-profile approach to avoid being seen as "capitalizing on disaster," and that approach backfired.

How can companies reconcile the public's desire for them to do good things for the community and the environment with the public's equally strong skepticism about corporate motives? Why do some companies' efforts to make their good deeds known meet with approval and others' with disdain? First, corporate philanthropy and social responsibility programs should be consistent with a company's vision to be perceived as credible, rather than as simply "check-the-box" activities or attempts to burnish a tarnished image.

Second, the means by which a company demonstrates its caring for the community should be carefully considered, using the communication framework provided in Chapter 2. If the company understands each of its constituencies—what the constituency members are concerned about, what is important to them, and what they already think about the company—it will be well positioned to structure the right kinds of programs and choose the right *channels* through which to communicate them. For instance, it may decide to describe its community outreach or environmental activities in its annual report or on its website rather than through advertising. It may decide that sponsoring a program that allows and encourages employees to volunteer their time in the community will be more effective than giving money to a local charity.

In the changing environment for business, corporate philanthropy and social responsibility are gaining visibility and importance in the eyes of many constituencies. A company that has a good understanding of its own constituencies and what is important to each group, and that gives thought to how to tie corporate responsibility programs into its corporate vision, will be well positioned to create programs that will enhance its reputation. As mentioned, Chapter 5 covers this topic in more detail.

What Is Corporate Advertising?

Now that we have an understanding of image, identity, and reputation, we will see how corporate advertising can be employed to shape an organization's image.

Corporate advertising can be defined as the paid use of media that seeks to benefit the image of the corporation as a whole rather than its products or services alone.

---

[36] Ibid.

Because all of a company's advertising contributes to its reputation, both product and corporate advertising should reflect a unified strategy. Corporate image advertising should "brand" a company the way product advertising brands a product.

A major difference between corporate and product advertising is who pays for each of the two types of advertising. A company's marketing department typically is responsible for all product-related advertising and pays for such ads out of its own budget. Costs are usually associated with a specific product or service. Corporate advertising, on the other hand, falls within the corporate communication area and either comes out of that budget or, in some cases, is paid for by the CEO's office.

Corporate advertising should present a clear identity for the organization based on a careful assessment of its overall communication strategy (see Chapter 2), and it generally falls into three broad categories: image advertising, financial advertising, and issue advocacy. Let's take a closer look at each of the three categories to understand what corporate advertising is all about.

Advertising to Reinforce Identity or Enhance Image

Many companies use corporate advertising to strengthen their identities following structural changes. As companies merge and enter new businesses, they need to explain their new vision, organization, and strategy to constituents who may have known them well in an earlier incarnation but are struggling to understand the new organization. These typically larger organizations often need to simplify their image to unify a group of disparate activities.

Tyco used corporate advertising to rehabilitate its image in the wake of corporate fraud by former CEO Dennis Kozlowski and former CFO Mark Swartz. Under Kozlowski, Tyco had become a confusing conglomerate of business units built by aggressive acquisitions. Even the company's own employees were unsure what businesses Tyco was in. Following operational improvements, new CEO Ed Breen hired Jim Harman from General Electric as Vice President of Corporate Advertising and Branding. Harman, who had overseen GE's "We bring good things to life" campaign, was tasked with demonstrating the breadth of Tyco's businesses, products, and services. Tyco used the tagline "a vital part of your world" in several print ads that portrayed the company's products and services as integral to daily life. The ads featured a background of more than 6,500 words listing Tyco products and services. The words formed a picture, such as a baby or a firefighter, demonstrating the importance and vitality of Tyco's offering. In 2005, Tyco won an award for best corporate advertising from *IR Magazine*.[37]

As discussed earlier in this chapter, identity audits are one way for organizations to manage their identity, image, and reputation with a variety of constituencies. When companies analyze their image with constituencies, they can then apply these findings to their corporate advertising strategy. If an organization's identity is very different from how it is perceived externally, for instance, it can use corporate advertising to close that gap. We saw how Burberry used a fresh print advertising campaign featuring model Kate Moss in a Burberry plaid bikini

---

[37] Suzanne Vranica, "Tyco Aims to Put Its Woes Behind It," *The Wall Street Journal*, June 15, 2004.

to change perceptions among consumers that the brand was (a) not for women and (b) stiff and stodgy. Corporate advertising can be an efficient mechanism for changing impressions about organizations if changes have really taken place. At Burberry, CEO Rose Marie Bravo was indeed expanding Burberry's women's clothing and accessory lines and working to raise Burberry's profile as a high-end retailer when the new ads appeared in print.

Effective image advertising also allows companies to differentiate themselves from rivals. For example, Nintendo won *Advertising Age* magazine's Marketer of the Year award in 2007 after a blitzkrieg of corporate advertising around its new product, the Nintendo Wii. After years of languishing behind competitors like Sony's PlayStation and Microsoft's Xbox, the company depended on this product to boost sales and reinstate the brand as a leader in the video gaming industry. In November 2006, with $200 million in marketing support, Nintendo's advertising campaign incorporated traditional media with word-of-mouth marketing and digital communications platforms. It appealed to nontraditional audiences, such as mothers, and empowered these groups by making them official Wii Ambassadors. The complementary TV and print ads (crafted by Leo Burnett USA, Chicago) all featured the signature phrase, "Wii would like to play." However, most important to the success of the marketing effort was its application of strategy (by looking toward audiences that represented the future success of the company) and its consistent messaging.

According to NPD Group analyst Anita Frazier, "Marketing played a huge role in the success of the Wii and DS, and I think the power of having a focused message executed throughout all the elements of the marketing campaign is evident. It's sort of like Marketing 101, but too many marketers forget that having a solid positioning and messaging is the most important thing to do before you spend the first dollar on executing the campaign."[38]

The campaign's success can also be attributed to the public relations (PR) behind it. As marketing/advertising and PR functions become increasingly integrated in many organizations, PR becomes a key partner in corporate advertising strategies.

"In our PR, we've always done outreach, but in this case, when we noticed something interesting happening online—like the weight loss using Wii Sports— we would draw it to the media's attention," said George Harrison, former senior VP, marketing and corporate communication, Nintendo North America, in *Advertising Age*. "The little things that kept showing up were picked up and blown out in marketing. . . . When we saw what people were doing or how they were getting creative, we would move on it."[39]

Advertising to Attract Investment

In Chapter 8, we will look at the importance of a strong investor relations function. One of the tools that companies use to enhance their images in the financial community is financial-relations corporate advertising. This kind of corporate advertising can stimulate interest in a company's stock among potential investors as

---

[38] http://www.iabcfortworth.com/emma_news/October_2007/nintendo_marketer2.html.
[39] Ibid.

well as buy-side and sell-side analysts (see Chapter 8 for more on analysts). Given the hundreds of companies that analysts cover, a good corporate advertising campaign can stimulate their interest to take a closer look at a particular one.

Although analysts focus heavily on company financials, in a survey of 200 research analysts—each of whom covered approximately 80 companies—"strength of management" was the number one factor influencing the decision to invest in a company.[40] Analysts place a high value on CEOs who express a coherent vision for their organizations, and as James Gregory of Corporate Branding LLC explains, "the CEO's ability to paint a picture of the company's future is the linchpin of a successful corporate advertising campaign."[41] For these reasons, companies' CEOs are often featured in corporate advertisements targeted at the financial community.

Some corporate advertisers assert that a strong, financially oriented corporate advertising campaign can actually increase the price of a company's stock. A W.R. Grace campaign that ran in the early 1980s is often cited as evidence. The television campaign, which ran as the company's "Look into Grace" series, highlighted the company's financial and business attributes and then asked, "Shouldn't you look into Grace?" Attitude and awareness studies of the ad campaign in test markets showed that its awareness and approval ratings were much higher after this campaign ran. In addition, the company's stock price increased significantly during the test campaign, though it did not go any higher with later campaigns. Corporate advertising expert Thomas Garbett, writing in the *Harvard Business Review*, stated that

> I interpret the relationship between corporate campaigns and stock pricing this way: advertising cannot drive up the price of a reasonably priced stock and, indeed, doing so might not be entirely legal; it can, however, work to ensure that a company's shares are not overlooked or undervalued.[42]

Professors at Northwestern University's Kellogg School of Business studied this trend using an econometric analysis of the link between corporate advertising and stock price. They determined that, indeed, corporate advertising has a statistically significant positive effect on stock prices. They further determined that the positive influence from such campaigns averaged 2 percent and was particularly strong during bull-market periods, such as in the mid- to late-1990s.

The implications of this study, if true, are exciting for companies. Even a one-point increase in the stock price can translate to the tens or hundreds of millions of dollars for large companies with many shares of stock outstanding. In addition, an improvement in stock price that improves the company's price-to-earnings ratio can present opportunities for stock options and dividends for employees, improving talent acquisition and retention.

Some companies view building their brand with investors as more important than doing so with customers. As Gary Patrick, founder of Patrick Marketing Group, observed, "There are business-to-business companies advertising during

---

[40] James R. Gregory, "The Impact of Advertising to the Financial Community," *BusinessWeek* special publication, 1999, p. 4.

[41] Ibid.

[42] Thomas F. Garbett, "When to Advertise Your Company," *Harvard Business Review*, March–April 1982, p. 104.

Friends or prime time baseball—clearly all they're doing is advertising to potential investors and Wall Street."[43]

Advertising to Influence Opinions

This kind of advertising often is called *issue* or *advocacy advertising* and is used by companies to respond to external threats from either government or special interest groups. Issue advertising typically deals with controversial subjects; it is a way for companies to respond to those who challenge the status quo.

Many companies started using issue advertising in the late 1970s and early 1980s to meet the challenges of what was perceived as the antibusiness media. By taking issues directly to the consumer, companies can compete with journalists for a share of the reader's mind. As a result, issue advertisements often are purposely placed on op-ed pages in prominent newspapers such as *The New York Times, The Wall Street Journal*, and *The Washington Post*. Perhaps the most famous example of this kind of advertising was Mobil Oil's series of issue advertisements, which ran for over 20 years. What began as a dialogue about the oil embargo in the early 1970s expanded to become a sort of bully pulpit for this powerful organization as it advocated positions on a wide variety of topics.

Many other organizations also have adopted the op-ed style for their advocacy ads, including Amway, whose approach typifies the more positive approach used by companies dealing with environmental issues. Amway ran a series of ads that positioned the company as environmentally aware. One had a photograph of five Amway distributors and the headline "Find the Environmental Activist." The copy goes on to explain that everyone in the ad is an environmental activist and that all Amway distributors are committed to the cause of environmental awareness. The tagline reads "And you thought you knew us."

This advertisement also reveals the problem, however, with much issue advertising. As David Kelley pointed out in an essay on the subject of issue advertising in *Harvard Business Review*, most companies "pay too much attention to the form and too little to the content of the message."[44] Does the tagline in the Amway ad, for example, imply "You thought we were a bunch of polluters because we specialize in detergents that come in huge containers"? Or does it mean "You thought we were just selling detergents when what we are really doing is protecting the environment"? Either way, the advertisement seems to be playing into the hands of critics rather than setting the agenda for the argument. Because the advertisement is so short, it never gets across the point that this company is trying to make. That is, it would like to argue directly with critics who charge Amway with environmental neglect.

Because companies typically are more conservative than their adversaries, particularly NGOs, their arguments often fall short of the mark. It is extremely difficult for a large corporation to take on a tough issue in the marketplace without offending someone. When companies try to please everyone, they ultimately dilute the power of their own messages.

[43] "Marketers Use TV Advertising to Attract Investment," *Investor Relations Business*, November 12, 2001, p. 17.

[44] David Kelley, "Critical Issues for Issue Ads," *Harvard Business Review*, July–August 1982, p. 81.

If a company decides to pursue an advocacy campaign, senior management must have the courage to argue forcefully for its ideas and must not be afraid to alienate certain constituencies in the process. For example, when the major book-sellers took on the conservative groups that called for a purging of all "dirty" books, they won the argument with advocates of First Amendment rights but lost with family-oriented fundamentalist groups. Organizations should thus proceed into the world of issue advertising with extreme caution and with a full under-standing of its inherent risks.

Who Uses Corporate Advertising and Why?

According to recent studies, over half of the largest industrial and nonindustrial companies in the United States have corporate advertising programs of one sort or another. Usually, a direct correlation exists between size and the use of corpo-rate advertising: The bigger the company, the more likely it is to have a corporate advertising program. Because large corporations tend to have more discretion-ary income, this correlation makes sense. In addition, larger companies tend to be more diversified and thus have a greater need to establish a coherent identity for a variety of activities, products, and services.

Corporate advertising also is used heavily by companies within more "controversial" industries: Cigarette companies, oil companies, pharmaceuticals, and other large industrial companies all have image problems to deal with, from concerns about health to drug recalls to pollution. Overall, heavy industry spends more on corporate advertising than consumer packaged-goods firms, which lead all other industries in product advertising. This ranking may be related to the presence in consumer product companies of a strong marketing focus that concentrates more on the four Ps of product, price, promotion, and place (distribution) than on developing a strong corporate image. Interestingly, recent government and public focus on the nutritional content of fast foods and snack foods is already leading to more corporate advertising from companies such as PepsiCo and McDonalds that are looking to show their companies in a more positive and healthy light.

A good corporate advertising program can clarify and enhance a company's image, and the absence of one can actually damage a company's reputation. Let's now take a closer look at some of the reasons companies invest in corporate advertising campaigns.

Increase Sales

The relationship between corporate advertising and sales is less clear than that between product advertising and sales, because corporate advertising is meant to do things that *eventually* boost sales but likely won't directly or immediately do so. This purpose creates a problem for managers trying to introduce corporate advertising into companies that have a heavy financial orientation. The numbers-oriented manager often will cite the lack of a direct connection between corporate advertising and sales as the best reason not to use corporate advertising.

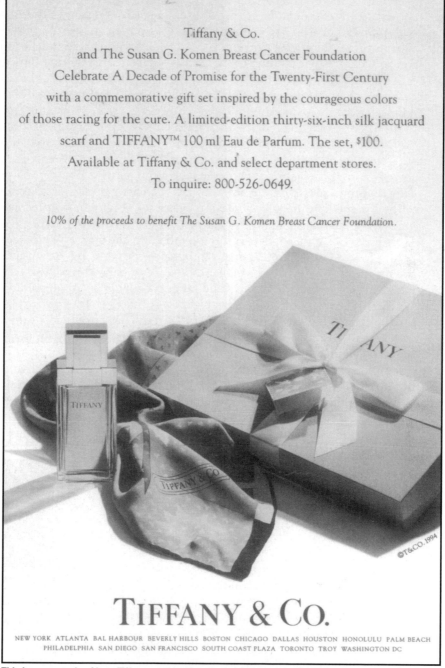

This is an example of how Tiffany combined product and corporate advertising. It represents the best of how philanthropic efforts can be used to foster goodwill with constituencies. Tiffany and Company Advertisement with The Susan G. Komen Breast Cancer Foundation. Photo credit: Tiffany and Company. Permission granted by Tiffany & Company.

Even so, there are growing efforts to identify a closer relationship between corporate advertising and sales. As a senior vice president of the Association of National Advertisers (ANA) remarked, "As has been seen in other marketing communications areas, corporate advertising managers are becoming more concerned with determining the Return-on-Investment [ROI] of their efforts."[45]

Although measuring the return on investment for individual marketing disciplines began 75 years ago with the monitoring of results from direct-mail campaigns, attempts to determine the ROI from integrated marketing campaigns are more recent. Several agencies, including Grey Global Group, McCann-Erickson WorldGroup, and J. Walter Thompson, are using new tools to better quantify results for clients, including measures such as cost per sale or cost per lead.[46] This sort of analysis can help companies make a stronger case for advertising budgets in difficult economic times and also may aid with their financial projections. The rise of the Internet has made it increasingly easy to measure newer forms of advertising, such as advertising banners placed on websites and in e-mail advertisements.

In 2000, AT&T Business Services took the unusual step of asking agencies competing for a $100 million business-to-business advertising assignment to project the return on investment of their proposed campaigns, and also to recommend which of the company's services should be most heavily advertised. In the future, "Advertising won't be treated as an expense, but as a strategic investment," said former marketing vice president Bill O'Brien.[47] Through corporate advertising, companies can draw out features about themselves that they think will appeal to the public and, as a result, make consumers want to buy products from them. For instance, S.C. Johnson & Son, a company established in 1886 and the maker of such brands as Glade, Pledge, Windex, and Ziploc, learned that 80 percent of consumers believed family-owned companies made products they could trust, versus only 43 percent who said the same of publicly owned companies. In response, the company rolled out a $450 million campaign highlighting the family heritage of S.C. Johnson, with the tagline, "S.C. Johnson—a Family Company."[48]

Create a Stronger Reputation

We have talked about the importance of reputation throughout this chapter. The best corporate advertising creates goodwill and enhances reputation by letting constituents in on what the organization is all about, particularly if the company does beneficial things that not everyone is aware of.

Amoco Chemical Company, acquired by BP in 1998, created a campaign that won an award from *BusinessWeek* in the late 1990s and is a classic example of this sort of advertising. One of the print ads for this campaign showed an airplane landing at night with the headline "Amoco Helps Make Coming Home a Little Safer." The ad went on to explain that the lighting masts use durable resin compounds based on material from Amoco Chemical. Although the advertisement was visually

[45] Association of National Advertisers' website, http://www.ana.net/news/1998/ 04_01_98.cfm.

[46] Laura Q. Hughes, "Measuring Up," *Advertising Age*, February 5, 2001, p. 1.

[47] Kathleen Sampey, "AT&T: Ads Are Investment; Shops Must Project ROI," *Adweek*, July 31, 2000, p. 6.

[48] Jack Neff, "S.C. Johnson Ads to Stress 'Family Owned,'" *Advertising Age*, November 13, 2001, p. 3.

appealing, another reason that this campaign made it into *BusinessWeek*'s "most memorable" list that year was the concept that chemicals are used for things most people don't even think about that make our lives better. The tagline, also memorable, read, "The Chemistry Is Right at Amoco." Learning more about the good things that come out of Amoco shifted some people's perceptions away from thinking of Amoco as another "big oil" company and a producer of environmental pollutants.

Companies also look to build credibility and enhance reputation by using endorsements from third-party organizations (TPOs).[49] Just as individuals rely on the *Zagat Survey* to confirm their choice in restaurants, many find this type of "seal-of-approval" advertising helpful in assessing companies, particularly lesser-known ones. An endorsement by a trusted and recognized TPO can inspire confidence in the consumer. Third parties can provide ratings or rankings of a company or its services, or they can be used as the subject of a story that illustrates how the company provided a service to them. Increasingly, the aggregated opinions of other consumers, as viewed on sites such as Amazon or Yelp, are serving as trusted third-party reviews. This presents interesting opportunities for companies to connect directly with the dedicated and respected customers that Malcolm Gladwell refers to as "mavens" in his book *The Tipping Point*. It also represents real risk for companies because nonexpert reviewers can share negative opinions that are based on misunderstanding and user error with millions of potential customers.

When companies receive positive reviews from a trusted TPO, they should consider referencing the endorsement in corporate advertisements. An example of this is the advertisement for a Van Kampen mutual fund that mentions the fund's five-star rating by the Morningstar investment guide. An example of a company that used TPOs to give it credibility for its lesser-known offerings is the campaign that Xerox Corp. launched for its consulting and operations services. One of these ads shows the Xerox name in large print with a car key sitting on top of it. The copy reads, "Enterprise Rent-A-Car wanted to reduce operational costs. Xerox found the key to success by moving 1.7 million documents onto their intranet every month." Another ad talks about how the company helped Honeywell lower its operational costs by millions of dollars. Although Xerox already had the name recognition that many smaller companies using TPO advertising do not, it was largely for photocopying equipment. This series of ads, with the tagline "There's a New Way to Look at It," revealed a much broader set of capabilities. Being able to talk about projects it had undertaken for large, well-known companies provided Xerox with more credibility as it attempted to boost its image as a more comprehensive service provider.

Corporate advertising also is widely used by companies to publicize their philanthropic activities, which, as discussed earlier in this chapter, also can lead to an enhanced reputation. These advertisements can create bizarre associations between otherwise diametrically opposed sectors of society, such as cigarette

---

[49] Dwane Hal Dean and Abhijit Biswas, "Third-Party Organization Endorsement of Products: An Advertising Cue Affecting Consumer Prepurchase Evaluation of Goods," *Journal of Advertising*, January 1, 2002, pp. 41–58.

manufacturers and the arts (The Altria Group, formerly Philip Morris), opera and oil (Texaco), and supertanker manufacturers and blue whales (Samsung).

Organizations using corporate advertising to enhance their reputations must be prepared for their opponents to respond negatively to what they may perceive as the company's attempt to smooth over a history of corporate wrongdoing or to apply a "quick fix" to a serious image problem.

As mentioned earlier in this chapter, Philip Morris, now Altria, has struggled to improve its reputation, despite allocating significant funds to the effort. The company's aggressive image advertising campaign touting its philanthropic activities, coupled with its identity program, actually worked to alienate critics further. Many viewed both as attempts to mask the company's true identity as a cigarette manufacturer responsible for thousands of cancer deaths. In fact, at Philip Morris's 2002 shareholder meeting, demonstrators waved a giant canvas banner depicting a skeletal Marlboro Man in a bandana marked "Altria."[50]

It is important, then, when using corporate advertising to enhance reputation that it be credible. Corporate advertising risks being perceived as not credible if, for instance, it ties closely to corporate vision, but that vision has not been properly communicated to the organization's constituencies through other channels as well. This requirement highlights the point made earlier that corporate advertising must be strategic and closely aligned with a company's overall communication strategy. In isolation, it will not have the power to change perceptions about the organization.

Recruit and Retain Employees

One of the most critical communication activities for any company is communicating with employees (see Chapter 7). If a corporate advertising campaign succeeds in explaining in simple terms what a large, complex organization is all about, it can be as helpful to employees as it is to the outside world. Corporate advertising is also an indirect way of building morale among employees. Trying to quantify this is very difficult, however. Garbett says that

> Putting a dollar figure on the savings attained by reducing employee turnover is difficult. Some say you should add recruitment and training costs, next multiply by the turnover rate, and then estimate the percentage of employees who might be persuaded to stay if they felt more positively about the company. Whatever the real figure, if corporate advertising can effect even a modest reduction in turnover, the savings to a large corporation is well worth the expense and effort of a campaign.[51]

Such advertising also helps companies attract the best and the brightest both at the entry level and for senior positions. A good corporate advertising campaign can create excitement among both potential and current employees. In 2002, GE launched a corporate print advertising campaign with four employee-related themes: diversity in leadership, the GE Fund, the GE Mentoring program, and

---

[50] "Philip Morris Annual Meeting Draws Most Extensive Protest in Corporation's History," *PR Newswire*, April 25, 2002, Online Lexis-Nexis Academic, April 2002.

[51] Thomas F. Garbett, *Corporate Advertising* (New York: McGraw-Hill, 1981), p. 120.

volunteerism at the company. Many of the ads show photographs of current GE employees as children. In one ad, a girl is pictured holding a globe; the text reads, "Introducing Eugenia Salinas who has traveled throughout the world as GE's General Manager, Americas Marketing for GE Medical Systems. She's part of the group of minority and women leaders across GE responsible for over $30 billion in annual revenues."[52]

Other ads include photos of GE employees who are involved in mentoring through the company-sponsored program or who participate in volunteer projects, along with members of their local community. Many such ads, ostensibly focused on employees, enhance a company's image with nonemployee constituencies as well. Consumers, for instance, may be impressed with GE's social responsibility programs or the caliber of its employees, which they read about in these print ads.

Conclusion

As we've seen in this chapter, identity, image, and reputation are integral to an organization's success and credibility, and using corporate advertising successfully can help with all three. Most managers who have not thought about corporate reputation tend to underestimate its value. This error is partly due to a lack of understanding about what corporate image, identity, and reputation are all about and what they do for an organization, but skeptics also should understand that an inappropriate or outdated identity can be as damaging to a firm as weak financial performance. Individuals seek consistency, and if perceptions about a corporation fail to mesh with reality, constituents take their business elsewhere.

Executives, then, need to be fully aware of the tremendous impact of identity, image, and reputation and must learn how to manage these critical resources. One way to do this, as illustrated by GE and Nintendo, is through corporate advertising. The decision to run a campaign should be based, above all else, on a firm's overall communication strategy. Whether the company is changing its image, is suffering from erroneous perceptions in the marketplace, or simply wants to continue a successful, well-received campaign that solidifies its identity, corporate advertising can be a tremendous resource in positioning the organization for future success.

No matter the strategy, an organization with a clear corporate identity that represents its underlying reality and is aligned with the images held by all of its constituencies will be rewarded with a strong reputation. Reputational success, in turn, matures into pride and commitment—among employees, consumers, and the general public—and these qualities are irreplaceable assets in an intensely competitive global business environment.

[52] GE company website, http://www.ge.com/campaign.htm.

JetBlue Airways: Regaining Altitude

New York–based JetBlue Airways had started 2007 on a roll; growth, in terms of both destinations and fleet size, was far outpacing even the most ambitious industry projections. And more important, the airline continued to enjoy a cult-like following among its loyal customers, thanks in large part to uncommonly attentive service, generous legroom, free satellite television feeds in every leather seat, and, of course, the company's signature Terra Blues potato chips. In fact, JetBlue ranked highest in customer satisfaction among low-cost airlines in 2006 and among all major airlines in the United States in 2005.[1]

Yet as a winter nor'easter barreled toward the New York metropolitan region on February 14, 2007, JetBlue's leaders were blissfully unaware that the next seven days would be by far the most trying in the company's eight-year history. Within five days, the company would have cancelled more than 1,000 flights, incurring tens of millions of dollars in losses in the process and tarnishing JetBlue's sterling reputation, thanks to a combination of bad luck, flawed decision making, and multiple systemic failures.

JetBlue founder and CEO David Neeleman encouraged his executive team to search for bold and inventive solutions to restore the company's public image, win back customers, and reassure employees and investors. If that meant parting with convention, then so be it, Neeleman said.

JETBLUE TAKES OFF

The 1999 launch of Jet Blue Airways was never supposed to work. After all, of the 58 startup jet airlines that had commenced operations since the U.S. government deregulated the industry

in 1978, only 2 survived. "It is a business whose margins are so razor thin that a couple of passengers on each plane can spell the difference between profit and loss and where a 1-cent rise in the price of jet fuel can cost the industry an added $180 million a year," wrote industry expert Barbara Peterson in 2004.[2]

Industry behemoths like Eastern Air Lines, Trans World Airlines, United Airlines, American Airlines, Braniff International Airways, Northwest Airlines, and Delta Air Lines reaped enormous profits and ruled the skies until Congress and President Jimmy Carter passed the Airline Deregulation Act of 1978. The primary purpose of the act was to eliminate government control over commercial aviation and encourage market forces to shape the industry's development.

Although the cut-throat competitive tactics used by the legacy airlines in the 1980s and 1990s caused most new companies to fail, competition persisted, and airfares dropped significantly into the twenty-first century, leading to the rise of low-cost carriers such as AirTran Airways, Southwest Airlines, and JetBlue Airways.

JetBlue was the brainchild of David Neeleman, an industry visionary who promised to "bring humanity back to air travel."[3] Neeleman, who was born in Brazil but grew up in Utah as part of a large Mormon family, was no stranger to startup airlines. He helped to build Morris Air, a Utah-based airline that Southwest acquired in 1993 for $129 million.

---

[1] B. Peterson, *Bluestreak: Inside JetBlue, the Upstart That Rocked an Industry* (New York: Portfolio, 2004).

[2] Ibid.

[3] Ibid.

Source: This case was prepared by Gregory G. Efthimiou of the University of North Carolina at Chapel Hill. It was published in 2008 as the winner of the Arthur W. Page Society's annual Case Study Competition Journal. Reprinted by permission.

Neeleman leveraged his industry experience and connections to create a company that would boast a fleet of brand new airplanes, low fares, and a host of customer-friendly embellishments that legacy carriers and start-ups alike would be hard pressed to match. Neeleman envisioned treating JetBlue's customers—never referred to as passengers—to comfy leather seats, paperless ticketing, and exceptional service by flight crew members. Every seat would come equipped with a television that featured dozens of free channels provided by satellite signal. Finally, to keep costs down, JetBlue would offer a virtually unlimited supply of appealing in-flight snacks instead of soggy meals that no one really wanted.[4]

Backed by an impressive capital reserve, Neeleman's plan worked far sooner than even the most optimistic industry observers predicted. With flights to and from previously underserved markets, JetBlue quickly shot to the top of J.D. Power and Associates' customer satisfaction surveys.[5] Based out of New York's John F. Kennedy International Airport, the startup soon expanded operations to Los Angeles, southern Florida, and a host of smaller markets, such as Buffalo, New York.

JetBlue's launch was particularly well timed. Despite frequent pricing skirmishes resulting from increased competition, the domestic commercial aviation industry started 2001 as the beneficiary of 24 consecutive quarters of profitability.[6] Passenger volume had risen at an average annual rate of 3.6 percent over the previous decade, and net profits for the industry totaled $7.9 billion in 2000.[7] Then the unthinkable happened.

The terrorist hijacking and downing of four U.S. jetliners in New York City, Washington, D.C., and rural Pennsylvania on September 11, 2001, crippled the industry. Consumer confidence in the safety and security of air travel plummeted, sending booking rates down by 70 percent when flights resumed after 9/11.[8] The industry, which generated 11 million jobs and constituted 9 percent of the U.S. gross domestic product, saw more than 80,000 jobs eliminated during the two months immediately following the attacks.[9] In fact, only three airlines managed to turn a profit in 2001—the low-cost carriers Southwest, AirTran, and JetBlue.[10]

Due in large part to its size and flexibility, JetBlue continued to impress in the years that followed. In 2002, *Advertising Age* crowned JetBlue the "Marketer of the Year," claiming that the company's branding efforts gave it a singular identity in a crowded and often confusing marketplace.[11] JetBlue flights were among the most on-time in the industry in 2003, the same year the airline filled most of its available seats on planes—two feats that rarely go hand-in-hand. By mid-2004, the company had turned a profit for more than 16 consecutive quarters.[12]

Although JetBlue reported a net loss of $1 million in 2006, primarily due to soaring jet fuel expenses, the company's operating revenue totaled $2.36 billion, which constituted growth of nearly 39 percent over the 2005 fiscal year.[13] By 2007, the airline's growing fleet of jets

[4] M. Cohn, "JetBlue Woes May Spur Wider Changes," *The Baltimore Sun*, February 20, 2007, p. A1.

[5] J. Bailey, "Long Delays Hurt Image of JetBlue," *The New York Times*, February 17, 2007, p. C1.

[6] S. Blunk, D. Clark, and J. McGibany, "Evaluating the Long-Run Impacts of the 9/11 Terrorist Attacks on U.S. Domestic Airline Travel," *Applied Economics* 38 (2006), pp. 363–70, see p. 363.

[7] Ibid.

[8] H. Kim and Z. Gu, "Impact of the 9/11 Terrorist Attacks on the Return and Risk of Airline Stocks," *Tourism and Hospitality Research* 5, no. 2 (2004), pp. 150–63, see p. 151.

[9] Ibid.

[10] T. Flouris and T. Walker, "The Financial Performance of Low-Cost and Full-Service Airlines in Times of Crisis," *Canadian Journal of Administrative Sciences* 22, no. 1 (2005), pp. 3–20.

[11] B. Tsui, "JetBlue Soars in First Months," *Advertising Age* 71, no. 38 (September 11, 2000). Retrieved November 10, 2007, from Business Source Premier database.

[12] Peterson, *Bluestreak*.

[13] "JetBlue Announces Fourth Quarter and Full Year 2006 Results," January 30, 2007, http://investor.jetblue.com/phoenix.zhtml?c=131045&p=irol-newsArticle_Print&ID=955585&highlight=.

served 52 destinations with more than 575 daily flights.[14] Even though an increasing number of critics forecasted growing pains for JetBlue after its meteoric rise, the love affair between the upstart airline and its faithful customers appeared to be as strong as ever.

THE PERFECT STORM

Valentine's Day 2007 got off to an inauspicious start in the New York metropolitan area. Bleak, gray skies blanketed the region, and weather forecasters warned of a wintry mix of precipitation. JetBlue officials at JFK International Airport gambled that temperatures would warm up enough to change the snowfall and icy slush into rain. Six JetBlue planes—four bound for domestic destinations, one headed for Aruba, and another for Cancun, Mexico— were loaded early in the day with passengers, luggage, and cargo. The planes pushed back from their respective gates and waited for word of a break in the storm. Meanwhile, several inbound flights landed, taxied, and filled most of the airline's dedicated gates.

With no end to the freezing rain in sight, JetBlue and airport officials hatched a plan to allow planes stranded on the tarmac to ferry back and forth to the few remaining open gates for offloading. This strategy failed, however, when the runway equipment used to tow the planes froze to the ground. As a JetBlue spokesperson would explain to a local newspaper: "We had planes on the runways, planes arriving, and planes at all our gates. . . . We ended up with gridlock."[15]

Meanwhile, almost all of the other airlines operating at JFK had called off their flights earlier in the day. Scores of JetBlue passengers in the terminal waited in vain to board flights that would inevitably be cancelled. "We thought

there would be these windows of opportunities to get planes off the ground, and we were relying on those weather forecasts," said Sebastian White, a corporate communications manager at JetBlue.[16] Freezing rain continued to fall, entombing hundreds of passengers inside JetBlue planes that were stranded on the runways at JFK. The worst, however, was yet to come.

ON THIN ICE

Deteriorating weather conditions at JFK and flaring tempers both inside JetBlue's terminal and aboard its planes exacerbated the company's crisis. Nine of the airline's jets sat idle on the tarmac for more than six hours before passengers were successfully offloaded and taken to the terminal.[17] Passengers aboard one JetBlue flight that landed at the airport were trapped inside the plane for a full nine hours.

Tensions inside the planes ran high during the seemingly interminable ground delays. The airline's pilots tried to provide frequent updates and apologies, while crew members in the cabins did their best to appease restless customers with snacks and beverages. It was not until 3:00 p.m. on Valentine's Day that JetBlue officials at JFK finally called the Port Authority of New York and New Jersey to request buses that the airline could use to shuttle passengers from the stranded planes back to the terminal.[18]

The crisis took a particularly troubling turn at Newark Liberty International Airport on February 15. Several passengers became unruly upon learning of additional flight cancellations, prompting JetBlue ticketing personnel to call in the police for protection.[19]

[14] "JetBlue Airways Names Dave Barger President and Chief Executive Officer," May 10, 2007, http://investor.jetblue.com/phoenix .zhtml?c=131045&p=irol-newsArticle&ID=998672&high- light=.

[15] A. Strickler, "Stormy Weather: Waiting til They're Blue; JetBlue Passengers Stranded on Planes for Hours Amid Icy Snarl at JFK Gates," *New York Newsday*, February 15, 2007, p. A5.

[16] S. White, personal interview, November 29, 2007.

[17] J. Bailey, "Long Delays Hurt Image of JetBlue," *The New York Times*, February 17, 2007, p. C1.

[18] J. Chung and A. Strickler, "A Labyrinth of Luggage as Travelers Search through Mounds of Baggage; JetBlue Cancels Hundreds of Weekend Flights," *Newsday*, February 18, 2007, p. A3.

[19] J. Lee, "JetBlue Flight Snarls Continue," *The New York Times*, February 16, 2007, p. 7.

JetBlue customers found little solace by calling the airline's reservations hotline or visiting JetBlue.com on the World Wide Web. By Friday, February 16, many callers who dialed the company's telephone number were greeted by a recorded voice that said, "We are experiencing extremely high call volume. . . . We are unable to take your call." Additionally, JetBlue's website listed flights as on schedule for departure, when the carrier had already cancelled many of those flights.[20] Widespread instances of lost baggage would only further infuriate JetBlue customers whose travel plans were disrupted by the Valentine's Day storm.

JetBlue soon discovered that many of its planes and flight crews scattered across the rest of the country were out of place due to the disruptions at its New York hub. The carrier was forced to cancel more than 250 of its 505 daily flights scheduled for Valentine's Day.[21] JetBlue called off 217 of its 562 scheduled departures for February 15 as well.[22]

"We had a problem matching aircraft with flight crews," said Jenny Dervin, JetBlue's director of corporate communications.[23] Company leaders quickly settled upon a strategy designed to "reset" the airline's operations. "Sometime in the afternoon [of February 16], it just fell apart," said Dervin.[24] "The folks running the operation [were] just exhausted. We said, 'Let's stop the madness.'" The plan to reset operations came at a steep price: JetBlue was forced to cancel approximately 1,200 flights between February 14 and February 19.

David Neeleman cited multiple operational failures that compounded the crisis. Among the primary culprits: inadequate communication protocols to direct the company's 11,000 pilots and flight attendants about where to go and when; an overwhelmed reservation system; and a lack of cross-trained employees who could work outside their primary area of expertise during an emergency.[25]

"We had so many people in the company who wanted to help who weren't trained to help," Neeleman said.[26] "We had an emergency control center full of people who didn't know what to do. I had flight attendants sitting in hotel rooms for three days who couldn't get a hold of us. I had pilots e-mailing me saying, 'I'm available, what do I do?'"

The cancellations during the five-day period cost the airline an estimated $20 million in revenue and an additional $24 million in flight vouchers given to customers who were impacted by the disruptions.[27] Within days of the storm, JetBlue lowered its operating margin forecast for the fiscal quarter and the year; investors immediately responded by selling off their shares of Jet Blue stock.[28] As the losses mounted, Neeleman became obsessed with finding a way to restore JetBlue's sterling reputation and win back disillusioned customers.

MISERY LOVES COVERAGE

"Call it the perfect storm, the imperfect storm, the Valentine's Day Massacre," said one JetBlue vice president.[29] Regardless of the label that the public affixed to the crisis, JetBlue officials knew the media interest in the story would be sky high. The company's corporate communications department fielded roughly 5,000 telephone

---

[20] M. Daly, "How Two Pilots Put Silver Lining in JetBlue Clouds," *New York Daily News*, February 18, 2007, p. 12.

[21] Bailey, "Long Delays Hurt Image of JetBlue."

[22] Strickler, "Stormy Weather."

[23] Lee, "JetBlue Flight Snarls Continue."

[24] J. Bailey, "JetBlue Cancels More Flights, Leading to Passenger Discord," *The New York Times*, February 18, 2007, p. A31.

[25] Ibid.

[26] Ibid.

[27] Ibid.

[28] P. Korkki, "Investors Mostly Glum in a Short Trading Week," *The New York Times*, February 27, 2007, p. C10.

[29] B. Capps, "Management's Misjudgment Gives JetBlue a Black Eye," *Advertising Age* 78, no. 18 (April 30, 2007). Retrieved November 10, 2007, from Business Source Premier database.

inquires from the media between February 14 and February 19.[30]

JetBlue's reputation as a successful and off-beat upstart airline only seemed to invite sensational newspaper headlines during the crisis. The *New York Post* published an article under the banner "Air Refugees in New JFKaos; Hordes Camp Overnight Before JetBlue Says: 'Tough Luck, No Flights.'" A *New York Times* story entitled "Long Delays Hurt Image of JetBlue" similarly predicted reputational damage for the carrier as a result of the crisis. The headline of a *Newsday* article asked the question virtually every industry observer wanted to know: "Can JetBlue Recover?" For their part, angry JetBlue customers provided plenty of material to reporters in search of a sound bite.

CONGRESS COMES CALLING

Just days after JetBlue's operational meltdown at JFK, members of Congress began calling for legislation designed to prevent air travelers from being held captive inside grounded airplanes for excessive amounts of time. Many suggested that the implementation of an industry-wide passenger bill of rights would be necessary to spur major airlines to action. Legislators argued that a bill of rights would entitle passengers to receive standardized compensation from carriers that fail to meet certain service levels, such as a flight that remains on the runway for hours after pushing back for departure.

With all eyes on the embattled company, JetBlue leaders knew they had to choose their public relations battles carefully. Leaders recognized that the company was at a crossroads. One option was to place a greater emphasis on the winter storm's role in the operational problems at JFK and across the country. The strategy of redirecting blame had certainly worked for other airlines in the past; after all, the public generally accepted that weather was a frequent cause of air travel disruptions. The corporate communications team at JetBlue's Queens-based headquarters also debated whether to put David Neeleman on the television news and talk show circuit, in addition to the YouTube mea culpa he had already issued.

But the biggest decision facing JetBlue's leadership team was a proposal set forth by Neeleman himself just days earlier. For JetBlue to regain its former prestige, Neeleman said that the airline had to do something novel, something impressive, something no competitor had ever done before to make amends to its customers. "I can flap my lips all I want," Neeleman said.[31] "Talk is cheap. Watch us."

Neeleman suggested a gambit that was likely to garner much needed positive attention for the beleaguered airline but would also commit the company indefinitely to millions of dollars in potential losses. Neeleman's idea was a JetBlue Airways Customer Bill of Rights that would specify in no uncertain terms how passengers would be compensated if the company failed to meet certain performance standards.

For example, customers would receive vouchers good toward future travel if their flight sat on the tarmac after landing for more than a certain number of minutes. The value of these credits would escalate the longer the passengers were forced to wait on board the plane. In essence, JetBlue would be putting its money in place of its mouth.

Members of Neeleman's executive team met the idea with skepticism. The ongoing costs associated with such a groundbreaking program would be unpredictable at best and staggering at worst. As the weekend progressed, Neeleman faced countless questions and staunch objections from the heads of Jet Blue's legal, finance, flight operations, government affairs, and

---

[30] J. Elsasser, "True Blue: After a Customer Relations Crisis, Lessons Learned at JetBlue," *Public Relations Strategist* 13, no. 3 (2007), pp. 14–19.

[31] J. Bailey, "Chief 'Mortified' by JetBlue Crisis," *The New York Times*, February 19, 2007, p. A11

marketing teams, to name a few.[32] No other airline has ever committed to something like this, they argued.

CONCLUSION

Neeleman, who was known for personally answering every customer letter or e-mail he received, viewed the Customer Bill of Rights as absolutely vital to restoring JetBlue's image. He contended that the bill of rights would reaffirm the public's perception that JetBlue viewed air travelers as human beings, not cattle to be shipped from Point A to Point B. "This is going to be a different company because of this," Neeleman said. "It's going to be expensive. But what's more important is to win back people's confidence."[33]

In numerous interviews over the weekend, Neeleman promised that he would reveal JetBlue's redemption plan to the world by Monday, February 19. If a customer bill of rights was going to be part of that plan, the CEO still had to convince many influential people inside the company.

CASE QUESTIONS

1. How could JetBlue have better communicated with its internal stakeholders across the country on Valentine's Day and during the days that followed to enhance its image with customers?

2. Should the corporate communications team at JetBlue have arranged for CEO David Neeleman to appear on the national television news and talk show circuit following the crisis? What might be the potential benefits and risks to the company's reputation?

3. Would you recommend a corporate advertising program for JetBlue?

4. If implemented, how would you market the JetBlue Airways Customer Bill of Rights to external and internal stakeholders? How would this affect JetBlue's reputation?

[32] White, personal interview.

[33] Bailey, "Chief 'Mortified.'"

CHAPTER FIVE

Corporate Responsibility

In the previous chapter, we discussed the importance of corporate reputation—the sum of an organization's constituency perceptions built up over time. An increasingly significant contributor to corporate reputation is the notion of corporate responsibility (CR), which is a corporation's social and environmental obligations to its constituencies and greater society. This new lens is increasingly being used by constituencies ranging from the general public to investors to analyze and critique modern-day corporate behavior.

When did society's expectations of corporations shift to include responsible and accountable behavior in addition to profit making? As recently as two decades ago, the general public viewed "do-gooding" as the primary domain of nonprofit organizations and good Samaritans. At the time, many considered businesses to be purely self-interested entities. Positioned in a corner directly opposite charities, the purpose of a corporation was profit maximization, with any efforts to give back to the community limited to check-writing and philanthropy at an arm's length. Milton Friedman, a University of Chicago economist, embodied the belief that businesses should be strictly economic, whereas governments and nonprofits should handle social and environmental issues. In the 1970s, Friedman's doctrines became famous through his *New York Times Magazine* article "The Social Responsibility of Business Is to Increase Its Profits," in which he declared: "What does it mean to say that 'business' has responsibilities? Only people can have responsibilities. A corporation is an artificial person and in this sense may have artificial responsibilities, but 'business' as a whole cannot be said to have responsibilities, even in this vague sense."[1]

In the 1970s, society began to more actively question the means by which corporations generate profits, acknowledging for the first time that corporate practices and society's well-being are closely linked.[2] Corporations became more environmentally aware once large-scale disasters such as Union Carbide's chemical leak in Bhopal, India, in 1984 and the Exxon *Valdez* oil spill in 1989 sparked wide-

---

[1] Milton Friedman, "The Social Responsibility of Business Is to Increase Its Profits," *The New York Times Magazine*, September 13, 1970.

[2] Joshua Daniel Margolis and James Patrick Walsh, *People and Profits? The Search for a Link between a Company's Social and Financial Performance* (London: Lawrence Erlbaum Associates, 2001).

spread uproar about the irresponsibility of big business.[3] In the 1990s, a series of exposés in the mainstream media revealed to many consumers for the first time the "sweatshop" labor conditions and child labor used in garment and footwear supply chains by companies such as Nike and the Kathie Lee line of clothing sold at Walmart. The exposés led to consumer outrage and boycotts, which prompted corporations to adopt codes of conduct to protect workers' rights.

Today, companies are increasingly aware of the impact that their operations have in their many communities, and beyond. We see companies forging into unprecedented territory by tackling issues ranging from income inequality and global pandemics to climate change and access to clean water—issues previously considered to be unrelated to their organizational mission. They are implementing community programs and partnerships with nongovernmental organizations (NGOs) and, most innovatively, are adapting their own business models to be more responsible and sustainable. In the new millennium, the for-profit and nonprofit sectors are no longer at odds; instead, the once distinct lines between them are blurring. The 2010 Edelman goodpurpose Study found that 86 percent of global consumers felt that "business needs to place at least equal weight on society's interests as on business' interests." Sixty-one percent of these consumers reported having a higher opinion of companies engaged in customer relations activities, regardless of the companies' motives for doing so. The study authors suggest that "purpose" is now the fifth "P" in the classic 4P's marketing model (product, price, placement and promotion).[4]

Many global executives today view corporate responsibility as being critical to their business strategy and operations. The 2011 PricewaterhouseCoopers CEO study revealed that nearly half of CEOs are planning to alter their products and practices in the next three years in response to increasing public pressure for more corporate responsibility.[5] A 2011 KPMG study suggested that companies and investors are increasingly looking at corporate responsibility initiatives as a proxy for "risk management, management quality, business reputation, and fundamental to the long-term performance of a company as well as enhanced long-term sustainable returns to investors."[6]

What Is Corporate Responsibility?

Corporate responsibility (often referred to as *corporate social responsibility*), *corporate citizenship*, *sustainability*, and even *conscious capitalism* are some of the terms bandied about in the news media and corporate marketing efforts as companies jockey to win the trust and loyalty of constituents around the world.[7] The acronym *ESG*, which stands for environment, social, and governance, is also used to describe corporate responsibility initiatives. The term *triple bottom line*, popularized

---

[3] "Just Good Business," *The Economist*, January 17, 2008.

[4] 2010 Edelman goodpurpose Study, Edelman, 2010, http://www.edelman.com/insights/special/ GoodPurpose2010globalPPT_WEBversion.pdf.

[5] 14th Annual Global CEO Study, PricewaterhouseCoopers, 2011, http://www.pwc.com/gx/en/ceo-survey/.

[6] International Survey of Corporate Responsibility Reporting, KPMG, 2011, http://www.kpmg.com/Global/en/ IssuesAndInsights/ArticlesPublications/corporate-responsibility/Documents/2011-survey.pdf.

[7] "Conscious Capitalism: Now *Creed* Is Good," BBC News, May 4, 2000.

in 1994 by John Elkington, founder of British consulting firm SustainAbility, is also used in corporate responsibility conversations and refers to profit, people, and planet.[8] With such a wide array of terms actively in use in business lexicon, we will define corporate responsibility at the outset of the chapter to guide the rest of our discussion on its effects on corporate reputation.

Corporate responsibility (CR) describes an organization's respect for society's interests, as demonstrated by taking ownership of the effect its activities have on key constituencies, including customers, employees, shareholders, communities, and the environment, in all parts of its operations. In short, CR prompts a corporation to look beyond its traditional bottom line (economic profit or loss) to consider the greater social implications of its business. This accountability often extends beyond baseline compliance with existing regulations to encompass voluntary and proactive efforts to improve the quality of life for employees and their families as well as for the local community and society at large. A responsible company makes a concerted attempt to reduce the negative social and environmental footprint of its operations through a thoughtfully developed strategy implemented over the long term and not merely through temporary, stopgap measures such as monetary contributions to charitable causes.[9] For example, ExxonMobil donating $250 million over 32 years to sponsor Masterpiece Theatre qualifies as philanthropy, but it cannot be categorized as CR as it makes no effort to mitigate the lasting impact of the company's operations. In contrast, Starbucks' efforts to minimize the negative effects of its coffee supply chain and retail operations by purchasing beans from fair trade growers and paying its employees wages higher than industry averages serve as cornerstones of its CR strategy.[10]

Many times, a company's corporate responsibility efforts involve donations of time and expertise, as opposed to cash. However, *Corporate Responsibility Magazine* observed in 2010 that the days of "checkbook philanthropy" were over and noted that contributions of in-kind products, services, and experience represented greater than 65 percent of corporate contributions.[11] In 2010, 64 percent of global consumers felt that it was no longer enough for a company to only donate money. Instead, consumers felt that companies must incorporate good works into the fabric of their business.

As companies focus on organizing and prioritizing their CR efforts, they are increasingly connecting with other organizations to develop guidelines and standards. As of December 2011, over 6,000 businesses in 135 countries around the world signaled their commitment to sustainability of human and environmental resources by participating in the "Global Compact" of the United Nations (UN).

To help guide companies in their corporate responsibility efforts, the UN Global Compact drew upon several diplomatic human rights and sustainability documents

---

[8] "Triple Bottom Line: It Consists of Three P's: Profit, People and Planet," *The Economist*, November 17, 2009, http://www .economist.com/node/14301663.

[9] "The ROI of CSR: Q&A with Geoffrey Heal," *Columbia Ideas at Work*, Columbia Business School, Spring 2008.

[10] Ibid.

[11] Jay Whitehead, "Black List Methodology," *Corporate Responsibility Magazine*, 2010, http://thecro.com/content/ bad-business-crs-black-list.

when crafting its list of Ten Principles. Companies can use this list to guide their corporate responsibility initiatives.

The UN Global Compact Ten Principles

Human Rights

Principle 1: Businesses should support and respect the protection of internationally proclaimed human rights; and

Principle 2: make sure that they are not complicit in human rights abuses.

Labour

Principle 3: Businesses should uphold the freedom of association and the effective recognition of the right to collective bargaining;

Principle 4: the elimination of all forms of forced and compulsory labour;

Principle 5: the effective abolition of child labour; and

Principle 6: the elimination of discrimination in respect of employment and occupation.

Environment

Principle 7: Businesses should support a precautionary approach to environmental challenges;

Principle 8: undertake initiatives to promote greater environmental responsibility; and

Principle 9: encourage the development and diffusion of environmentally friendly technologies.

Anti-Corruption

Principle 10: Businesses should work against corruption in all its forms, including extortion and bribery.[12]

The UN also launched the Principles for Responsible Investment initiative in 2006 to help investors identify and invest in companies with strong corporate responsibility practices. By December 2011, over 915 investor groups with combined assets of $US 30 trillion under management had joined the initiative.[13]

In shaping a CR strategy, a corporation ideally acknowledges and integrates the full spectrum of constituencies' "extra-financial" concerns—social, environmental, governance, and others—into its strategy and operations. The Global Reporting Initiative (GRI) describes five interdependent capital asset classes: financial, human, natural, social and technological.[14] *The Economist* has described CR as "part of what businesses need to do to keep up with (or, if possible, stay slightly ahead of) society's fast-changing expectations."[15] Developing an authentic CR strategy

[12] "The Ten Principles," United Nations Global Impact, http://www.unglobalcompact.org/AboutTheGC/TheTenPrinciples/index.html.

[13] http://www.unpri.org/about.

[14] "Carrots and Sticks: Promoting Transparency and Sustainability," Global Reporting Initiatives, 2010, https://www.globalreporting.org/resourcelibrary/Carrots-And-Sticks-Promoting-Transparency-And-Sustainbability.pdf.

[15] "Do It Right," *The Economist*, Special Report: Corporate Social Responsibility, January 17, 2008.

signals a corporation's intent to look beyond short-term financial returns and focus on long-term success and sustainability by managing those expectations. This consideration often requires the executives of public companies to fight prevailing pressures to achieve strong quarterly results at the expense of longer-term, often less tangible benefits.[16]

Despite these challenges, Harvard Business School guru Michael Porter and consultant Mark Kramer argue that CR is a strategy that, if implemented thoughtfully and thoroughly, can enhance a corporation's competitiveness. They analyze the interdependence of a company and society by using the same tools used to analyze overall competitive positioning and strategy development. In this way, CR can be used strategically to set an "affirmative [CR] agenda that produces maximum social benefit as well as gains for the business."[17] A CR strategy should not be reactive but should *proactively* identify the social consequences of a company's entire value chain—the full spectrum of all of the activities it engages in when doing business—to pinpoint potential problems and opportunities wherever business and society intersect.[18]

In 2011, almost half of the 250 largest global companies reported gaining financial value from their CR initiatives. KPMG has identified the two principle drivers of increased value from corporate responsibility as: cost savings and improved reputation.[19] A recent IBM survey found that 87 percent of executives were focusing on CR activities that would help them to improve efficiency, and 69 percent were focusing on CR activities that would help with new ideas for revenue generation.[20]

To help companies with strong corporate responsibility platforms gain more credibility and recognition, an American non profit, B Lab, has created the concept of the B corporation certification, with the "B" standing for "Benefit." B corporations are companies that meet "rigorous and independent standards of social and environmental performance, accountability, and transparency." Companies can apply to B Lab for B corporation status much in the same way that companies can apply to certifying bodies to achieve fair trade, organic or LEED certification. At the end of 2011, B Lab reported that 502 companies with a combined $US 2.5 billion in revenue had become B corporations. These companies represent 60 industries and include investment groups and construction firms. Consumer product companies such as King Arthur Flour, Dansko, Method, Seventh Generation, Guayaki, EO Products, and Numi Organic Tea have achieved B corporation status.

On its website, B Lab lists the following reasons that companies may want to become a B corporation: "Differentiate your brand, maintain mission, save money [particularly via partnerships and discounts negotiated for members by

[16] Henry Mintzberg, Robert Simons, and Kunal Basu, "Beyond Selfishness," *MIT Sloan Management Review* 44, no. 1 (Fall 2002).

[17] Michael E. Porter and Mark R. Kramer, "Strategy & Society: The Link between Competitive Advantage and Corporate Social Responsibility," *Harvard Business Review*, December 2006.

[18] Ibid.

[19] "International Survey of Corporate Responsibility Reporting 2011," KPMG, 2011, http://www.kpmg.com/Global/en/ IssuesAndInsights/ArticlesPublications/corporate-responsibility/Documents/2011-survey.pdf.

[20] "Corporate Social Responsibility: Leading a Sustainable Enterprise," IBM Institute for Business Value, 2009, http://public .dhe.ibm.com/common/ssi/ecm/en/gbe03226usen/GBE03226USEN.PDF.

B Lab], generate press, attract investors, improve and benchmark performance, and build a movement."[21]

In addition to its own certification, B Lab has taken its mission a step further and is working with state governments to legitimize the B corporation as a legal incorporation option. Similar to the C corporation, S corporation, limited liability company (LLC), and limited liability partnership (LLP), B corporation status reflects the organizational structure of a company, as well as the tax laws that affect it. By the end of 2011, California, Hawaii, Virgina, Maryland, Vermont, and New Jersey had passed legislation recognizing B corporations.[22] B Lab is pursuing this agenda because it believes that "current corporate law makes it difficult for businesses to consider employee, community, and environmental interests when making decisions."

With more states, including New York, reviewing B corporation legislation, and with consumers agitating for more corporate responsibility, it seems increasingly likely that corporations that do not make an effort to carve out their own CR niche will be left trailing their competition.

The Twenty-First Century's CR Surge

Corporations are increasingly aware that as they look out for society's best interests, they are actually looking out of their own interests, too, particularly in the long run. As Charles Handy notes, "business needs a sustainable planet for its own survival, for few companies are short-term entities; they want to do business again and again, over decades."[23] Businesses do not exist in a vacuum—they inevitably intersect with society and are mutually dependent for their survival. As *Financial Times* assistant editor Michael Skapinker argues, "companies cannot thrive in collapsing societies. Without political stability, the future of business is grim. . . . Even in the most stable countries, companies need the community's approval to function. Opinion can turn against them fast: witness European consumers' distaste for genetically modified food, or the attacks on pharmaceutical companies over the pricing of AIDS drugs in Africa."[24] This argument includes corporations' need for an environmentally stable context in which to operate. Pressing environmental and social issues today—from climate change to income inequality—pose serious threats to "business-as-usual" operations. Sal Palmisano, Chairman of the Board and former CEO of IBM, describes the new expectations corporations must meet to survive in light of these risks: "All businesses today face a new reality. . . . Businesses now operate in an environment in which long-term societal concerns—in areas from diversity to equal opportunity, the environment and workforce policies—have been raised to the same level of public expectation as accounting practices and financial performances."[25]

---

[21] http://www.bcorporation.net/.

[22] http://www.bcorporation.net/.

[23] Charles Handy, "What's a Business For?" *Harvard Business Review*, December 2002.

[24] Michael Skapinker, "Corporate Responsibility Is Not Quite Dead," *Financial Times*, February 12, 2008.

[25] Daniel Yankelovich, *Profit with Honor: The New Stage of Market Capitalism* (New Haven: Yale University Press, 2006), p. 9.

Corporations slow to adapt to this new reality pay a price. An often-cited example is Walmart's 2004 discovery of a report prepared by McKinsey & Co. subsequently made public by walmartwatch.com, a public education campaign devoted to challenging Walmart to become a better corporate citizen revealing that up to 8 percent of Walmart consumers surveyed at the time had ceased shopping at the chain because of its reputation, which at the time included a perceived CR deficit.[26] Walmart's CEO at the time, Lee Scott reacted with the comment: "We thought we could sit in Bentonville, take care of customers, take care of associates—and the world would leave us alone. It doesn't work that way anymore."[27] In a published statement, Scott also admitted Walmart had been caught off-guard by its entanglement in social and environmental issues: "To be honest, most of us at Walmart have been so busy minding the store that the way our critics have tried to turn us into a political symbol has taken us by surprise. But one thing we've learned from our critics . . . is that Walmart's size and industry leadership mean that people expect more from us. They're right to, and when it comes to playing our part . . . we intend to deliver."[28]

People today are expecting more. *The Economist* has described CR as "a do-gooding sideshow" that has now turned mainstream.[29] In 2010, IBM surveyed 1,500 global CEOs and found that 76 percent of the CEOs believed that they are expected by customers to increase their focus on social responsibility initiatives.[30] In today's world of heightened awareness of climate change, human rights, and scarcer resources, a corporation's "extra-financial" behavior—how well it treats its stakeholders and the world in which it operates—contributes greatly to its trustworthiness. Trust is not an abstract notion; it can have a significant impact on a company's bottom line. For example, the 2011 Trust Barometer published by the international public relations firm Edelman revealed that 73 percent of people have refused to buy the products or use the services of a corporation they do not trust.[31]

Large corporations started the new millennium on a precarious note, the effects of which still linger today. Enron's and WorldCom's respective scandals shocked the world and undermined the average person's trust in the motives and operations of big business. Enron's now famous Code of Conduct—last published in July 2000, prior to the company's downfall—described such fundamental values as respect, integrity, communication, and excellence. Belief in the altruistic motives of big business subsequently crashed; by 2002, a *BusinessWeek*/Harris survey reported that 79 percent believed that "most corporate executives put their own personal interests ahead of employees and shareholders."[32] Over a decade later, the general public still demonstrates low levels of faith in corporations.

[26] Marc Gunther, "The Green Machine," *Fortune*, August 7, 2006.

[27] "The Debate over Doing Good," *BusinessWeek*, August 15, 2005.

[28] Yankelovich, *Profit with Honor*, p. 10.

[29] "Just Good Business."

[30] "Capitalizing on Complexity: Insights from the Global Chief Executive Officer Study, IBM," 2010, http://www-935.ibm.com/services/c-suite/series-download.html.

[31] Edelman Trust Barometer 2007, http://www.edelman.com.

[32] Yankelovich, *Profit with Honor*, p. 24.

RESPECT: We treat others as we would like to be treated ourselves. We do not tolerate abusive or disrespectful treatment. Ruthlessness, callousness, and arrogance don't belong here.

INTEGRITY: We work with customers and prospects openly, honestly and sincerely. When we say we will do something, we will do it; when we say we cannot or will not do something, we won't do it.

COMMUNICATION: We have an obligation to communicate. Here, we take the time to talk with one another . . . and to listen. We believe that information is meant to move and that information moves people.

EXCELLENCE: We are satisfied with nothing less than the very best in everything we do. We will continue to raise the bar for everyone. The great fun here will be for all of us to discover just how good we can really be.

Our Values

RESPECT We treat others as we would like to be treated ourselves. We do not tolerate abusive or disrespectful treatment. Ruthlessness, callousness and arrogance don't belong here.

COMMUNICATION We have an obligation to communicate. Here, we take the time to talk with one another... and to listen. We believe that information is meant to move and that information moves people.

INTEGRITY We work with customers and prospects openly, honestly and sincerely. When we say we will do something, we will do it; when we say we cannot or will not do something, then we won't do it.

EXCELLENCE We are satisfied with nothing less than the very best in everything we do. We will continue to raise the bar for everyone. The great fun here will be for all of us to discover just how good we can really be.

In June 2011, Gallup reported that only 19 percent of Americans say they have a "great deal" or "quite a lot" of confidence in big business, compared to 78 percent for the military.[33]

At the same time, widespread Internet access—with an estimated 2.1 billion people online as of March 2011[34]—has redefined the notion of transparency for corporations. The Internet and social media sites such as Twitter and Facebook now serve as powerful forums for like-minded people to educate and organize themselves. Individuals also now have a powerful tool for spreading once proprietary company information. It is easier than ever for constituents to monitor companies and to criticize them for everything from human rights violations that take place in a distant corner of a company's supply chain to carbon emissions that are in excess of local regulatory limitations. Even traditional media is benefiting from the Internet when reporting on corporate responsibility issues. In 2010, *Corporate Responsibility Magazine* published a "black list" of the 30 companies that it rated worst in terms of transparency regarding their corporate responsibility practices. *Corporate Responsibility Magazine* made a point of noting that the 30 "black list" companies had underperformed both the S&P 500 and the magazine's list of "100 Top Corporate Citizens," based on the three-year total return.[35] This controversial list became fodder for many online blogs and websites, and so reached a much wider audience than its magazine counterpart alone.

Further attention is paid to corporations' CR efforts through a proliferation of socially responsible indices and rankings, such as the "Best in Social Responsibility" category on *Fortune*'s Most Admired Companies list. In 2009, Newsweek began publishing its annual Green Rankings to evaluate corporate responsibility initiatives. Many corporations today vie for inclusion on widely admired indices including the FTSE4Good Index—an index created by benchmarking company FTSE and designed specifically to help measure the performance of companies meeting globally recognized corporate responsibility standards.[36] Another index of such companies is the Dow Jones Sustainability World Index, which is comprised of the top 10 percent of 2,500 companies worldwide according to longterm economic, environmental, and social sustainability criteria.[37]

These communication channels and points of engagement directly influence constituencies' impressions of a corporation. A corporation lacking a CR strategy and a clear communication plan for its CR strategy runs the risk of losing control of its reputation in today's highly-networked and highly-scrutinized business environment.

The Upside of CR

Although CR is taking center stage thanks to a business environment of proliferating risks, adopting a socially responsible strategy can offer a compelling upside

---

[33] "Confidence in Institutions," Gallup poll, June 9–12, 2008, http://www.gallup.com.

[34] "Internet Usage and World Population Statistics," Internet World Stats, http://www.internetworldstats.com, March 31, 2008.

[35] Dirk Olin and Mark Bateman, "Bad Business—CR's Black List," *Corporate Responsibility Magazine*, March 2010.

[36] http://www.ftse.com/Indices/FTSE4Good_Index_Series/index.jsp.

[37] http://www.sustainability-index.com.

to corporations. Contrary to Friedman's claims, responsible business practices do not necessarily undermine a corporation's profit motive. In fact, many CEOs today describe acting responsibly as pragmatic—it makes good business sense. A well-executed CR strategy can translate into an array of benefits, including attracting and retaining customers, identifying and managing reputational risks, attracting the best-quality employees, and reducing costs. Walmart—top company on the *Fortune* 100, driven by a fierce cost-cutting mantra—explains the value of CR from a strategic perspective. In former CEO Lee Scott's words: "By thinking about sustainability from our standpoint, it is really about how do you take the cost out, which is waste, whether it's through recycling, through less energy use in the store, through construction techniques we're using, through the supply chain. All of those things are simply the creation of waste."[38] Cutting costs allows Walmart to charge even lower prices, which supports its mission of helping customers to, "Save Money. Live better." General Electric (GE) has also achieved significant cost savings through eco-friendly action, investing in alternative energy technologies in 2002 when oil was priced at $25 per barrel. In 2011, with oil prices six times higher and rising, GE is reaping the benefits of the demand it predicted nine years ago.[39]

The scale and nature of the benefits from CR activities for an organization can vary depending on the business and are often difficult to quantify, though increased efforts are being made to link CR initiatives directly to financial performance. In the meantime, a strong business case exists that CR makes good business sense and positively affects the bottom line.

Reputation risk management

Managing reputational risk is a central part of any robust corporate communication strategy. As Berkshire Hathaway CEO Warren Buffett once famously noted: "It takes 20 years to build a reputation and five minutes to ruin it. If you think about that, you'll do things differently." Corruption scandals or environmental accidents can devastate a carefully honed corporate reputation in a matter of days. These events can also draw unwanted attention from regulators, courts, governments, and media. Building a genuine culture of "doing the right thing" within a corporation— the foundation of any genuine CR strategy—can help offset these risks.

Brand differentiation

In crowded marketplaces, companies strive for a unique selling proposition that can separate them from the competition in consumers' minds. Corporate responsibility can help build customer loyalty based on distinctive ethical values. Several major brands, such as Stonyfield, Seventh Generation, TOMS shoes, and The Body Shop, are built on such ethical values. GE CEO Jeffrey Immelt emphasized the importance of differentiating a brand by staying ahead of issues and evolving with ever-changing constituency concerns: "When society changes its mind, you better be in front of it and not behind it, and [sustainability] is an issue on which society

[38] "Alan Murray," *The Wall Street Journal*, March 24, 2008.
[39] Morgen Witzel, "A Case for More Sustainability," *Financial Times*, July 2, 2008.

has changed its mind. As CEO, my job is to get out in front of it because if you're not out in front of it, you're going to get [ploughed] under."[40]

Talent attraction and retention

As we will discuss in more detail later in the chapter, a CR program can aid in employee recruitment and retention. It can also help improve the image of a company among employees, particularly when they become involved through fundraising activities, community volunteering, or helping shape the company's CR strategy itself. Using these tactics to strengthen goodwill and trust among present and future employees can translate into reduced costs and greater worker productivity. In 2010, an estimated 34 percent of employees said that they would take a pay cut to work for a socially responsible firm.

Once a company decides to implement corporate responsibility practices, it should be sure to communicate them to its employees and other key constituencies to maximize the return on its efforts. That same 2010 study found that a full 53 percent of employees were not sure if their company had any CR practices in place.

License to operate

Corporations want to avoid interference in their business through taxation or regulations. By taking substantive voluntary steps, they may be able to persuade governments and the wider public that they are taking current issues such as health and safety, diversity, or the environment seriously and thus avoid intervention. Expenses today can result in future cost savings or increased revenue streams from new, socially responsible products and services. Consider DuPont saving more than $2 billion from energy use reductions since 1990—an upfront investment that, years later, continues to pay in spades.[41] In the words of Abby Joseph Cohen, senior U.S. investment strategist at Goldman Sachs, "Companies are taking a broader view that allows them to see that a cost today may reduce future liabilities, and the reduction of those future liabilities in turn has a positive impact on their cost of capital."[42] Acting before regulations force them to can position corporations as well-respected leaders in responsibility and sustainability.

CR critics

Despite mounting evidence in support of CR's benefits, followers of Milton Friedman and others continue to argue there is no place for social responsibility in business. These critics rail against CR as detracting from a corporation's commercial purpose and effectiveness, thereby inhibiting free markets. In this view, responsibility and profitability constitute a zero-sum game; corporations are for-profit institutions whose primary purpose is profit and who lose competitiveness through altruistic, profit-diminishing behavior. Some critics claim CR is little more than a public relations strategy, in which companies cherry-pick their good

---

[40] Ibid.

[41] Porter and Kramer, "Strategy & Society."

[42] Stephanie Strom, "Make Money, Save the World," *The New York Times,* May 6, 2007.

activities to showcase and ignore the others, creating an inaccurate image of a socially or environmentally responsible company. Others contest that CR programs are often undertaken in an effort to distract the public from the ethical questions posed by their core operations. In general, however, constituencies are increasingly calling for more corporate responsibility and demanding that companies rise to the occasion.

CR and Corporate Reputation

As we move on from our overview of CR and its general benefits, we will focus our discussion on one central question: What kind of contribution can responsible business practices make to the strengthening of a corporation's reputation? Research indicates its effect is growing, with more than one-half of business executives believing that a recognized commitment to corporate responsibility contributes "a lot" to a company's overall reputation.[43] Studies have also confirmed that the average person's decisions about what to buy and whom to do business with are influenced by a company's reputation for social responsibility. In turn, CR has become a critical means to build trust with corporate constituents. College-educated Americans today have begun viewing social responsibility as more important than an overall corporate brand or financial performance, second only to the quality of their products and services when deciding which companies to trust. In 2010, a study revealed that, despite the recession, 70 percent of consumers were willing to pay more for a product from a socially responsible company.[44] Interestingly, the 2010 goodpurpose Edelman study found that consumers in emerging markets were more likely to purchase products from brands that support a good cause than those in developed markets. Eighty percent of consumers in Brazil, China, Mexico, and India expect companies to donate a portion of profits to a good cause.[45]

Although the evidence for engaging in corporate responsibility initiatives is compelling, companies should be aware that instituting and publicizing CR activities in an effort to boost reputation can backfire. A *McKinsey Quarterly* article presented research that found that companies with poor perceived product quality can actually be *hurt* by publicizing their CR efforts because consumers then assume that the company is focusing on the wrong issues. The study's authors recommend that companies follow these three steps: (1) "don't hide market motives" (consumers understand a need for profit), (2) "serve stakeholders' true needs," and (3) "test your progress."[46] A recent study indicated that sustainability factors were still very rarely the primary purchase driver.[47] Research published in 2011 in the *Journal of Service Research* found that, particularly in the case of an educated consumer,

---

[43] *Safeguarding Reputation*, No. 2, Weber Shandwick/KRC Research, 2006.

[44] Corporate Social Responsibility Branding Survey, 2010.

[45] 2010 Edelman goodpurpose Study.

[46] CB Bhattacharya, Daniel Korschun, and Sankar Sen, "What Really Drives Value in Corporate Responsibility," *McKinsey Quarterly*, December 2011, https://www.mckinseyquarterly.com/What_really_drives_value_in_corporate_responsibility_2895.

[47] The Futures Company, "How to Sustain Sustainability," 2011, http://futuresco.thefuturescompany.com/file_depot/0-10000000/0-10000/1/conman/How+to+Sustain+sustainability+1.pdf.

being known for high-quality products and services more effectively protects a company's reputation than being known for good corporate responsibility does.[48]

Despite the risks, the acknowledgment of CR's significant effect on reputation, constituency trust, and even the bottom line, many corporations are not capitalizing on these trends. A significant gap exists between executives recognizing the importance of CR and companies taking action to implement a thoughtful and effective CR strategy. Although almost three-quarters of CEOs say companies should integrate environmental, social, and governance issues into their strategies and operations, only half say that their respective companies are actually doing so.[49] This gap presents an opportunity—many corporations still have the chance to take responsible action and, in doing so, differentiate themselves from the competition and build valuable goodwill.

We will now examine several key corporate constituencies—customers, investors, employees, NGOs, and environmentalists—to take a closer look at each group's evolving expectations of corporate responsibility and what corporations can and should be doing in response to strengthen their reputations.

Consumer Values and Expectations: Taking Matters into Their Own Hands

Millions of everyday consumers possess the unprecedented power to determine the fates of corporations. In his book *Supercapitalism*, former U.S. Labor Secretary Robert Reich contends that the average person had an easier time expressing values as citizens through democracy 30 years ago.[50] He argues that today many use their role as consumers to communicate those same values at the cash register.[51] We also see more individuals taking matters of philanthropy into their own hands. Founded in 2005, Kiva—a California-based website that connects lenders and borrowers from around the world—has amassed some 650,000 lenders who make online donations in $25 increments. These lenders have funded more than 695,000 borrowers from Tanzania to Tajikistan with $US 26 million lent thus far.[52] On its website, Kiva boasts a 98.96 percent repayment rate. The site has sparked a new model of philanthropy, connecting donors and recipients, eliminating the middleman, and, in the process, empowering individuals. At the same time, consumers' personal values are reflecting a greater individual commitment to responsibility. In 2010, Americans gave almost $US 291 billion to charities, down from a historic high of more than $US 300 billion in 2007, but up almost 4 percent from the 2009 giving level.[53]

Compelling evidence exists that consumers are willing to use this individual empowerment to act on their values—reaching into their pocketbooks to pay

---

[48] Andreas B. Eisingerich and Gunjan Bhardwaj, "Does Social Responsibility Help Protect a Company's Reputation?" *MIT Sloan Management Review*, March 23, 2011, http://sloanreview.mit.edu/the-magazine/2011-spring/52313/does-social-responsibility-help-protect-a-companys-reputation/.

[49] Debby Bielak, Sheila M. J. Bonini, and Jeremy M. Oppenheim, "CEOs on Strategy and Social Issues," *The McKinsey Quarterly*, October 2007.

[50] Matt Woolsey, "Supercapitalism: Transforming Business," *Forbes*, September 6, 2007.

[51] Ibid.

[52] "When Small Loans Make a Big Difference," *Knowledge@Wharton*, June 3, 2008.

[53] "The Annual Report on Philanthropy for the Year 2010," Giving USA Foundation, 2011, http://www.givingusareports.org/products/GivingUSA_2011_ExecSummary_Print.pdf.

more for the sake of corporate responsibility. Consider fair trade coffee, which is priced at a premium compared with regular, noncertified coffee. The International Fair Trade Association defines fair trade as incorporating "concern for the social, economic and environmental well-being of marginalized small producers . . . not maximiz[ing] profit at their expense."[54] Despite the higher price tag, demand for fair trade products continues to grow; Fairtrade International reported that global consumers spent approximately $4.36 billion euros ($US 5.66 billion) on certified products in 2010, a 28 percent increase over 2009.[55]

Organic food—viewed by many as better for their health, for the health of farm workers, and for the environment—has experienced a similar explosion, benefiting organic food retailers such as Whole Foods. The U.S. sales of organic food and beverages have grown from $US 1 billion in 1990 to $US 26.7 billion in 2010.[56] The United States had an estimated 14,500 certified organic farms in 2011, up from 10,000 in 2006, but analysts worry that demand is still outstripping supply.

In addition to going the extra mile to pay more for socially responsible products, consumers have been willing to punish corporations for their lack of corporate responsibility. Research reveals that 35 percent of all Americans have avoided a product because they perceived a company as not socially or environmentally responsible.[57] The practice dates back to 1830, if not earlier, when the National Negro Convention called for a boycott of slave-produced goods and has prompted shifts in corporate behavior throughout the past two centuries.[58] The Rainforest Action Network established its influence as an NGO by orchestrating a boycott of Burger King in 1987 for importing beef from countries where rainforests are destroyed to provide pasture for cattle.[59] Burger King's sales subsequently declined by 12 percent, prompting Burger King to cancel $35 million worth of beef contracts in Central America and announce an end to rainforest beef imports.[60] Or consider Shell reeling from its poor handling of its 1995 decision to dispose of an oil storage platform that it no longer needed, the Brent Spar, by simply sinking it in the Atlantic Ocean. Environmental NGO Greenpeace staged protests, using vivid and emotional language, prompting a widespread boycott of Shell stations in northern Europe, with sales volumes in Germany dropping as much as 40 percent in June 1995.[61]

Not only must corporations be aware of the changing values and behavior of consumers, they also must remember that expectations of corporate responsibility are far from homogeneous across the globe, differing to a sometimes large degree across countries, regions, and hemispheres. Monsanto, the world's largest

---

[54] Andrew Downie, "Fair Trade in Bloom," *The New York Times*, October 2, 2007.

[55] "High Trust and Global Recognition Levels Make Fairtrade an Enabler of Ethical Consumer Choice: Global Poll," GlobeScan for Fairtrade International, October 11, 2011, http://www.globescan.com/news_archives/flo_business/.

[56] "Industry Statistics and Projected Growth," Organic Trade Association, June 2011, http://www.ota.com/organic/mt/business.html.

[57] Andrew Winston, "Conflicted Consumers," *Harvard Business Online*, June 2008.

[58] Leo Hickman, "Should I Support a Consumer Boycott?" *The Guardian* (UK), October 4, 2005.

[59] Rainforest Action Network, http://www.ran.org.

[60] Ibid.

[61] David P. Baron, "Facing-Off in Public," *Stanford Business*, August 2003.

seed company, suffered for failing to recognize the strong opposition to genetically modified (GM) foods prevailing in Europe, based on fears that GM crops can create animal and human health issues and take a negative environmental toll. Despite soaring food prices, Europeans appear to be particularly apprehensive about the use of such crop technologies,[62] whereas similar concerns are far less prevalent in Monsanto's home base in the United States.

Investor Pressures: The Growth of Socially Responsible Investing

Earlier in the chapter, we discussed Milton Friedman's argument that there is no place for social responsibility in for-profit entities. In this view, executives are merely agents of investors, the individuals who own the corporation.[63] How does Friedman's argument reconcile with investors' increased use of capital markets as a mechanism to encourage socially responsible corporate behavior? Investors today are demonstrating an increased interest in socially responsible companies, rewarding them by using CR more frequently as part of their criteria to invest. Almost two-thirds of Americans cite a company's record of social responsibility as "an influential factor when making a decision to purchase stock or invest in a company."[64] As a result, between 10 and 12 percent of money professionally managed in the United States today can be categorized as socially responsible investing (SRI),[65] including social or environmental goals and an investment strategy that employs screening, divesting, or shareholder activism. As *Fortune* magazine describes, this development constitutes a multitrillion dollar "wager that socially responsible companies will outperform companies that don't engage a wide array of stakeholders, from shareholders and customers to employees and activists, in an ongoing conversation about what can be done better."[66] High-profile investment management shops driven by socially responsible missions are springing up, such as Generation Investment Management, founded in 2004 by former U.S. vice president and Nobel Peace Prize winner Al Gore and former Goldman Sachs Asset Management head David Blood, with an investment philosophy that integrates sustainability research with rigorous fundamental equity analysis. The investment strategy is by no means altruistic; it is based on a conviction that sustainability will be an important contributor to long-term business performance. In Blood's own words:

> For long-term investors, assessing a company's ability to create enduring value depends on both financial analysis and analysis of other material, yet often non-traditional, factors. Climate change is one such factor that is a concrete business risk *and* opportunity for some industries, and an emerging issue for others. Researching how well companies are prepared for a lower-carbon world is plain common sense when looking at a long-term investment horizon. Investors of all stripes are beginning to price this into their investment decisions.[67]

[62] Sam Cage, "High Food Prices May Cut Opposition to Genetically Modified Food," Reuters, July 8, 2008.

[63] Friedman, "The Social Responsibility of Business."

[64] "Rethinking Corporate Social Responsibility."

[65] "The ROI of CSR: Q&A with Geoffrey Heal," Columbia Ideas at Work.

[66] Telis Demos, "Beyond the Bottom Line," *Fortune*, October 23, 2006.

[67] David Blood, "A New Climate for Investment," *Financial Times*, September 23, 2006.

Even China—a country with a less-than-pristine environmental and human rights record—announced its $US 200 billion sovereign wealth fund's intentions to seek profits in socially responsible companies by avoiding investments in such industries as gambling, tobacco, and arms manufacturing.[68]

Several large financial institutions, such as Goldman Sachs and UBS, are adapting their research departments to meet the growing demand for equity research that integrates environmental, social, and governance considerations. With a specialized team operating out of London, Goldman Sachs' "intangibles research" incorporates social and environmental risks into its corporate research and analysis, stating that "risks and opportunities arising from climate change and its regulation will be of increasing interest to investors in coming years."[69] Goldman Sachs also launched its GS SUSTAINS initiative in the United States. Executive director of the program Marc Fox explains the economic rationale behind paying attention to extra-financial risks: "We are looking to develop a new way of measuring companies that does not necessarily fit in to the accounting framework. The ultimate goal is to identify long-term investment drivers."[70] Goldman's research is yielding proof that corporate responsibility and sustainability could serve as proxies for good management. What started decades ago as a highly specialized type of investing—gaining momentum through the Vietnam War and the anti-apartheid movement in South Africa—has become a mainstream investment strategy used regularly by pension funds and insurance companies around the world. Active socially responsible investment is on the rise. In the United States, assets that are considered to be socially responsible investments have been growing at a faster rate (34 percent) since 2005 than the total universe of investment assets under professional management (3 percent). By the start of 2010, an estimated $US 3.07 trillion, nearly one-eighth of professionally managed assets, was part of a socially responsible investment.[71]

Responsibility Inside and Out: Employee Involvement in CR

In our chapter on Internal Communications, we underscore the essential role employees play as brand ambassadors for a corporation. The same holds true in the implementation of a CR strategy. The next generation of corporate leaders is actively searching for responsible practices in corporate track records as they recruit and pick a place to start their careers. In 2011, *Harvard Business Review* shared research finding that 88.3 percent of MBA graduates from top programs would take a pay cut to work for an ethically responsible company and would be willing to forgo an average of $US 8,087 in compensation.[72] Top business schools around the world are offering a greater number of corporate responsibility,

[68] "China Fund Shuns Guns and Gambling," *Financial Times*, June 13, 2008.

[69] "Goldman Sachs Nears $1 Billion in Renewable, Clean Energy Investments," *Energy Washington Week* 3, no. 24 (June 14, 2006).

[70] Ibid.

[71] "The 2011 Socially Responsible Investing Report, TIAA-CREF (a member of the Social Investment Forum)." TIAA-CREF, 2010, http://www.tiaa-cref.org/ucm/groups/content/@ap_ucm_p_tcp/documents/document/tiaa01007775.pdf.

[72] "New MBAs Would Sacrifice Pay for Ethics," *Harvard Business Review*'s "The Daily Stat," http://web.hbr.org/email/archive/dailystat.php?date=051711.

values-based leadership, and sustainable enterprise courses and programs, addressing business students' desire not just to work hard but to do some good at the same time.[73] Net Impact, an association of 20,000 business professionals working to improve corporate responsibility, reports that it has active chapters at 90 percent of top MBA campuses.

Once a corporation has attracted top talent, engaging those employees from all levels of the organization in a company's CR efforts is imperative. Employees are often the primary spokespeople for a corporation, responsible for much word-of-mouth information shared and impressions formed. Furthermore, making employees central to a CR strategy can boost employee goodwill and morale, decrease turnover, and increase operational efficiencies by encouraging employees to identify opportunities for sustainability and cost-savings.[74] Many corporations are missing out on this upside—though more than three-quarters of executives say corporate citizenship fits their companies' traditions and values, only 36 percent report talking to their employees about corporate citizenship.[75] IBM serves as an example of a company successfully engaging its employees in CR issues, hosting regular brainstorming sessions focused on corporate responsibility and sustainability. It often refers to its now famous, and still largest, first "InnovationJam" held in 2006.[76] During this InnovationJam, more than 150,000 IBM employees, family members, clients, and partners in 104 countries joined in an online conversation on IBM's global intranet (www.ibm.com/ibm/think-2007). Driven primarily by IBM employees, more than 46,000 observations and ideas were posted on how to translate IBM's technologies into economic and broader societal value. IBM allocated $100 million to explore 10 promising business opportunities suggested, including creating access to branchless banking for the underprivileged masses around the world and working with utility companies to increase power grid and infrastructure efficiency.

Stanley Litow, IBM's vice president of corporate citizenship and corporate affairs and president of IBM's Foundation, further explains IBM's approach to corporate responsibility: "In the *Harvard Business Review*, Rosabeth M. Kanter described the IBM approach as going from "spare change to real change. With the spare change approach, the company makes X amount of dollars and it gives its spare change back to the community, with the goal being generosity. But with the real change approach, you take what is most valuable to the company—in our case, our innovation technology, and the skill and talent of our people—and contribute it into the community. The real change approach is strategic, it's a systemic part of the way we operate as a company, and that is the case for tie-in to business strategy. In the end, it's even more generous to do it that way."[77]

[73] Strom, "Make Money, Save the World."

[74] Gabrielle McDonald, "In-house Climate Change: Use Communication to Engage Employees in Environmental Initiatives. (The Green Revolution)," *Communication World*, November–December 2007.

[75] "Time to Get Real: Closing the Gap between Rhetoric and Reality, The State of Corporate Citizenship 2007," Boston College Center for Corporate Citizenship, December 2007.

[76] http://www.collaborationjam.com/.

[77] "From Spare Change to Real Change: An Interview with Stanley Litow," *LEADERS*, April 2010, http://www.leadersmag.com/issues/2010.2_Apr/Making%20a%20Difference/Litow.html.

Strong evidence exists that the general public now views genuinely responsible behavior as starting *inside* the four walls of an organization. As Fleishman-Hillard posted on its website following the 2011 Fortune Green Brainstorming Conference, "A company cannot meet its sustainability or reputation goals without a smart strategy that incorporates employees."[78] After conducting research with Fleishman-Hillard, the National Consumers League reports that 76 percent of American consumers agree that for a company to be socially responsible, it should prioritize salary and wage increases for employees over making charitable contributions.[79] Observers credit the Google workplace environment for its strong social responsibility reputation, because the company does not traditionally score in the top five for its environmental or community, cause involvement on the major rankings.[80] As Robert Fronk of Harris Interactive explains: "corporate responsibility, in the minds of consumers, starts with your own employees first."[81] Corporations have an excellent opportunity to differentiate themselves based on such *internally* responsible behavior. There is a sharp contrast today between executive talk and action pertaining to the treatment of employees. Although four out of five senior executives "see the importance of valuing employees and treating them well," only half of companies surveyed offer health insurance to employees, and less than one-third provide either training or career development to low-wage employees.[82]

Building a Values-Based Culture

A critical element of valuing employees is codifying corporate beliefs—including those pertaining to employees and other constituencies—in a set of corporate values for each employee to embody. A clear and prominent set of values or code of ethics instilled in employees should ideally serve as a navigational compass for everyday work activities. Employees who live and breathe their company's values are far less likely to engage in legal or ethical breaches. A strong, values-based culture can also contribute to an organization's competitive edge, increasing employee pride, loyalty, and willingness to go the extra mile for the sake of the corporation's mission.[83] Former IBM chairman Thomas J. Watson described the importance of corporate values and strong employee faith in them in this way:

> Consider any great organization—one that has lasted over the years—and I think you will find that it owes its resiliency, not to its form of organization or administrative skills, but to the power of what we call beliefs and the appeal these beliefs have

[78] "Trends in Corporate Sustainability," Fleishman-Hillard blog, April 6, 2011, http://sustainability.fleishmanhillard.com/2011/04/06/trends-in-corporate-sustainability/.

[79] "Worker's Rights: Social Responsibility All about Worker Welfare, Survey Says," http://www.nclnet.org/worker-rights/107-corporate-social-responsibility/303-social-responsibility-all-about-worker-welfare-survey-says.

[80] Harris Interactive, "The Annual RQ 2007–2008."

[81] Ibid.

[82] "Time to Get Real."

[83] Francis Joseph Aguilar, *General Managers in Action: Policies and Strategies*, 2nd ed. (Oxford: Oxford University Press, 1992).

for its people. This, then, is my thesis: I firmly believe that any organization, in order to survive and achieve success, must have a sound set of beliefs on which it premises all its policies and actions. Next, I believe that the most important single factor in corporate success is faithful adherence to those beliefs. And finally, I believe that, if an organization is to meet the challenges of a changing world, it must be prepared to change everything about itself except those beliefs as it moves through corporate life.[84]

For a values-based corporate culture to take root and thrive, the tone must be set from the top. Warren Buffett, CEO of Berkshire Hathaway and noted philanthropist, is adamant about this, taking an active role in clearly communicating his ethical expectations to his employees. Using blunt, everyday language—and analogies that any employee can easily identify with—he explicitly states intolerance for ethical wrongdoings, citing ethics as more important than profits. Most important, Buffett creates a clear connection between the individual actions of employees and corporate culture, in turn shaping the organization's overall reputation. Buffett emphasized this personal accountability in a now legendary September 2006 memo to Berkshire Hathaway employees: "Your attitude on such matters, expressed by behavior as well as words, will be the most important factor in how the culture of your business develops. And culture, more than rule books, determines how an organization behaves. Thanks for your help on this. Berkshire's reputaion is in your hands."

Research underscores the enormous impact corporate leaders have on the atmosphere of a workplace and the values and behavior encouraged within it. Deloitte has found that 75 percent of employees identify either their senior or middle management as the primary source of pressure they feel to compromise the standards of their organizations.[85]

Ensuring that employees are striking a healthy balance in their lives is another important piece of building an ethical culture. Deloitte's Ethics & Workplace survey also found that an overwhelming 91 percent of employed adults polled claim they are more likely to behave ethically in the workplace when they maintain a good work-life balance.[86] A positive working environment reduces stress and frustration levels, thereby diminishing the likelihood of cutting corners to meet unrealistic demands. It is disturbing to consider research by corporate trend-tracking service DYG SCAN pointing to a pattern of employees no longer believing in employer loyalty, concern, and personal commitment.[87] Investing in employees to foster a sense of mutual accountability and encouraging the free airing of issues without fear of reprimand or retaliation can go a long way toward strengthening an ethical culture.[88] Taking another step to provide employees with resources— such as ethics training to prepare them for dilemmas or a hotline to call if one occurs—can be critical to keep a corporate culture aligned with the strong values that must underpin all successful corporate citizenship efforts.

---

[84] Ibid.

[85] "Trust in the Workplace," Ethics & Workplace Survey, Deloitte, 2010, http://www.deloitte.com/assets/Dcom-UnitedStates/Local%20Assets/Documents/us_2010_Ethics_and_Workplace_Survey_report_071910.pdf.

[86] Allen, "Creating a Culture of Values."

[87] Yankelovich, *Profit with Honor*, p. 43.

[88] David Gebler, "Is Your Culture a Risk Factor?" *Working Values Ltd.*, September 2005.

Memorandum

To: Berkshire Hathaway Managers ("The All-Stars")
From: Warren E. Buffett
Date: September 27, 2006

The five most dangerous words in business may be "Everybody else is doing it." A lot of banks and insurance companies have suffered earnings disasters after relying on that rationale.

Even worse have been the consequences from using that phrase to justify the morality of proposed actions. More than 100 companies so far have been drawn into the stock option backdating scandal and the number is sure to go higher. My guess is that a great many of the people involved would not have behaved in the manner they did except for the fact that they felt others were doing so as well. The same goes for all of the accounting gimmicks to manipulate earnings—and deceive investors—that has taken place in recent years.

You would have been happy to have as an executor of your will or your son-in-law most of the people who engaged in these ill-conceived activities. But somewhere along the line they picked up the notion—perhaps suggested to them by their auditor or consultant—that a number of well-respected managers were engaging in such practices and therefore it must be OK to do so. It's a seductive argument.

But it couldn't be more wrong. In fact, every time you hear the phrase "Everybody else is doing it" it should raise a huge red flag. Why would somebody offer such a rationale for an act if there were a good reason available? Clearly the advocate harbors at least a small doubt about the act if he utilizes this verbal crutch.

So, at Berkshire, let's start with what is legal, but always go on to what we would feel comfortable about being printed on the front page of our local paper, and never proceed forward simply on the basis of the fact that other people are doing it.

A final note: Somebody is doing something today at Berkshire that you and I would be unhappy about if we knew of it. That's inevitable: We now employ well over 200,000 people and the chances of that number getting through the day without any bad behavior occurring is nil. But we can have a huge effect in minimizing such activities by jumping on anything immediately when there is the slightest odor of impropriety. Your attitude on such matters, expressed by behavior as well as words, will be the most important factor in how the culture of your business develops. And culture, more than rule books, determines how an organization behaves.

Thanks for your help on this. Berkshire's reputation is in your hands.

Strategic Engagement: The Continued Influence of NGOs

In today's business context of precarious trust in corporations, nongovernmental organizations (NGOs) continue to rank among the most trusted institutions in the world. Edelman's 2011 Trust Barometer reveals the gap between the level of global trust in NGOs (61 percent) and trust in government (52 percent).[89] As Edelman describes: "NGOs have moved quickly into the 'trust-void' and have taken advantage of the downward spiral in public perceptions of government, media, [and] corporations . . . 'thought-leaders' are two to three times as likely to trust an NGO to do what is right compared to large companies because they are seen as being motivated by morals rather than just profit."[90] Making active use of social networking sites including Facebook and video-sharing sites such as YouTube has enabled NGOs to recruit thousands if not millions of new supporters on the Internet in recent years to spread their agendas with unprecedented speed and volume.

[89] 2011 Edelman Trust Barometer, http://www.edelman.com/trust/2011/uploads/Edelman%20Trust%20Barometer%20 Global%20Deck.pdf.

[90] "Non-Government Organizations More Trusted Than the Media, Most-Respected Corporations or Government," http://developmentgateway.org, p. 2.

In recent years, the public has come to trust NGOs more than they trust corporations or governments, particularly pertaining to CR issues such as health, the environment, and human rights.[91] According to one global advocacy group focused on sustainability: "[NGOs] are the moral compass and ethical watchdogs against the forces of government and capitalism that seek to despoil the planet and crush the faceless majority."[92] These voluntary organizations rally around a particular cause and hold corporations accountable for their behavior, launching campaigns against them whenever they fall short of expectations. In turn, NGOs have come to realize that such anticorporate campaigns can be far more powerful than antigovernment campaigns, particularly when targeting well-known, global brands. They possess several characteristics that enable them to catch public attention and approval. First, NGO communications are often sophisticated and controversial, and thus more likely to receive media attention. Second, the smaller size and agility of NGOs enable them to act faster than more bureaucratic corporations with layers of legal protocol.[93] The pervasive use of the Internet and its communication tools such as Twitter and Facebook has only strengthened and lengthened the reach of NGO communications, enabling local organizations to voice their messages to a global audience and pose an even greater threat to corporate reputations. Organizations including Amnesty International, Greenpeace, the National Wildlife Federation, Oxfam, the Rainforest Action Network, Friends of the Earth, Sierra Club, and Public Citizen maintain active blogs to spread messages and engage new supporters on CR and sustainability.[94] As a result, the chairman of both Goldman Sachs International and former chairman of BP, Peter Sutherland, places NGOs alongside multinational corporations in terms of world influence: "The only organizations now capable of global thought and action—the ones who will conduct the most important dialogues of the 21st century—are the multinational corporations and the NGOs."[95]

For the most part, NGOs have achieved great success in controlling this dialogue, proactively engaging corporations to effect the change they seek. As Randall Hayes, founder of the Rainforest Action Network (RAN), explains: "If you [as an NGO] are not talking to business, you are just preaching to the choir. The real change to protect the environment is going to come from the business sector; we can't depend on government regulation to solve our problems."[96] Therefore, NGOs pick their targets with great care and in recent years have recognized the rising power of the capital markets to influence responsible corporate behavior. In 2000, RAN launched an ongoing advocacy campaign against Citigroup and its environmental record, staging visible demonstrations at corporate headquarters in New York, organizing consumer boycotts, and running a full-page ad in the

[91] Edelman Trust Barometer 2007.

[92] International Foundation for the Conservation of Natural Resources Fisheries Committee, "IFCNR Special Report: How NGOs Became So Powerful," February 20, 2002.

[93] Paul A. Argenti, "Collaborating with Activists: How Starbucks Works with NGOs," *California Management Review*, November 1, 2004.

[94] "Corporate Social Responsibility and Sustainability in the Blogosphere," Edelman, 2006.

[95] Speech by David Grayson, "The Public Affairs of Civil Society," January 26, 2001.

[96] SustainAbility, Global Compact, and United National Environment Programme, "The 21st Century NGO: In the Market for Change," June 2003, p. 30.

International Herald Tribune that depicted then CEO Sandy Weill as an environmental villain.[97] In 2003, Citigroup agreed to meet with RAN to discuss environmental strategy, resulting in what RAN dubs the most far-reaching commitment made by any bank to date.[98] Mounting pressure from RAN and other NGOs has served as the catalyst for more than 40 financial institutions to sign on to the Equator Principles—a clear and consistent standard for assessing and managing environmental and social risk in project financing—since June 2003.[99] Various NGOs targeting a number of banks simultaneously was a highly effective tactic, breeding a sense of CR competition among leading financial institutions to trump one another's commitments, thereby making it much more difficult for any socially or environmentally damaging project to slip through the cracks and gain financing.

The widespread influence of NGOs reaffirms the need for corporations to think strategically about relationships with these organizations when building and executing a CR strategy. NGOs have the power to wreak havoc with eye-catching, direct, and powerful communication campaigns. As a result, corporations must identify opportunities to collaborate with NGOs and establish relationships before a crisis strikes or negative press airs. The CR team within a corporation should attempt to anticipate the mind-set of NGOs, pinpointing issues of mutual concern from their critical vantage point. Companies must not only be poised to discuss these issues with their NGO critics but should also foster ongoing dialogues on CR topics with all constituents to gauge existing concerns and communicate the efforts they are making to address them. Finally, and perhaps most critically, a company's corporate communication team should be actively involved in crafting the NGO and overall CR communication strategy to ensure consistency across all messages shared with internal and external constituents.

Being Green: The Corporation's Responsibility to the Environment

Earlier in the chapter, we described corporations as requiring a healthy and prosperous society to exist. In 2006, former U.S. vice president and Nobel Peace Prize recipient Al Gore's highly acclaimed documentary *An Inconvenient Truth* revealed that environmental stability is not something to be taken for granted. The film vividly depicted the environmental concerns of this century and bred increasing anxiety over climate change among millions of consumers. In turn, increasing evidence exists that constituents are rewarding companies that are environmentally responsible, doing their bit to preserve the planet. A recent study found that 72 percent of global consumers expect corporations to take action to preserve and sustain the environment.[100] Companies are scrambling to meet this demand and position themselves in an environmentally friendly light. In 2008, Chris Hunter, then vice president at environmental consulting firm GreenOrder and a former energy manager at Johnson & Johnson, observed: "Ten years ago, companies would call up and say 'I need a digital strategy.' Now, it's 'I need a green strategy'."[101]

[97] Matthew Yeomans, "Taking the Earth into Account," *Time Europe* 165, no. 19 (May 9, 2005).

[98] Christopher Wright, "For Citigroup, Greening Starts with Listening," *Ecosystem Marketplace*, April 4, 2006.

[99] Fiona Harvey, "Committed to the Business of the Environment," *Financial Times*, September 17, 2006.

[100] 2010 Edelman goodpurpose Study.

[101] Ben Elgin, "Little Green Lies," *BusinessWeek*, October 29, 2007.

Environmentally responsible behavior can not only attract consumers, it can also offer enormous cost savings for companies willing to make the upfront investment. Many companies—including Whole Foods, Microsoft Corp., Macy's Inc., and Target Corp.—have partnered with solar developers to reduce carbon output and reliance on utility-supplied power, setting 15- to 20-year electricity fixed costs, which are often less expensive than retail prices.[102] Walmart has become an unexpected leader in environmentally responsible practices, surprising countless skeptics who voiced doubt when it unveiled its green plan in the fall of 2005. Working with the Rocky Mountain Institute (RMI), a sustainability and energy efficiency "think-and-do" tank based in Snowmass, Colorado, Walmart conducted an "efficiency overhaul," auditing energy use across its supply chain.[103] Switching freezer case lighting from incandescent to cooler LED bulbs is projected to save Walmart $US 2.6 million per year, and installing diesel backup generators in the company's truck fleet to reduce engine idling will save $25 million annually.[104] The green plan has cost Walmart $US 500 million, but it anticipates substantial savings on the horizon, including $300 million from trucking fleet fuel efficiency alone by 2015.[105] RMI senior consultant Lionel Bony describes such efforts in terms of corporations staying ahead of the curve to stay competitive in a future business context that will be defined by dwindling resources and stringent environmental regulations: "you can either do it now and take some time to do it incrementally, or you can do it when it's right in your face."[106] In 2010, 64 percent of global consumers indicated that they would support increased government regulation to protect the environment, even if it would mean a negative impact on corporate profits.[107]

Those companies that act as environmental leaders can seize an opportunity to truly differentiate their brands in a sea of corporations claiming corporate responsibility. Coca-Cola is assuming such a leadership position in the water conversation arena and in the process is protecting the resource most critical to its production process and future profitability. In June 2007, Coca-Cola announced its investment of $US 20 million over five years to improve global water conservation, partnering with the World Wildlife Fund to preserve seven of the world's major rivers.[108] In 2006, Coca-Cola used 290 billion liters of water to produce its beverages—approximately one-fifth of daily water consumption in the United States and an increasingly precious commodity, particularly in the developing countries that represent some of Coca-Cola's most lucrative markets.[109] By 2025, the World Wildlife Fund anticipates two-thirds of the world's population will face water shortages,[110] a dangerous prospect for Coke. As Neville Isdell then CEO of Coca-Cola, stated bluntly during his announcement of the Company's long-term goal to become "water

---

[102] Ibid.

[103] Chris Turner, "Getting It into Your System," *Access Review* (FedEx), Volume 2, 2008.

[104] Ibid.

[105] Ibid.

[106] Ibid.

[107] Ibid. "2010 Edelman goodpurpose Study."

[108] Dune Lawrence, "Coca-Cola to Spend $20 Million on Water Conservation," *International Herald Tribune*, June 6, 2007.

[109] Ling Woo Liu, "Water Pressure," *Time*, June 12, 2008.

[110] Ibid.

neutral" by returning all water used in its beverages and their production to nature and communities through conservation, recycling, and other programs: "Water is the main ingredient in nearly every beverage that we make. Without access to safe water supply, our business simply cannot exist."[111]

Coca-Cola is also making efforts to repair the trust of certain stakeholders damaged by prior water controversies, most notably when the NGO Centre for Science and Environment alleged that certain Company products contained trace amounts of pesticide residues in India in 2003. Jeff Seabright, Coca-Cola's vice president of environment and water resources, admits that the Company's initial public relations during the crisis was ineffective, and that the company has used the episode as a valuable lesson that perception is just important as reality in successfully running a responsible and sustainable operation. In Seabright's words: "If people are perceiving that we're using water at their expense, that's not a sustainable operation. We sell a brand. For us, having goodwill in the community is an important thing."[112] Looking to the future, Chinese consumption represents significant revenue opportunities for Coca-Cola and countless other multinational corporations. To protect the opportunities in this market—riddled with environmental problems—Coke has partnered with NGOs to boost environmental education and encourage river conservation and rainwater harvesting in the country.

With increased attention to corporations' environmental efforts, NGOs and the general public will be watching to see whether companies deliver on their lofty promises. Consider FedEx—an environmental leader awarded a Clean Air Excellence prize by the Environmental Protection Agency in 2004 for its plans to replace its 30,000 medium-duty trucks over the next 10 years with clean-burning hybrid trucks, sparing the atmosphere 250,000 tons of greenhouse gases annually.[113] The overnight shipping leader has not fully delivered on its commitment; at the end of 2011, FedEx reported that only 408 all-electric and hybrid vehicles were in use, representing a tiny percentage of its fleet.[114] As FedEx VP for environmental affairs and sustainability Mitch Jackson explained, profitability is the underlying reason for the lack of follow-through: "We do have a fiduciary responsibility to our shareholders. We can't subsidize the development of this technology for our competitors."[115]

As we move on to discussing best practices in CR communications, we will highlight the importance of backing promises and claims of responsibility with bona fide action to maintain and strengthen trust and goodwill with increasingly conscious constituents.

Communicating about Corporate Responsibility

The strongest of CR strategies are lacking if they do not include a clear communications component. Although CR may be housed in varying areas of an organization—within the human resources, business development, or corporate communication departments of an organization, for example—corporate communicators must be actively engaged in CR messaging to ensure consistency and integration with the

---

[112] Ibid.

[113] Ibid.

[113] Elgin, "Little Green Lies."

[114] http://about.van.fedex.com/corporate_responsibility/the_environment/alternative_energy/cleaner_vehicles.

[115] Ibid.

overall communication and reputation management strategy. We will now review a number of key considerations a corporation should keep in mind when building and communicating its CR strategy.

A Two-Way Street: Creating an Ongoing Dialogue

As we discussed earlier in the chapter, tracking and responding to constituency expectations is a key component of a CR strategy's success, enabling a company to stay ahead of the changing ethos and, in the process, strengthen its reputation. A primary way to monitor constituency expectations is by fostering an ongoing and active dialogue with consumers, shareholders, and the general public about the social and environmental role companies should play. The 2010 Edelman good-purpose study found that 63 percent of consumers want brands to make it easier for consumers to make a positive difference.[116] A lack of dialogue can lead to a lack of awareness of external opinions on issues of corporate responsibility. A recent IBM study reveals that customers are the chief stakeholders driving corporate social responsibility, yet 76 percent of businesses surveyed admit they don't understand their customers' CR concerns, and only 17 percent of businesses are even asking about them.[117]

The Dangers of Empty Boasting

Given the array of potential benefits CR brings to the table—including loyal workers and customers—many companies are eager to position themselves as responsible. Unfortunately, this eagerness has produced a wave of companies trumpeting actions that are not necessarily backed by substance. In its most recent State of Corporate Citizenship report, the Center for Corporate Citizenship at Boston College's Carroll School of Management found that although most executives believe that CR is important, and 54 percent believe that CR is even more important during a recession, only 40 percent of all companies (and 65 percent of large companies) have a person or team dedicated to corporate citizenship issues.[118] Corporations make false or hollow claims at their own risk: Vigilant NGOs and other corporate critics are quick to pinpoint any inaccuracies. For example, "greenwashing" is a popular term used to describe the act of misleading consumers regarding the environmental practices of a company or the environmental benefits of a product or service.[119] The Rainforest Action Network offers a "Greenwash of the week" feature on its official blog "Understory," exposing the reality behind eco-friendly marketing with videos posted on YouTube.[120] In 2010, TerraChoice Environmental Marketing reported in its annual "Sins of Greenwashing" study that a staggering 95 percent of consumer

[116] Ibid.

[117] George Pohle and Jeff Hittner, "The Right Corporate Karma," *Forbes*, May 16, 2008.

[118] "Weathering the Storm: The State of Corporate Citizenship in the United States 2009," Center for Corporate Citizenship at Boston College Carroll School of Management.

[119] TerraChoice, http://www.terrachoice.com.

[120] http://understory.ran.org.

products claiming to be "greener" were guilty of one or more of the seven sins of greenwashing.[121] It is no wonder that constituents continue to be skeptical about companies' motivations and the realities behind CR claims. Additional suspicion stems from the high percentage of advertising dollars currently streaming into CR-related content. General Electric unveiled its highly acclaimed $1 billion "Ecomagination" campaign in 2005, featuring a collection of environmentally friendly products that now constitute 7 percent of GE's sales.[122] In 2007, GE reported spending nearly all of its corporate advertising budget on Ecomagination, advertisements. Journalists noted that more than 90 percent of GE sales was actually derived from its standard, less eco-friendly products.[123]

In this environment of watchfulness and skepticism, corporations need to work hard to bridge the divide between rhetoric and reality. As the Natural Marketing Institute (NMI) explains regarding corporate environmentalism: "the future of the green movement will require a new level of sophistication and clarity as consumers increasingly discern between those companies that are truly sincere versus those that are perceived as participating for superficial reasons."[124] Even the leading hybrid car manufacturer Toyota has not been immune to the criticism of false advertising. The environmental community has expressed dissatisfaction with Toyota over its efforts to block legislation before Congress to boost fuel economy for all new vehicles from 25 to 35 miles per gallon by 2020, claiming the target is technologically unrealistic. In response, a "How Green is Toyota?" campaign was launched by a number of environmental groups, resulting in more than 100,000 e-mails sent to Toyota's top U.S. executive.[125] In October 2007, protesters draped a Toyota dealership in Detroit with images of flag-wrapped coffins and the tagline "Driving War and Warming."[126] Toyota answered by launching its biggest advertising campaign to date, featuring a commercial of the Prius constructed using grass, twigs, and earth and asking: "Can a car company grow in harmony with the environment? Why not? At Toyota, we're not only working toward cars with zero emissions. We're also striving for zero waste in everything else we do." The ads unleashed new criticism of Toyota's eco-awareness as little more than a slick PR effort, sparking a new round of protests using Toyota's own slogan, "Why not?"[127]

The Transparency Imperative

With the Internet granting NGOs and average consumers unprecedented corporate access, pressure is mounting for companies to reveal proactively both the good and the bad elements of their operations. Companies have achieved success in

[121] "The Sins of Greenwashing," TerraChoice with Underwriters Laboratories, 2010, http://sinsofgreenwashing.org/findings/greenwashing-report-2010/.

[122] Strom, "Make Money, Save the World."

[123] Elgin, "Little Green Lies."

[124] Lifestyles of Health and Sustainability 2007, Consumer Trends Database, http://www.nmisolutions.com.

[125] Keith Naughton, "Toyota's Green Problem," *Newsweek*, November 19, 2007.

[126] Ibid.

[127] Ibid.

turning a critical eye on themselves and citing where they can be doing better on the CR front. For example, Nike and The Gap became the first to reveal the names of their overseas factories—acknowledging the shortcomings in their respective global supply chains—and set a new benchmark several years ago.[128] In much the same manner as mystery shoppers gauge the quality of in-store customer service, Nike conducts surprise inspections of its global manufacturers to ensure they meet worldwide standards.[129] Patagonia has taken transparency a step further through its interactive website "The Footprint Chronicles," tracking the negative impact of its products on the environment "from design to delivery."[130] The site highlights Patagonia's environmental shortcomings and emphasizes its willingness to keep improving: "We're keenly aware that everything we do as a business—or have done in our name—leaves its mark on the environment. As yet, there is no such thing as a sustainable business but every day we take steps to lighten our footprint and do less harm."[131] Such self-critical transparency can help build trust with constituencies. Transparency is also important for companies seeking favorable media coverage for their CR reporting and CR initiative updates. In a recent blog post entitled, "I Published My Sustainability Report . . . So, Where's My Media Coverage?" on the Fleishman-Hillard website, the highly regarded PR and marketing communications consulting firm provides the following advice to companies: (1) "don't sugarcoat your story," (2) "be real," and (3) "be detailed." The blog posting goes on to quote a sustainability reporter for *Fortune*, Marc Gunther, who shares the following insight: "Sustainability reports arrive almost daily now. A company's report needs to stand out from the crowd. Ideally, by pushing the envelope in some way—being forward-thinking, or unusually transparent, or willing to talk honestly about setbacks and frustrations as well as accomplishments and points of pride."[132] Positioning oneself as fallible—and determined to do better—can go a long way in winning the hearts of story-seeking journalists and skeptical consumers who are now armed with unprecedented insights into companies' business practices via the Internet and the widespread voices of critical NGOs. Consumers today are savvy enough to recognize that identifying a problem is the first necessary step toward solving it.[133]

Getting It Measured and Done: CR Reporting

Providing metrics—hard-and-fast evidence of CR efforts and results—will become increasingly important as more stakeholders pay close attention to the claims and realities of corporate behavior.

---

[128] Jeffrey Hollender and Bill Breen, "Can You Be Green without Also Being Transparent?" *Harvard Business Online*, June 19, 2008.

[129] William J. Amelio, "Worldsource or Perish," *Forbes*, August 17, 2007.

[130] "The Footprint Chronicles," http://www.patagonia.com.

[131] Ibid.

[132] Elin Nozewski, "I Published My Sustainability Report . . . So, Where's My Media Coverage?" Fleishman-Hillard Sustainability blog, December 15, 2011, http://sustainability.fleishmanhillard.com/2011/12/15/i-published-my-sustainability-report-so-where%E2%80%99s-my-media-coverage/#.

[133] Hollender and Breen, "Can You Be Green without Also Being Transparent?"

In such a metrics-conscious environment, the demand for corporations to issue crisp and clear CR reporting will only continue to increase. During 2010, 5,200 CR reports were published, growing exponentially from the mere 27 reports produced in 1992,[134] with Europe producing more than three times as many CR reports as either Asia or North and Central America, which tied for second place.[135] A 2011 study by KPMG found that a full 95 percent of the 250 largest global companies now report on their corporate responsibility efforts, though to varying degrees. This number represents an increase of more than 14 percent from 2008. Smaller companies also report on their corporate responsibility efforts, but in the 2011 study, only 48 percent of companies with revenues less than $US 1 billion in revenue were reporting.[136] What makes a CR report effective? First, it should appeal to the full range of a corporation's constituencies, providing both quantitative and qualitative evidence of CR efforts. Other important traits of a CR report include the disclosure of the bad as well as the good, acknowledging room for improvement, relevant and direct content (without burying truths in dense reports), creative and engaging delivery of facts, and the engagement of employees and other constituencies in the creation of the reports.[137] Starbucks uses its CR reporting as an opportunity to engage with groups ranging from environmentalists and coffee suppliers to academics and board members to ensure content is relevant and meaningful.[138] More creative formatting, such as storytelling through video, can also be an effective way for organizations to humanize their CR reporting. Finally, much like an audited annual report, an external verification statement from a neutral third party can increase a CR report's credibility. In 2009, however, only 25 percent of CR reports included such external verification.[139]

Quantifying CR efforts can pose significant challenges given the many intangible aftereffects of responsible behavior, many of which take years to accrue. Environmental impact is much easier to quantify than social impact, but companies should make efforts to measure wherever and however possible. Columbia Business School Professor Geoffrey Heal argues that environmental reporting could also be improved by standardizing it in a format similar to the Generally Accepted Accounting Principles (GAAP) to enable cross-company comparisons[140] and possibly even fuel CR competitiveness among companies. The Global Reporting Initiative (GRI) produces a global standard in sustainability reporting, which more than 1,000 organizations from 60 countries use, but its guidelines do not include quantification.[141] Although critics can easily ignore or discount anecdotal evidence of CR successes, crisp numbers—backed by a clear methodology and explanation—are far more difficult to dispute.

[134] "CR Reporting Awards 07 Official Report: Global Winners & Reporting Trends," CorporateRegister.com, March 2008.

[135] Ibid.

[136] International Survey of Corporate Responsibility Reporting, KPMG, 2011. http://www.kpmg.com/Global/en/IssuesAndInsights/ArticlesPublications/corporate-responsibility/Documents/2011-survey.pdf.

[137] Ibid.

[138] Amy Anderson, Starbucks Communications Program Manager, 2008 International Association of Business Communicators International Conference, June 23, 2008.

[139] "CR Reporting Awards 07."

Conclusion With a record number of companies devoting significant budgets and human capital to CR efforts, there is more CR chatter to compete with, which makes it difficult to differentiate a company as responsible. By July 2008 an article in the *Environmental Leader* had dubbed this effect "green fatigue" or "green noise."[142] In this environment, responsibility is no longer an option; it is a necessary condition that a corporation must meet to maintain positive relationships with its constituents and ensure its ongoing survival. The following list of key takeaways can ensure a thoughtful communication strategy is properly integrated to fuel the success of a corporation's CR program.

1. It Starts on the Inside

Throughout the chapter, we have emphasized the importance of engaging employees in a CR strategy. Walmart cites employee engagement in its CR efforts as a critical part of its green plan's success. Each employee is encouraged to make voluntary changes in his or her life to make a positive individual contribution to the environment—from using compact fluorescent lights to riding a bike to work—which helps them rally more personally around Walmart's corporate environmental efforts and share those messages in-store with consumers.[143] At Walmart Canada, vice presidents draw from the lower ranks of the company's 75,000-employee pool to pull together 14 "Sustainability Value Networks," teams that submit proposals and action plans on topics including greenhouse gas reduction and operational waste reduction.[144] Ensuring a CR strategy resonates strongly with employees can help drive greater efficiencies and positive feelings of ownership and membership in a company that stands for something greater than profits alone.

2. Collaborate with Friends and Foes

The old adage holds true in CR communications: Keep your friends close and your enemies even closer. The continued influence of NGOs presents an opportunity for corporations to forge partnerships to defend against attacks and build credibility with the millions of consumers who hold these cause-driven organizations in high regard. McDonald's, for example, worked closely with the Environmental Defense Fund in the early 1990s to change from plastic, foam packaging to paper through a collaborative effort. Earlier in the chapter, we discussed Citigroup's strategic discussions with RAN as a first critical step in its staking a trailblazing position of environmental leadership among fellow financial services firms that then followed suit by signing on to the Equator Principles.

---

[140] "The ROI of CSR."

[141] Ibid.; Global Reporting Initiative, http://www.globalreporting.org.

[142] Valerie Davis, "Are Consumers Falling Off the Green Wagon and Should We Care?" *Environmental Leader*, July 10, 2008.

[143] David Dias, "Giant Steps," *Financial Post Business* (Canada), July/August 2008.

[144] Ibid.

3. *Present the Bad with the Good*

The importance of transparency cannot be overstated in the implementation of a CR strategy. Companies that do not disclose or downplay the negative attributes or effects of their operations do so at their own peril. Given the sophistication and vigilance of NGOs and the average consumer with a Twitter account, today, a company's constituents will likely find out the truth whether or not the company proactively tells them. Being transparent means being clear in CR communication and not clouding realities with vague or verbose prose. Admitting mistakes and missteps is the first, necessary step to correcting them. Constituents will be more forgiving and trustful of a company that openly discusses its challenges in implementing CR initiatives than they will be of companies that attempt to mask or misrepresent shortcomings. Clarity also means using metrics and quantifying CR efforts wherever possible and, just as important, explaining the methodology. Constituencies will only appreciate and engage with a company's CR strategy if they are able to understand what it is and how the results are being measured.

4. *Stay One Step Ahead of Antagonists*

Corporations should keep a finger on the pulse of influencers, critics, and all constituents to gauge existing opinions and spot potential trouble brewing well in advance of a CR crisis erupting. This monitoring will enable a company to tell its own story and maintain a strong grasp on its reputation. In the words of Mary Jane Klocke, Director of North American Shareholder Marketing at BP: "Engagement raises brand awareness, offers valuable insights and perspectives from key stakeholders and gives us avenues of influence and opportunity to get the facts out . . . rather than have the [socially responsible investment or SRI] community receive its information from the media or other third parties."[145]

5. *Match Rhetoric with Action*

Constituencies today have little patience for self-aggrandizing corporations that inaccurately inflate their CR efforts or do not deliver on promises made. The greater the number of corporations that vie to win approval through CR efforts, the more savvy and discerning constituencies will be in separating hollow rhetoric from bona fide results. Companies should also be careful to never express complacency in their efforts to be responsible. Just as the business environment—and a corporation's intersection with social, environmental, and governance issues—is constantly in flux, so a CR strategy must be continually reshaped.[146] David Douglas, vice president of eco responsibility at Sun Microsystems, explains: "A big mistake is to send the message that your company believes it has done all it can do. There is always room for improvement when it comes to developing business practices that create social and business value. To indicate otherwise brings the credibility of your company's entire [CR] program into question."[147]

[145] Garrett Glaser, "Lessons Learned in Promoting CSR," *Corporate Responsibility Officer*, 2007.

[146] Porter and Kramer, "Strategy & Society."

[147] Glaser, "Lessons Learned in Promoting CSR."

Starbucks Coffee Company

On an overcast February afternoon in 2000, Starbucks CEO Orin Smith gazed out of his office window in Seattle and contemplated what had just occurred at his company's annual shareholder meeting. In prior years, the meeting had always been a fun, all-day affair where shareholders from around the country gathered to celebrate the company's success. This year, however, Smith and other senior Starbucks executives heard an earful from the activist group Global Exchange at the meeting. Global Exchange, a human rights organization dedicated to promoting environmental, political, and social justice around the world, criticized Starbucks for profiting at the farmers' expense by paying too little for beans and not buying "fair trade" beans. Not only did the activists disrupt the company's annual meeting to the point that the convention hall security police asked the activists to leave, but they also threatened a national boycott if the company refused to sell and promote fair trade coffee. Although Smith strongly disagreed with using the shareholders meeting as a public forum, he knew there was a strong likelihood his company could face serious reprisals if it did not address the issues raised by Global Exchange.

FAIR TRADE COFFEE

Fair trade began after World War II as religiously affiliated non profit organizations purchased handmade products for resale from European producers. Fair trade was an economic model based on fair labor compensation and mutual respect between producers and consumers. By the late 1990s, the fair trade movement had gained a foothold in the United States, and in early 1999, TransFair USA, a third-party certification agency, launched its Fair Trade Certified coffee label. During that summer, Global Exchange began a campaign to educate consumers and the media about labor conditions in the coffee industry, focusing on getting the message out to specialty coffee consumers. Although the activists were successful in educating pockets of consumers, they knew their effectiveness was limited without directing blame for the farmers' woes. Global Exchange decided to take an anti corporation stance and focused its attention on the most visible brand in specialty coffee: Starbucks.

At this time, fair trade coffees were coffees that were purchased directly from cooperatives of small farmers at a guaranteed floor price. Unlike shade and organic coffees, fair trade coffee focused on the workers' economic sustainability. Fair trade coffee attempted to cut out or limit the middlemen and provided much-needed credit to small farmers so that they could end their poverty cycle. Licensing organizations in individual importing countries certified fair trade coffee from farmers listed on the Fair Trade Registry. Consequently, there was a host of different certifying agencies, and fair trade coffee accounted for different market share in each country.[1]

STARBUCKS' ISSUES WITH FAIR TRADE COFFEE

For Starbucks, the real issues were brand perception and the consumer proposition. Starbucks hesitated to sign a fair trade license, not wanting to commit until it had carefully weighed all of the implications.[2] According to Starbucks

[1] Rice & McLean, p. 78.

[2] Smith, personal interview.

Source: This case was sponsored by the Allwin Initiative for Corporate Citizenship and prepared by Alison Stanley, T'02, under the direction of Professor Paul A. Argenti, with the cooperation of Starbucks Coffee Company. © 2002 Trustees of Dartmouth College. All rights reserved.

executives, their chief concern with fair trade coffee was finding top-quality beans from cooperatives that had not demonstrated an ability to produce quality beans to Starbucks standards. From earlier cupping analyses, Starbucks had little evidence that fair trade coffee met its quality standards. Starbucks was beginning to move toward purchasing more of its coffee through direct relationships with exporters or farmers and negotiated a price based on quality. The company was willing to pay higher prices for great-quality beans and had developed long-term contracts with many of its suppliers.

Mary Williams, senior vice president (VP) of the coffee department, was known throughout the coffee industry as a "tough cupper" who would not settle for anything less than top-quality beans and explained, "the relationships I have with farmers were built over the last 20 years. It's taken some of them years before I would use their beans consistently and pay them $US 1.26 or more. Now I was being asked to use another farmer who I didn't know and pay him the same price without the same quality standards?"[3] On average, farmers sent samples and met with Starbucks coffee buyers at their farms for at least 2 years before Starbucks accepted their beans. In weighing the fair trade coffee issue, Williams had secondary concerns with how the farmers she worked with would react when they discovered that other farmers received the same price without being held to the Starbucks quality standards. This was not a trivial issue because it was more expensive to grow high-quality beans. Further, she feared that the smaller cooperatives would not be able to guarantee that they could take back a low-quality shipment and replace it based on Starbucks' volume and quality needs.

Starbucks was also concerned about its brand exposure if the quality of fair trade coffee turned out to be very different from the rest of its 30 whole bean coffee line. Coffee quality was a critical component of the Starbucks brand, and if it was compromised the value of the brand could be seriously diminished. "Honestly, we didn't want to put our brand at risk," said Tom Ehlers, VP of the Whole Bean department. "This was an uncharted category and as marketers, we were concerned about endorsing a product that didn't meet our quality standards."[4] The Whole Bean department would face several challenges in introducing fair trade coffee to 3,200 stores in the United States. First, it would have to come up with a good story for fair trade coffee. "A lot of our business is about the romance of coffee—where it comes from and how to make it come alive for the customer. We weren't really sure where fair trade beans would be coming from because of the quality," explained Tim Kern, Whole Bean product manager.[5]

And how would fair trade coffee be priced? Starbucks coffee was a high-margin business, but if the company were to charge a premium for fair trade, how would customers perceive this? Although pricing was a secondary issue to consider, it was not a reason for Starbucks to abandon fair trade coffee. Orin Smith recalled, "In fact, a number of people believed that the sale of low quality Fair Trade coffee undermined their entire business proposition with customers: Starbucks and other specialty coffee companies had persuaded customers to pay high prices for quality coffee. This enabled roasters to pay the highest prices in the industry to coffee sellers." If quality was reduced, specialty coffee would be no different than mass market coffee and the consumer would be unwilling to pay premium prices. This would destroy the industry's ability to pay price premiums to producers.

THE STARBUCKS CULTURE

In 1990, Starbucks' senior executive team drafted a mission statement laying out the guiding principles behind the company. The

[3] Mary Williams, Starbucks SVP Coffee Department, July 24, 2002.

[4] Tom Elhers, personal interview, Starbucks VP Whole Bean, July 25, 2002.

[5] Tim Kern, personal interview, Starbucks Whole Bean product manager, July 25, 2002.

EXHIBIT 5.1 Starbucks Mission Statement

Establish Starbucks as the premier purveyor of the finest coffee in the world while maintaining our uncompromising principles as we grow. The following six guiding principles will help us measure the appropriateness of our decisions:

- Provide a great work environment and treat each other with respect and dignity.
- Embrace diversity as an essential component in the way we do business.
- Apply the highest standards of excellence to the purchasing, roasting and fresh delivery of our coffee.
- Develop enthusiastically satisfied customers all the time.
- Contribute positively to our communities and our environment.
- Recognize that profitability is essential to our future success.

team hoped that the principles included in this mission statement would help partners (Starbucks' term for employees) to gauge the appropriateness of their decisions and actions. As Orin Smith explained, "Those guidelines are part of our culture and we try to live by them every day." After drafting the mission statement, the executive team asked all Starbucks partners to review and comment on the document. Based on their feedback, the final statement (see Exhibit 5.1), put "people first and profits last." In fact, the number one guiding principle in Starbucks' mission statement was to "provide a great work environment and treat each other with respect and dignity."

Going forward, Starbucks did three things to keep the mission and guiding principles alive. First, it provided all new partners with a copy of the mission statement and comment cards during orientation. Second, when making presentations, Starbucks' leadership continually related decisions back to the appropriate guiding principle or principles they supported. And third, the company developed a "Mission Review" system through which any partner could comment on a decision or action relative to its consistency with one of the six principles. The partner most knowledgeable on the comment had to respond directly to such a submission within two weeks, or if the comment was anonymous, the response appeared in a monthly report. As a result of this continual emphasis, the guiding principles and their underlying values had become the cornerstones of a very strong culture.

After buying Starbucks, CEO Howard Schultz had worked to develop a benefits program that would attract top people who were eager to work for the company and committed to excellence. One of Schultz's key philosophies was to "treat people like family, and they will be loyal and give their all." Accordingly, Starbucks paid more than the going wage in the restaurant and retail industries, granted stock options to both full- and part-time partners in proportion to their level of base pay, and offered health benefits for both full- and part-time partners. As a result of its commitment to its employees, Starbucks enjoyed a low annual employee turnover (60 percent verus the restaurant industry average of 200 percent) and employees reported high job satisfaction. All of this satisfaction had fostered a strong culture that employed a predominately young and educated workforce of individuals who were extremely proud to work for Starbucks. Their pride came from working for a very visible and successful company that tried to act in accordance with the values they shared. According to Smith, "It's extremely valuable to have people proud to work for Starbucks and we make decisions that are consistent with what our partners expect of us."

CORPORATE RESPONSIBILITY AT STARBUCKS

Just as treating partners well was one of the pillars of Starbucks' culture, so too was contributing positively to the communities that it served, and to the environment. Starbucks made this commitment not only because it was the right thing to do but also because its workforce was aware and concerned with global environmental and poverty issues. In addition to sustaining and growing its business, Starbucks supported causes "in both the communities where Starbucks stores were located and the countries where Starbucks coffee was grown."

On the local level, store managers were granted discretion to donate to local causes and provide coffee for local fundraisers. One Seattle store donated more than $500,000 to Zion Preparatory Academy, an African-American school for inner-city youth. CEO Howard Schultz used his own money to start the Starbucks Foundation, which provided "opportunity grants" to nonprofit literacy groups, sponsored young writers programs, and partnered with Jumpstart, an organization helping Headstart children. Although the Starbucks Foundation was technically separate from the company, Starbucks made an annual donation to the foundation.

On the international level, in 1991, Starbucks began contributing to CARE, a worldwide relief and development foundation, as a way to give back to coffee-origin countries. By 1995, Starbucks was CARE's largest corporate donor, pledging more than $100,000 a year and specifying that its support go to coffee-producing countries. The company's donations helped with projects such as clean-water systems, health and sanitation training, and literacy efforts. By 2001, Starbucks had contributed more than $1.8 million to CARE.

In 1998, Starbucks partnered with Conservation International (CI), a nonprofit organization that helped promote biodiversity in coffee-growing regions, to support producers of shade-grown coffee. The coffee came from cooperatives in Chiapas, Mexico, and was introduced as a limited edition in 1999. The cooperatives' land bordered the El Triunfo Biosphere Reserve, an area designated by CI as one of the 25 "hot spots" that were home to over half of the world's known plants and animals. Since 1999, Starbucks had funded seasonal promotions of the coffee every year, with the hope of adding it to its lineup of year-round offerings. The results of the partnership had proven positive for both the environment and the Mexican farmers. Shade acreage increased by 220 percent, and farmers received a price premium of 65 percent above the market price and increased exports by 50 percent. Since the beginning of the partnership, Starbucks made loan guarantees that helped provide over $750,000 in loans to farmers. This financial support enabled these farmers to nearly double their income.

In 1992 Starbucks developed an environmental mission statement to articulate more clearly how the company interacted with its environment, eventually creating an Environmental Affairs team tasked with developing environmentally responsible policies and minimizing the company's "footprint." Additionally, Starbucks was active in using environmental purchasing guidelines, reducing waste through recycling and energy conservation, and continually educating partners through the company's "Green Team" initiatives. In 1994, Starbucks hired Sue Mecklenburg as the first director of Environmental Affairs. Although Starbucks had supported responsible business practices virtually since its inception, as the company grew, it felt more pressure to protect its image. It was Mecklenburg who developed the idea of using paper sleeves instead of double cupping.

At the end of 1999, Starbucks created a Corporate Social Responsibility department, and Dave Olsen was named the department's first senior vice president. According to Sue Mecklenburg, "Dave really is the heart and soul of the company and is acknowledged by others as a leader. By having Dave be the first Corporate Responsibility SVP, the department had instant credibility within the company." Between 1994 and 2001, Starbucks' CSR department grew from only one person to fourteen.

THE FAIR TRADE DECISION

Starbucks had defined being a socially responsible corporation "as conducting our business in ways that produce social, environmental and economic benefits to the communities in which we operate." Starbucks knew that consumers were increasingly demanding more than just a "product," at the same time that employees were increasingly electing to work for companies with strong values. In a 1999 survey by Cone Communications, 62 percent of respondents said they would switch brands or retailers to support causes they cared about. Another survey conducted in 2001 showed that 75 to 80 percent of consumers were likely to reward companies for being "good corporate citizens," and 20 percent said they'd punish those who weren't.[6] The company cared about being a responsible corporation for a variety of reasons: increasing employee satisfaction, maintaining quality supply sources, obtaining a competitive advantage through a strong reputation, and increasing shareholder value.[7]

As he looked out over the busy port in Seattle's South of Downtown district, Orin Smith pondered all of these issues. Although offering fair trade coffee was a good objective and consistent with the company's aims of being a socially responsible organization, Smith knew he could not base his decision on this factor alone. Even though Smith had a rough idea of which issues his executive team would bring up during the discussion, as the CEO he had to consider the larger picture. He drummed his fingers on the desk and asked himself how Starbucks could support fair trade coffee given that the company had limited resources, a strong reputation to protect, and shareholders who were willing to support causes only so much.

CASE QUESTIONS

1. What are the key issues for Starbucks?
2. What are the problems associated with the decision to offer fair trade coffee from a communication perspective? What are the problems associated with not offering fair trade coffee?
3. What should Smith do?

---

[6] Alison Maitland, "Bitter Taste of Success," *Financial Times*, March 11, 2002, p. 2.
[7] Packard, "Sustainability Practices Presentation."

CHAPTER SIX

Media Relations

One of the most critical areas within any corporate communication function is the media relations department. The media are both a constituency and a conduit through which investors, employees, customers, and community members receive information about and form images of a company. Consumers, for instance, might see a *"Dateline NBC"* segment on a particular firm or read an article about it in Bloomberg *BusinessWeek* or *The Wall Street Journal Online*. With the growth of digital communications platforms, the media's role as disseminator of information to a firm's key constituencies, including the general public, has changed dramatically in recent years. Virtually every company now has some kind of media relations department that covers both traditional and new media, whether it is one part-time consultant or a large staff of professionals.

In this chapter, we look at what media relations professionals do, and also how companies should approach the different members of the now expanded "press." We examine who the media are, how firms communicate with the media through relationship building, and what constitutes a successful media relations program in today's changing business environment.

The Evolution of the News Media

The news media are omnipresent in our society. With the advent of television in the late 1940s and early 1950s and the tremendous growth of the Internet since the 1990s, what had once been the exclusive domain of the print medium in newspapers as well as radio increasingly has become part of the visual realm through televisions, connected devices, and computers.

The arrival of television moved the "headline news" that had formerly been found in newspapers to a new, nearly instantaneous medium. Newspapers adapted by taking over the kind of analysis that had previously appeared in weekly news magazines such as *Time* and *Newsweek*. The news magazines, in turn, took over the feature writing and photo journalism that used to appear in older monthlies such as *Life Magazine* and the *Saturday Evening Post*.

Referred to as "the press" in earlier times, the expanded media are a powerful part of American society. The First Amendment of the Constitution guarantees the right of free speech in the United States, and over the years, the media have helped shape attitudes in this country on issues as diverse as gun control and hemlines, abortion and corporate pay. A free press also makes politicians accountable for their actions in both public and private life. Even politicians would argue that the media bring the distant world of politics into the home of the average citizen.

Today, with the rise and adaptation of digital communications platforms, including blogs, social media networks, virtual worlds, "mash-ups," and wikis, the average citizen has actually *become* part of the media. These digital platforms have empowered anyone with an Internet connection to act as a journalist of sorts, hence the term *citizen journalism*. These stakeholders can have an incredible impact on businesses, politicians, and global political events, as we shall discuss later.

Although most Americans feel strongly about the rights of a free press to say or print whatever it likes, business has always had a more antagonistic relationship with the press. This relationship stems in part from the privacy that corporations enjoyed in the early part of the last century. Unaccustomed to dealing with the news media, most companies simply acted as if the news media didn't matter. Later in the twentieth century, companies were forced to rethink this isolationist approach due to a number of developments, including laws governing the disclosure of certain information by public companies at regular intervals, a Supreme Court ruling in 1964 that required proof of malicious intent to win libel cases against the media, more public interest in business (see Chapter 1), and more media interest in business.

These last two events in particular—increased public and media interest—had a profound effect on business and its dealings with the media. Which came first? Although it is difficult to determine whether the media generated heightened interest in business or were simply responding to changes in public attitudes, what is certain is that sometime in the 1970s, business coverage started to change. Since then, the private sector has become much more public.

Part of what perpetuated this shift in attitudes was the public's realization that business had a tremendous effect on their lives. Incidents like the oil embargo, environmental problems at Love Canal, and questionable advertising on children's television programs all became enmeshed in other controversies in the 1970s, such as Watergate and the Vietnam War. People began to see companies as entities controlling important parts of their lives that did not have to answer to anyone in the way that government did to voters. Special interest watchdog groups emerged to deal with this problem and to make business more accountable.

Business leaders were used to the privacy that they had maintained for decades and were reluctant to admit that times had changed. Even today, some older business professionals resist accepting the importance of communicating through the media, and are especially wary of embracing social media. This kind of attitude is increasingly risky and less common, however, as each industry—from oil and gas, to financial services, to pharmaceuticals—has found itself the subject of some level of public scrutiny, and many companies have learned the hard way that when a crisis hits, having poor or nonexistent relationships with the media, or lacking an effective online communications strategy, will only make the situation worse.

Companies have also had to adapt to a 24/7 news cycle, and to media that include an expanded set of voices. The news media, once a powerful conduit of information exchange between businesses and stakeholders, now find themselves competing not only with other news outlets, but also with companies and their stakeholders. The ability to create content has been democratized by emerging digital communications channels, thus redefining media relations in a profound way. This new media reality doesn't just challenge news outlets; businesses that once depended on their outbound press releases to be interpreted by reporters and then communicated to stakeholders are now forced to engage directly with audiences in two-way conversations rather than one-sided monologues.

The Growth of Business Coverage in the Media

Before the 1970s, business news was relegated to a few pages toward the back of the newspaper (consisting mostly of stock quotations) and to a handful of business magazines; it received virtually no coverage at all in national and local television news broadcasts. As public attitudes changed, however, the business news sections in newspapers gained recognition and began to expand. Because the media are interested in satisfying the needs of readers and viewers, they had to meet the public's growing interest in the private sector and its participants.

Around the same time that *The New York Times* developed Business Day, a separate section published every day devoted to business issues, *The Wall Street Journal* became the number-one selling newspaper in the United States. Business magazines started to become profitable, and television networks and their local affiliates began to devote segments to business news.

Today, so many magazines and websites are devoted to business news that it is nearly impossible to find a topic not thoroughly covered by one media outlet or another. In recent years, news of corporations, the stock market, and business personalities has often become the lead story on national news television and radio broadcasts. With 24-hour business networks such as CNBC and Fox Business Network, business coverage on TV news channels and countless business articles published on the Internet, corporate news is now virtually impossible to ignore.

Compared to decades past, business news today is actually exciting. The large-format *Fortune* magazine found in doctors' offices in the 1950s and 1960s was basically a dull vehicle for companies to express their points of view. *Fortune* was more successful than others, however, because it allowed executives to check its quotes—a practice then unknown anywhere other than this one magazine. Today its cover stories appeal to a wider audience. *Forbes* gains attention from a broad readership by publishing salaries of top entertainers, whereas Bloomberg *BusinessWeek* attracts an audience through features such as its widely read rankings of business schools and corporate boards.

As coverage of business increased, however, the media industry was consolidating. Fifty corporations controlled the vast majority of all news media in the United States in 1983. By 2004, only five corporations owned and operated 90 percent of this country's "mass media."[1] Thus, economics plays a big part in what gets covered, as major industrial companies worry more about the bottom line at their media subsidiary (for instance, ComCast and its NBC network). Media conglomerates, including the new discredited Rupert Murdoch's News Corp, have also increasingly been accused of allowing politics to shape their coverage. According

---

[1] Media Reform Information Center website, http://www.corporations.org/media.

to its critics, News Corp's Fox News has changed the old model of television news, which was pitched toward a mass audience across the political spectrum and aspired to standards of fairness in reporting, and replaced it with an aggressive drive toward partisan coverage.[2]

Most executives today recognize that the media are more likely to cover the mistakes rather than the companies. In general, the worse the news is about a company or its CEO, the more likely it is to become a major news story that will capture the media's (and the public's) attention, if only briefly. A 1997 study conducted by the Pew Research Center revealed that the public wanted more reporting on corrupt business practices by a margin of 60 to 28 percent.[3] By 2005, this impulse for more transparency and greater reporting on business had morphed into a growing movement toward making journalism more transparent and treating the public as a partner in the process rather than a passive participant. At the same time, the public became increasingly distrustful of businesses and the traditional news media that covered them. With the rise in corporate scandals, beginning most notably in 2001 with the infamous dissolution of the energy company Enron, the public developed a stronger appetite for corporate exposés. The traditional model of media as mouthpiece for corporate news seemed outdated in a time when people craved authenticity and engagement. Digital media, with its emphasis on transparancy and fast reaction time, gained influence. This trend is reflected in the results of the 2011 Trust Barometer: respondents indicated that search engines and online news sources were their first and second choices for news about a company, followed by print and broadcast. This rise in news coverage from nontraditional sources in the form of online news and blogs lends further support to the need for corporations to develop a thoughtful approach to their media relations with both traditional and nontraditional media.

Building Better Relations with the Media

To build better relationships with members of the media, organizations must take the time to cultivate relationships with the right people. This task might be handled by employees within the company's media relations department (if one exists) or given to a public relations firm to handle. Either way, companies should be sure to avoid falling into some of the common pitfalls of what has historically been media relations "standard practice."

For example, most old-style public relations experts rely on a system of communication with the media that no longer works. That system sends out press releases (or video news releases) to a mass audience and hopes that someone will pick up the story and write about it. Why is this system no longer valid? The vast majority of press releases go unread by reporters in the United States—due to both the massive quantities of releases these reporters receive daily and the time constraints

---

[2] "Fox Most Trusted News Channel in US, Poll Shows," *The Guardian Online*, January 27, 2010.

[3] "Bad News: Another Study Finds Media Really Has Problems," *PR Reporter*, April 7, 1997, p. 1.

under which reporters work. The same is true for many of the mailings, e-mails, and voicemails from public relations agencies. Overall, journalists prefer receiving emails. Writes one respondent in the PWE New Media 2011 Journalist Survey of his or her communication preferences: "Email only. I throw or lose 100 percent of all hard copy communication. Sending a huge press kit in a folder announces to the world: 'I'm living in the year 1978!' "[4] However, any e-mail that isn't personally addressed to the recipient is likely to be deleted or programmed to be automatically sent to the spam folder. Just as with snail mail, people are now almost programmed to jettison anything in their mailbox that doesn't seem personalized or relevant.

Part of the problem is that the measure of success in the media relations business has for years been the amount of "ink" (or coverage) that a company gets, whether aided by in-house professionals or an outside consultant. Yet few companies try to figure out what value a "hit" (as it is called in the business) in a relatively unimportant publication has in terms of a firm's overall communication strategy. Getting lots of ink, which means lots of articles written about a company, may not have any value if it doesn't tie back to the company's strategic communication objectives (see Chapter 2) it started out with in the first place.

As discussed later in this chapter, most communication measures to date have focused on the quantity or efficiency of communication output. Although the majority of communications professionals still judge success on their ability to place material in the media rather than on the impact such coverage might have on shifting opinion, awareness, or moving markets, there is evidence that this is changing. Senior executives within companies are increasingly interested in demonstrating return on investment (ROI) from their public relations (PR) and marketing efforts. The most recent BenchPoint Global Survey of Communications Measurement found that the overwhelming majority (88 percent) of communications professionals believe that measurement is an integral part of the PR process.[5] Organizations such as Interbrand and the Reputation Institute are at the forefront of this trend.

The message to companies about press releases is thus: Use mass-mailed releases sparingly. Organizations should reserve this method for stories that they are sure will have a wide audience. In such cases, the same result can be achieved by placing the story on the *Public Relations Newswire* (*PR Newswire*) or convincing The Associated Press to put the story out on its wire, if it is a major story that will have mass appeal. Most of the time, what works best is to find out who the right journalists are for a given story and then to develop a tailored game plan for connecting with them. In the era of digital information, there is no excuse for not knowing an editor or reporter's background and interests. Martha Groves, staff reporter with the *LA Times*, writes, "At a time when PR 'professionals' have access to abundant online resources and could theoretically target *just* the right reporter, they continue to pitch ideas indiscriminately. I cover local news in Los Angeles but receive countless

[4] http://www.pwrnewmedia.com/2011/powerlines/february/downloads/2011JournoSurveyReleaseResponses.pdf.

[5] BenchPoint Global Survey of Communications Measurement 2009, http://www.benchpoint.com.

e-mails that do not relate even remotely to my beat. Publicists do their clients and themselves no favors by wasting reporters' and editors' time."

Many PR agencies credit themselves with successfully developing media relationships by researching both personal and professional details, such as birthdays or children's birthdays, or following the practice of targeted "gifting." Companies seldom use this tactic, however, because it takes more time to conduct such research. In a field cluttered with information coming from a variety of sources, however, this is actually the best approach. The responses in the PWR New Media 2011 Journalist Survey illustrate this point: "don't bother me if you don't know my market, my interests or my needs. Don't pretend to be my buddy if you've never met me; above all be competent."[6]

Conducting Research for Targeting Traditional Media

The way a typical media research operation might unfold for a company is as follows: First, senior managers working with the members of the corporate communication department determine what objectives they have for a certain story. Let's assume, for example, that the story is about a major company that is moving into a new foreign market. The managers' objective might be to create awareness about the move into the new market and also discuss how the firm has changed its global strategy. Thus, this story is part of an overall trend at the company rather than a one-shot, tactical move. Given these considerations, the company would begin to search for the right place to pitch the story.

To do this, the corporate communications professionals conduct research to find out who covers their industry and the company specifically. Identifying print, radio, and TV reporters is relatively easy for most companies because the same reporters typically cover the same beat for a period of time and have established relationships with the company either directly or indirectly in that process. Some of these reporters—typically those from print journalism—would definitely be interested in the story. If the company is maintaining its records properly, it can determine at a glance which reporters will most likely cover the story and, more important, who will be likely to write a "balanced story" (code words for a positive piece) about this strategic move.

How do companies determine who is going to write a positive piece before rather than after pitching the piece? This point is where ongoing research pays off. Each time a journalist covers a firm in the industry, the corporate communications professionals need to determine what *angle* the reporter has taken. To continue with our example, suppose a look at the records shows that *The Wall Street Journal* reporter who covered the company's beat has recently written a piece about a competitor firm moving into a different market as part of its new global strategy. Chances are, this reporter will not be interested in writing the same story again about another company.

By conducting this kind of research, companies can avoid giving reporters information that they are not interested in, and communications need only occur

---

[6] PWR New Media 2011 Journalist Survey.

when a company's media audience is most likely to be receptive. Although this system is not foolproof, it generally yields better results than sending out a story to 300 reporters hoping that four or five may pick it up, with no idea who they are or what angle they are likely to take on the story.

Today, companies can easily access information about the journalists who cover them. Consultants generate computer analyses of reporters' articles, ask industry sources to provide critiques of writers they know, and find out personal information about targeted journalists. Earlier generations of PR professionals worked hard to get such information at long lunches with reporters, but new technology allows corporate communications professionals to access such information through electronic databases, such as Cision's MediaSource or the Bulldog Reporter's MediaBase.

In addition to figuring out who is covering a company's beat, the firm's corporate communication team needs to determine what kind of a reporter it is dealing with. For a television network, such as CNN, this determination means knowing who the producer for the piece will be. Then a communications professional from the company can call the head office in Atlanta and purchase the producer's last two or three stories. For a business magazine such as *Forbes*, electronic databases—such as LexisNexis or Factiva—contain stories that reporters have written over a period of time. Those written in the last two years are most likely to be useful to your company.

What can corporate communications professionals learn by looking at previous stories the producer at CNN has filed and earlier stories that the *Forbes* reporter has written? An individual tends to write about things or put together reports in a particular way. Very few reporters change their style from one story to the next. They have found an approach that works for them—a formula, so to speak—and they tend to stick with formulas that work.

What this kind of analysis usually reveals is that the journalist tends to write or present stories with a particular point of view. One such analysis performed for a company on a *Forbes* reporter's work showed that he liked to write "turnaround" stories. That is, he liked to present the opposite point of view from what everyone else had written about. So, if a company, for example, is trying to make a case for such a turnaround, this reporter would be more likely to write the kind of article that would be helpful for the company despite his negative tone.

Watching the CNN producer's work could help determine how this individual conducts interviews, how the stories are edited, whether he or she likes to use charts and graphs as part of the story, and so on. Let's say that the producer, for example, seems to present balanced interviews, as opposed to antagonistic ones, and likes to use charts and graphs. Again, this makes it seem as if such a producer could easily turn out a positive story for the company—a goal that should be pursued.

Corporate communication departments should perform this type of analysis for each call that comes in. Many executives complain about the amount of time such analysis takes, but the benefits of handling an interview with this kind of preparation make the effort involved well worthwhile.

Researching and Engaging the Expanded "Press"

Imagine, as in the story above, that the company preparing for its launch into a new foreign market must go about identifying influential media in the online space. Identifying these members of the expanded "press," namely bloggers, online communities, social media outlets and citizen journalists, is more complicated. Bloggers, unlike print journalists, comment on everything from politics to business, from entertainment to the environment. They are not governed by formal training, long-established standards, or official affiliations. While many bloggers are merely exercising their right to free speech, others offer insightful commentary about issues and, in turn, have influence over large audiences. Arianna Huffington, for example, is an incredibly influential media representative who built an empire that lives and works solely in the online space. The Huffington Post, her left-wing online news and commentary site, is the top US internet news provider, receiving more monthly unique visitors than the websites of the *New York Times*, the *Wall Street Journal*, the *LA Times* and the *Washington Post*.[7] To still be caught up in the argument of "Is a blogger a journalist?" is in many ways to have missed the boat.

Understanding which bloggers are influential to a particular company's stakeholders is the first step in research and engagement. There are strategies for identifying the bloggers who influence a specific audience, and most, not surprisingly, come in the form of online monitoring and tracking devices. The ever-growing list of applications that help executives wrap their arms around the massive volume of online conversations includes Technorati, Google Alerts, BlogPulse, and Compete. But more than being reactive and simply waiting for a blogger to praise or pummel their company, corporate communications executives must court bloggers proactively for coverage. Best in class companies, such as HP, successfully identify the bloggers who exercise significant influence over their stakeholders by having a stable of employees charged with actively surveying cyberspace (HP has over 50 corporate blogs alone).

HP learned early on that bloggers are the greatest sales rep they never had on payroll: in September 2008, HP made an announcement that it had increased personal computer sales by 10 percent the previous May simply by leveraging the blogging community to promote HP's recently released HDX Dragon computer system. Skeptics immediately assumed that the company had paid off bloggers to build buzz around the product, but according to HP's vice president and general manager for the Personal Systems Group, bloggers weren't given a cent. All HP executives did was send 31 new computer systems to 31 influential bloggers in the tech space, offering to let them give them away in competitions to their readers.[8]

Investing in web-based communications platforms is another way to bring bloggers and online commentators directly to you. There are many examples of companies that are successfully harnessing the power of digital media to reach the expanded media. For example, Microsoft has built an online newsroom called "PressPass" within its main website that brings corporate information, news, fast

[7] Jay Yarow, "*Huffington Post* Traffic Zooms Past *The New York Times*," *Business Insider*, June 9, 2011.

[8] Matt Marshall, "HP Announces an (Almost) Unbelievable Blogger Campaign," *VentureBeat Digital Media*, September 24, 2008.

facts, PR contact information, image galleries, and broadcasts into one central location for journalists to access. Likewise, General Motors' European arm built a social media newsroom to archive news, aggregate recommended blogs, offer multimedia downloads, and consolidate RSS feeds.

Blogger outreach is not only for the biggest, most tech-savvy corporate players; on the contrary, organizations of any size can identify influential blogger communities and effectively engage with them. The first thing any communications executive must do is resist the urge to send press releases and pitches to the bloggers they identify. True to almost all "digerati," bloggers are extremely conscious of maintaining authenticity, which means that communications executives must learn to listen if they have any hope of successfully targeting them.

As when working with traditional print journalists and TV producers, communications executives must review a substantial backlog of the blogger's posts along with the comment threads before reaching out and pitching a story. Katie Paine, CEO of communications firm KDPaine & Partners, recommends "reading as much as six months' worth of posts before ever engaging the blogger. Listen until you understand the tone and nature of the conversation." This level of reconnaissance gives executives a thorough understanding of the blogger's interests, as well as the audience's level of engagement with the blog.

Pitching bloggers is in many ways more complicated than pitching their traditional counterparts: because online media are so much about customization and individual engagement, bloggers can be flippant with communications executives who "corrupt" the code of authenticity with generic press releases and hand-spooned media pitches. Beyond making sure that the approach is entirely germane to a blogger's audience, the pitch itself must be as concise as possible. "You certainly can't spin your way to a blogger's heart," claims Sir Martin Sorrell, chief executive of advertising at marketing giant WPP. "Respect and engagement are essential. Handing product over to bloggers wouldn't be enough. If, however, you invite bloggers in to get their ideas on a brand, you might succeed. Get them involved; give them something of value. The prize for getting it right? The stakeholder becomes a brand loyalist and tells other people."[9]

Responding to Media Calls

In addition to doing their homework on reporters, companies can strengthen their relationships with the media through the way they handle requests for information. Many companies willingly spend millions of dollars on advertising but are unwilling to staff a media relations department with enough personnel to handle incoming calls from the media.

This refusal can be a costly mistake, as responding to such requests carefully, and timely, can make a powerful difference in how the company appears in the story. Let's say that a company has gotten negative press over the last couple of years because it has not kept up with the times, but it is now working on a campaign to change its image. A call comes in from a reporter at CNN, and another

[9] Sir Martin Sorrell, "Public Relations: The Story Behind a Remarkable Renaissance," Institute for Public Relations Annual Distinguished Lecture, New York, November 5, 2008.

call comes in from a reporter at *Forbes*. What should the communications staff do to ensure that both of these requests are met in a timely manner and one that will reflect best on the company?

To begin with, calls should come into a central office that deals with all requests for information from important national media. Although this sounds like common sense, calls are often answered by an administrative assistant who cannot distinguish between important and unimportant calls from the media. Many an opportunity has been lost because someone failed to get the right message to a media relations expert in the corporate communication department.

Next, the person who takes the call should try to find out what angle the reporter is taking on the story. In our example, the CNN reporter may or may not have a particular point of view, but the *Forbes* reporter probably does, as that publication prides itself on taking a particular approach to its stories. The company needs to find out what that approach is before responding to the request. Let's assume that the CNN reporter wants to look at the company's activities as part of an industry trend toward more upscale positioning. The *Forbes* reporter, on the other hand, seems to imply from the conversation that she sees the company's new approach in a less-than-positive light.

The person responsible for that telephone call should try to get as much information as possible while being careful not to give in return any information that is not already public knowledge. The tone of the conversation should be as friendly as possible, and the media relations professional should communicate honestly about the possibilities of arranging an interview or meeting other requests. At the same time, he or she should find out what kind of deadline the reporter is working under.

This issue is often a point of contention between business and the media. Particularly with senior executives who are accustomed to arranging schedules at their own convenience, a call from the media at an inconvenient time can be an annoyance. But all reporters must meet deadlines. They have to file their stories—whether on television or radio, in print, or on the web—on a certain date, by a certain time. These deadlines usually have little flexibility, so knowing in advance what the deadline is allows you to respond within the allotted time. Being aware of deadlines is similarly critical when proactively pitching a story to avoid irritating reporters under deadline crunches and, by doing so, leaving them with a negative impression of the company. Responding to media calls in a timely manner is especially critical today given the speed at which news can travel online. Politicians from Eliot Spitzer to Congressman Anthony Weiner and sports personalities such as Tiger Woods have learned that in the age of texting and Twitter, not communicating is communicating. Any brand that does not have an online communications strategy in place when called on for comment, or worse, when a crisis hits, risks its reputation. (See Chapter 10 for more on crisis communication.)

Preparing for Media Interviews

Once the research and analysis are complete, the employee who will be interviewed, whether at the executive level or more junior, needs to be prepared for the actual meeting with the reporter. If the interview is to be conducted by phone, as is often the case for print articles, a media relations professional should plan to sit in on the interview. The following approach works best.

First, the employee should be given a short briefing on the reporter's or producer's prior work, using examples gathered in the research phase discussed earlier, so that he or she develops a clear understanding of the reporter's point of view. For example, if the reporter tends to write turnaround pieces, the appropriate passages from relevant stories should be shown to the employee.

One *Fortune* 500 CEO prepared for an interview with CNN by watching the last two or three major stories the producer had filed. Having done so, he was able to begin the conversation with the producer by saying how much he liked one of the stories. This positive beginning set the tone for the rest of the interview. Additionally, after learning that the producer always used a list of bullet points as part of each story, the CEO developed a list of points he wanted to communicate about the company in bullet-point form and handed it to him before he left. When the story was broadcast, it was positive about the company, and the list of bullet points was right up there on the television screen, which delighted the CEO, who had worried for days about the interview.

Once the employee has been briefed on the reporter's background and likely angle, he or she should be given a set of questions that the reporter is likely to ask. These questions can be developed from what the communications staff member working on this interview has gleaned in previous conversations with the reporter, from an analysis of the reporter's work, and from what seem to be the critical issues on the subject. If possible, the communications specialist should arrange a trial run with the employee to go over answers to possible questions. The employee also should understand that the agenda for a news story is hard to change—once the reporter has decided to write or produce a particular kind of story, it is difficult to introduce a new topic into the discussion.

In preparing for a television interview or webcast, a full dress rehearsal is absolutely essential. The interview should look as if it is totally natural and unrehearsed when it actually occurs, but the employee should be prepared well in advance. This requirement means thinking about what to communicate to the reporter, no matter what he or she asks during the interview. While the employee cannot change the agenda for the interview, as discussed earlier, he or she can get certain points across as the dialogue moves from one idea to the next.

In addition to thinking about what to say, the employee needs to think about the most interesting approach to expressing these messages. Using statistics and anecdotes can help bring ideas alive in an interview. What is interesting, however, depends on the audience. Many people mistakenly assume that the reporter is their audience, but it is the people who will watch the interview with whom they are really communicating. Communications professionals and executives must keep this in mind in determining the best approach for a television or online interview. Know your audience: the tone and tenor for an appearance on an evening news hour are different from those of a webchat with a blog for new mothers. (See Chapter 2 for more on communication strategy, especially analyzing constituencies.)

INTERVIEW TIPS

Communications expert Mary Munter suggests the following tips when preparing for a media interview:*

- Keep answers short; think in 10-second sound bites.

- Avoid saying "no comment"; explain why you can't answer and promise to get back to the reporter when you can.

- Listen carefully to each question; think about your response; only answer the question you were asked.

- Use "bridging" to move the interviewer from his or her question to your communication objective.

- Use anecdotes, analogies, and simple statistics to make your point.

- Keep your body language in mind throughout the interview.

*Adapted from Mary Munter, "How to Conduct a Successful Media Interview," *California Management Review*, Summer 1983, pp. 143–150.

Finally, the employee needs to be prepared to state key ideas as clearly as possible at the beginning of the interview. Answers to questions need to be as succinct as possible. Especially in television, where sound bites of three or four seconds are the rule rather than the exception, executives need training to get complicated ideas into a compact form that the general public can easily understand. Andrew Grant, head of Tulchan Communications and a longtime veteran of Brunswick Public Relations, advises that a spokesperson must be able to: "distill the company into a story he or she can tell over lunch and a journalist should be able to walk away and write it down on the back of a cigarette packet."[10]

Gauging Success

Measurement has been transformed thanks to the ease with which communications campaigns conducted using digital platforms can be tracked. Although most communications professionals still believe that press clippings are the strongest indicators of a campaign's success, various other deliverables, including advertising value equivalent (AVEs), internal reviews, benchmarking, opinion polling, and specialist media evaluation tools, have emerged to help measure success and in turn, ROI. Digital campaigns, in particular, now come with specialized measurement tools, including Klout for measuring levels of online influence on Twitter and Facebook, and Google Analytics. As digital communications platforms grow, so, too, will the tools developed to measure their reach. The debate on how to measure campaign results is continuously evolving alongside social media. The amount of ink that a company gets, however, does not necessarily indicate whether it is achieving its communication objectives. Verizon, for example, keeps records of all of its media hits, and looks not only at where the ink has landed but also at how well the company's key messages are communicated. Nancy Bavec, former director of media relations at Verizon, explains, "The entire department's compensation

[10] Dan Bilefsky, "Join the Sultans of Spin Media Relations," *Financial Times*, July 13, 2000, p. 19.

is tied to its ability to elevate Verizon's media scores."[11] Part of elevating a media score is finding out where the media hits have landed (with which constituencies), not just determining that the media carried a story on the company.

Microsoft also actively tracks its media scores. To prepare for its May 2006 debut of Windows Live One Care, Microsoft executives teamed up with PR firm Waggener Edstrom to shape messaging around the product launch. Having secured a retail partnership with Best Buy, the team created two separate brand message campaigns—"One Care is like a 'pit crew' for the PC" and "One Care is the all-in-one PC care service"—and tested each to see which would get the most traction with the media. The messages took very different approaches, and were both tested over a three-month period that straddled the launch. Using Waggener Edstrom's proprietary measurement system, the team collected metrics that analyzed the volume, tone, depth, and key message pickup of each brand message with media coverage prior to and following the retail launch. The metrics revealed that the all-in-one campaign was far more successful, having been picked up 77 times in media mentions compared to the pit crew's 24 mentions. Based on the results, Microsoft abandoned the pit-crew approach for the all-in-one campaign.

In addition to this sort of media monitoring and analysis, more sophisticated approaches to the measurement of media relations, referenced earlier, have the power to:

- Identify which communications activities create the most value in terms of a specific business outcome.
- Evaluate how well an organization's various communications functions perform against an industry average.
- Demonstrate the total value created by a communications department in terms of one or more business outcomes.
- Drive strategic and tactical decision making in the communications function, hedging reputational risk and managing major events such as mergers and top management changes.
- Highlight actual corporate value created by communications activities.[12]

Maintaining Ongoing Relationships

By far the most critical component in media relations is developing and maintaining a network of contacts with the media. Building and maintaining close relationships is a prerequisite for generating coverage. A company cannot simply turn the relationship on and off when a crisis strikes or when it has something it would like to communicate to the public. Instead, firms need to work to develop long-term relationships with the right journalists for their specific industry. This effort usually means meeting with reporters just to build goodwill and credibility. The media relations director should meet regularly with journalists who cover the

[11] Benchmark Global Survey of Communications Measurement, 2009.

[12] Quoted in "How Do Your PR Efforts Measure Up in the Wired World?" *Interactive PR and Marketing News*, November 26, 1999, p. 1.

industry and also should arrange yearly meetings between key reporters and the CEO. The more private and privileged these sessions are, the better the long-term relationship is likely to be. Most communications professionals now recognize that nothing can match face-to-face meetings when it comes to building relationships with key editors and journalists. Bloggers, as well, need to be proactively courted through press days and meetings where the company focuses on listening and engagement.

One example of a company's successful efforts to build strong media relations is Matalan Clothing Retailers in the United Kingdom. The company offers journalists tours of its headquarters, including opportunities to try on its clothing in changing rooms and, most surprising, to fully analyze its distribution network. Chris Lynch formerly of Ludgate Communications, a representative of Matalan, explains, "We tactically avoid granting phone interviews in order to get journalists to meet us face-to-face. Otherwise it ends up being just about the numbers."[13] By taking such a personalized approach, Matalan quickly became a favorite company among journalists. This success has continued, with Matalan recently winning "Home Retailer of the Year" in the National Home Awards sponsored by the *Daily Telegraph*.[14]

Many companies take a less "integrated" approach than Matalan and use the more typical venue of a meeting between a member of the media and a company executive. Because these meetings often have no specific agenda, they can be awkward for all but the most skilled communicators. Within organizations, people assigned to handle media relations should enjoy "meeting and greeting," should be tapped into the company's top-line strategic agenda, and should be able to think creatively.

Often these kinds of meetings occur at lunch or breakfast. They should be thought of as a time to share information about what is going on at the company, but with no expectation that a story will necessarily appear anytime soon. In the course of such a conversation, the skillful media relations professional will determine what is most likely to interest the reporter later as a possible story. Without being blatant about it, he or she can then follow up at the appropriate time with the information or interviews that the reporter wants.

Media relations professionals should expect to be rebuffed from time to time. They may get turned down for lunch several times by reporters who are particularly busy, only to find them very receptive to a long telephone conversation. As is true with personal relationships, media relations professionals will find that they simply do not get along with every journalist they come into contact with. Unless the reporter is the only one covering a company's beat at an important national media outlet, this awkwardness should not be an insurmountable problem. When personality conflicts do occur, professionals can and should work around them to ensure that the overall relationship of the company with that media outlet is not jeopardized and media opportunities are not missed.

---

[13] Paul Argenti and Courtey Barnes, *Digital Strategies for Powerful Corporate Communications*, p. 71.

[14] Paul Argenti, "Demonstrating the Value of Communications through Measurement," July 2005.

One hotel executive at a major chain didn't think that he needed to have any sort of relationship with the reporter covering his beat at *The Wall Street Journal*. After almost two years of being left out of nearly every major story on the industry, a consultant persuaded him to try again to establish a relationship with this reporter. The reporter was only too happy to make amends as well, as she needed the company's cooperation as much as it needed her. Nonetheless, that attitude cost the company nearly two years of possible coverage that it would not get back.

Building a Successful Media Relations Program

What does it take, then, to create a successful media relations program? First, organizations must be willing to devote resources to the effort. This rule does not necessarily have to mean huge outlays of money; an executive's time can be just as valuable.

Jim Koch, brewmaster and president of the company that makes Sam Adams beer, brought his beer into the national limelight through the skillful use of media relations with the help of one outside consultant at a fraction of the cost of a national advertising program. More recently, on a much smaller scale, two sisters who started a greeting card company that specialized in cards that targeted a gay audience were interested in building a relationship with the media. Through their own efforts, writing letters and reading the newspapers to find out who the best reporters would be for their message, they were able to get hits in both *The New York Times* and *The Wall Street Journal*. In both cases, the media relations effort paid off in sales, which was the ultimate goal.

For many larger companies, the media relations effort will involve more personnel and often the use of outside counsel. What follows is what is needed, at a minimum, for the effort.

Involve Media Relations Personnel in Strategy

As one public relations executive at a large company put it, "They like to keep us in the dark, like mushrooms, and then they expect us to get positive publicity, usually at the last minute." Instead, companies need to involve someone, preferably the most senior corporate communication executive, in the decision-making process. Once a decision has been made, it is much more difficult to talk management out of it because of potential problems with communications.

Although the communications point of view will not always win in the discussions that take place at top management meetings, having these individuals involved will at least allow everyone to be familiar with the pros and cons of each situation and decision. Communications professionals who are involved in the decision-making process also feel more ownership for the ideas that they need to present to the media.

---

[15] "Join the Sultans of Spin Media Relations," p. 19.

[16] Matalan website, http://www.matalan.co.uk/.

Develop In-House Capabilities

Using consultants and public relations firms may be beneficial in some cases, but most companies choose to develop an in-house media relations staff. Companies can save thousands of dollars a month by using internal staff and investing in the right tools to conduct research for analyzing the media. Often, the best communicator for a brand is an employee; firsthand knowledge is difficult to outsource.

One problem for many companies, however, is that they do not consider media relations to be important enough to hire professional staff in this area. Lawyers, executive assistants, and even accountants often handle communications because of the unfortunate assumption that, because "anyone can communicate," it doesn't matter whom you put on this assignment. Companies must recognize that building relations with the media is a skill and that individuals with certain personalities and backgrounds are better suited to the task than others.

Use Outside Counsel Strategically

For many companies, including those with limited budgets and resources, outsourcing media relations often makes more sense than hiring an internal communications team. Companies that do decide to look externally for media relations support have a wealth of options, from industry-specific consultancies to global communications agencies such as Edelman that have in-house digital capabilities. One of the chief arguments for hiring external agencies is that PR professionals have influence with press in a given space: they specialize in maintaining relationships with specific journalists and editors and are in constant dialogue with the media.

Many companies choose to outsource only part of their media relations strategy. eReleases, for example, is an online company that focuses on mass press release distribution. eReleases pulls from its database of over 100,000 journalists to target and submit a company's press release; it will also write the release for an additional fee. Public relations agencies charge thousands of dollars for the same kind of service. Other companies choose to outsource aspects of their digital media needs, including search engine optimization (SEO), social media measurement, and content production. Even the most experienced press release writer may struggle to capture the clipped, casual tone of a Twitter account. As media relations evolves under the influence of social media, large, integrated campaigns will become more popular, and they will require the skillsets of many different kinds of media agencies, from a team to manage the Facebook account to a production house that can produce video content.

Developing an Online Media Strategy

Until recently, media coverage—newspaper headlines or more in-depth profiles on television news shows like "60 Minutes"—has been the primary means for exposing corporate flaws. Accordingly, companies with well-managed media relations programs have had some leverage to get their own side of the story communicated to the public. Over the last two decades, however, wireless communication and the Internet have transferred an enormous amount of power into the hands of individuals. As Patricia Sturdevant, general counsel to the Washington-based National Association of Consumer Advocates, explains, "The Internet is a very effective new weapon for the consumer. Before the Internet, unless you had a lot of time or money, there wasn't any way to get the public's attention to a problem. Now, you can broadcast it to the entire world in an instant."[17] We will see in Chapter 10 that one disgruntled Dunkin' Donuts customer created a crisis situation for the company by launching his own anti-company website. This same occurrence has happened on a massive level for retail behemoth Walmart, which is the target of countless sites and blogs created solely to trash its reputation. Digital communications platforms, have enabled consumers to seize control of corporate messages and reputations and, in effect, have their way with them.

Thus, the Digital Age has many implications for business, including an expansion to individuals of powers that were previously concentrated in the hands of the organized media. Accordingly, companies' media strategies need to be augmented with tactics for dealing with this new dimension of coverage, including, for instance, establishing a forum for constituencies to share opinions, concerns, and complaints about the company, and a proactive effort to monitor information circulating about the company in various media channels, including blogs.

As much as digital communications platforms present the threat of the unknown, so, too can they provide incredible opportunities to communicate messaging in new, more creative and effective ways. Media relations has transformed into an almost unrecognizable rendition of its former self, with much of the messaging about a brand coming not from the one-way pipeline of company to press to shareholder, but from content created directly by the consumer. Social media and instant reporting by "on-scene" witnesses with handheld devices means that much of what is being communicated about companies is actually the product of individuals in addition to the mainstream media. By creating integrated media campaigns that tie together strong online and offline channels, companies can increase visibility and brand awareness. Most already have: over 84 percent of *Fortune* Global 100 companies use social media and blogging as part of their broader marketing efforts—77 percent have an active Twitter feed; 61 percent have Facebook pages.

One of the best integrated media campaigns of recent years is Nike's "Write the Future" campaign. Although not an official sponsor of the 2010 World Cup, Nike wanted to be a part of the global conversation around what would be one of the most watched sporting events in history. Nike Football's Facebook page was the cornerstone of the campaign strategy. Fans had the chance to see the brand's three-minute World Cup ad spot—starring the world's greatest soccer stars—before it aired. By the morning of the early release date, over 107,000 fans had signed up for the sneak peek. The day of the release, the film was viewed online 12,000,000 times and Nike Football Facebook fans tripled from 1.1 to

3.1 million. Nike went a step further with the "Write the Future" campaign, inviting fans to participate in "The Chance" contest, in which they could make http://www.facebook.com/pages/NIKE-WRITE-THE-FUTURE-JASON-CHOI-LIKE-LIKE-LIKE/136549343026199?v=wall"\t"_blank their own Facebook World Cup campaign. Nike would choose the best of the user-generated campaigns and send the creators to the Nike Academy soccer camp, where they could be scouted. Over-all, Nike garnered 1.9 billion online campaign impressions and over 40 million online views of the film. Nike reported that global sales grew 7 percent following the campaign.

Another example of a company that uses social media to its benefit is Southwest Airlines. The company has invested significant resources to build a completely integrated digital communications strategy into its overall communications plan. This strategy includes a blog (www.blogsouthwest.com), on which every employee—from executives to pilots to airplane mechanics—is encouraged to post. Thanks to its genuine content and its pledge to transparency, the blog has become a viable way for Southwest executives to communicate with employees and consumers alike and to strengthen stakeholders' relationship with the brand.

In addition to the blog, Southwest executives also count the "Wanna Get Away" microsite, YouTube videos, a Facebook page, a Twitter application, and a presence on LinkedIn among their branded digital platforms.

Twitter and Facebook have proven to be useful channels for connecting with both customers and media, but there are many others. LinkedIn was launched in May 2003 as a "grown-up" version of Facebook, appealing to a more business-minded audience by focusing on professional rather than personal relationships. By 10 years later, LinkedIn had more than 100 million users. LinkedIn gives executives the ability to follow news-related web activity among members of their organizations, competitors, and their industries as a whole via a widget that can be downloaded. This widget automatically aggregates the links, news stories, and media sources that are most relevant to the user's profile, allowing individuals to track the news that their connections are consuming.[18]

Extend Your Media Relations Strategy to the Blogosphere

Studies have shown that the public is often far more trusting of other consumers than it is of traditional institutions, including corporations. According to the 2011 Edelman Trust Barometer, nearly half of all respondents trust "a person like yourself" to relay credible information about a company.[19] This response helps explain the phenomenal rise of blogs from the 1990s to the present, as blogs can be "owned" and operated by any average individual—or company, as seen by the previous Southwest Airlines example. There are more than 150 million blogs in the United States, and because of their speed, bloggers can and do alter the volume and tone of any conversation.[20] For example, bloggers played a key role in the 2008

---

[18] "Global Social Media Checkup Report 2011," Burson Marsteller, http://www.slideshare.net/BMGlobalNews/bursonmarsteller-2011-global-social-media-checkup.

[19] Case Studies, http://www.mindshareworld.com.

[20] Edelman Trust Barometer 2011.

presidential election, with candidates Hillary Clinton, Barack Obama and John McCain all maintaining blogs to stay connected with voters.

In addition, blogs are an important tool for corporations and journalists alike to track consumer points of view and concerns. However, not all corporate executives took kindly to the blogosphere at first, and their companies' brands and reputations suffered accordingly. Failing to embrace the Internet as a viable and potentially violent communications tool has serious implications. In addition to the aforementioned critical sites created by consumers, huge multinational companies across the globe have been forced to go up against single individuals to protect their brands and get the true story out to media. Dell's reputation was thrown for a loop when, in June 2005, an irate blogger by the name of Jeff Jarvis lambasted the company for poor customer service. Within hours, hoards of consumers who were in agreement with his claims posted comments, thus creating a maelstrom of negativity throughout the blogosphere. The company remained in the doghouse for months after failing to address the discontent properly in cyberspace; however, beginning with the launch of its own blog (Direct2Dell) in July 2006, executives finally joined the online conversation and slowly began to rebuild its tarnished image.

The blog was put to good use when another potential crisis—a widespread battery recall—hit. Dell's chief blogger Lionel Menchaca addressed the issue in a human voice and enabled customers to comment freely. The blog also offered information on how customers could get a replacement battery. Michael Dell even launched IdeaStorm.com and implored customers to give the company advice. New metrics show that the customer service rating has risen significantly

The following are some guidelines on blogs:

- *Take blogs seriously.* Find those that seem to be most interesting for your industry, bookmark them, and read them regularly.

- *Act fast.* If you need to respond to something on a blog, do so quickly and honestly. Similarly, if you are writing a corporate blog, make sure it is transparent and very up to date. Some corporations, such as McDonald's and Mazda, have tried to tap into the power of the blogosphere by creating fake blogs, only to have this tactic backfire when real bloggers exposed them.

- *Don't dismiss requests for interviews and information from bloggers.* Many are also established journalists, and if they are unhappy with your attitude, you may find your e-mail exchange published in full on their site.[21]

Tapping into the information circulating on the Internet can give companies extraordinary access to information about customer needs and complaints. Monitoring Internet "chats" and blogs can enable companies to learn about current constituency needs and tailor actions to meet those that are most vital to the company's reputation and bottom line. By using the Internet proactively, companies can glean valuable insights about constituency attitudes, sentiments, and reactions to which they might otherwise not have access. In many ways, a company should

---

[21] Edelman and Intelliseek, "Trust 'MeDIA': How Real People Are Finally Being Heard," White Paper, Spring 2005.

view the Internet as an unprecedented and ideal survey group. Without a doubt, online monitoring can help companies gauge the sentiments of constituencies, allow them to respond effectively, and help them stay on top of today's information surge. However, companies should not become so consumed by the power of the Internet that they neglect other important media channels.

Handle Negative News Effectively

Although negative news reporting has existed as long as the press itself, its frequency and severity has increased exponentially in the context of the changing business environment for all of the reasons discussed in this chapter. Now more than ever, organizations' reputations are vulnerable to the innumerable unknowns that infiltrate the press due to digital communications platforms. Corporate leaders have seen their peers crumble under the glare of unanticipated negative press and are changing their business strategies and response mechanisms accordingly.

Because of the widespread reach of the Internet, companies must keep their proverbial fingers on the pulse of the ever-changing news landscape and constantly work to build the goodwill that will help them to weather any negative press. (See Chapter 4 for more on reputation management.) Bad publicity online can legitimately threaten the bottom line. Many companies assign employees or obtain external specialists to search for negative stories about the company. Others take advantage of low-cost tools, including search engines such as Google that can be used to set up free alerts. The "consumer opinion" section on Yahoo!'s site alone (http://dir.yahoo.com) lists over 200 consumer opinion sites—criticizing companies such as American Express, Ford, Nike, Walmart, and even Yahoo! itself.

When companies carefully monitor web activity, they are better positioned to respond in a timely manner to negative comments. FedEx recently won praise for issuing a prompt video apology in response to a customer's video of a FedEx courier delivering a box clearly labeled as a computer monitor by tossing it over the customer's fence instead of carefully placing it on the ground, or delivering it to the front door of the customer, who claimed to be home at the time. After opening the box to discover a broken monitor, the customer reviewed footage of the delivery that was captured by his closed-circuit security camera and posted what he discovered to YouTube with the title: "FedEx Guy Throwing My Computer Monitor." The video, posted on Monday, December 19, 2011, quickly went viral. On Thursday, December 22, 2011, FedEx responded with a heartfelt video apology that it uploaded to YouTube. The original video received more than 4 million views by January 22, 2011, the date that FedEx posted its apology. By January 2012, the video had received 8 million views, meaning that for 4 million viewers, the company's apology was just a click away when they watched the offending video.

Many companies recognize that the best way to prevent negative news is to ensure that inside information does not leak. Companies are getting smarter about creating social media guidelines and policies that regulate the flow of sensitive information from within. Any corporate employee, whether a CEO with a blog or a recent MBA hire with a LinkedIn profile, can potentially make "news" without proper guidance. Ever the innovator, IBM was one of the first companies to introduce guidelines for social platforms. In the spring of 2005, the company created a wiki and asked employees to contribute advice and fair restrictions in terms of authoring blogs. This wiki became the first iteration of the company's blogging policy, and evolved into what is now IBM's comprehensive "Social Computing Guidelines," authored by the employees themselves.

When a company does stumble upon bad news circulating about itself—be it a condemning attack in the blogosphere or a hostile op-ed article in a daily newspaper—the communications department should quickly assess the potential damage that the news might cause. Who is the person who has issued the complaint? Are the comments valid? Is the person speaking only as an individual, or does she or he represent a broader constituency, such as investors or employees? If a broader constituency, how widespread are the complaints? If a rogue website has been constructed, how many hits per day has it received, and how have people generally responded to the negative message? If an unflattering newspaper article has been printed, how wide is the paper's circulation?

Once these questions are answered, a company's task force or permanent crisis communication team—including members of senior management—must brainstorm some potential actions. Company lawyers should be consulted to discuss what legal stance the company might need to take. Lawyers will be able to offer advice about whether newspaper articles, websites, or blogs are defamatory, warranting a lawsuit against the perpetrator.

Conclusion

As technology develops new mechanisms for disseminating information, the media relations function will continue to evolve away from the old PR flak model into a professional group that can help organizations get their message out quickly, honestly, and to the right media.

Companies today are under constant scrutiny from many of their constituencies. A demand for instantaneous information accompanies this public watchfulness, and the pressure is increasing with each new technological innovation. Managers must be prepared to answer this demand by considering all constituencies in dealing with the media agents who inform them. By crafting messages with care and then using proper media channels, companies can tap into this powerful "conduit constituency," the media, to ensure that the best possible message is disseminated and heard.

[22] Maja Pawinska Sims, "Monitoring the Web—Blogging the Great Untapped Resource," *PRWeek*, June 10, 2002.

[23] YouTube.com and http://www.digitaltrends.com/web/fedex-responds-to-computer-monitor-chucking-delivery-man-debacle/

[24] IBM Social Computing Guidelines, http://www.ibm.com/blogs/zz/en/guidelines.html.

Adolph Coors Company

Shirley Richard returned from lunch one April afternoon in 1982 and found a message on her desk that Allan Maraynes from CBS had phoned while she was out. "God, what's this?" was all she could say as she picked up the phone to discuss the call with her boss, John McCarty, vice president for corporate public affairs. In her second year as head of corporate communication for what was then the nation's fifth-largest brewer, Richard was well aware of the Adolph Coors Company's declining popularity—a decline that she partially blamed on an ongoing conflict with organized labor. But the conflict was hardly breaking news, and she was almost afraid to ask why CBS was interested in the company.

Richard found out from her boss that Maraynes was a producer for the network's news program "60 Minutes." Reporter Mike Wallace had already phoned McCarty to announce plans for a "60 Minutes" report about the company. Program executives at CBS were aware of accusations of unfair employment practices that the AFL-CIO had raised against Coors and wanted to investigate the five-year battle between the brewery and organized labor.

Once McCarty explained the message from Maraynes, Shirley Richard sank into her chair. She had spent the last year working hard to understand organized labor and its nationwide boycott of Coors beer, and she was convinced that the company was being treated unfairly. She believed the union represented only a small subset of Coors's otherwise satisfied workforce. But Richard also doubted whether the facts could speak for themselves and was wary of the AFL-CIO's ability to win over the media. She was well aware of Mike Wallace's reputation for shrewd investigative reporting.

On the other hand, "60 Minutes" was considered by many corporations as anti-big-business, and Richard had no idea how corporate officials would respond under the pressure of lights, camera, and the reporter's grilling questions. McCarty and Richard met with the two Coors brothers to discuss the network's proposal and to determine whether producer Maraynes should even be allowed to visit the Coors facility. Company president Joseph ("Joe") Coors and chairman William ("Bill") Coors were skeptical of the prospect of airing the company's "dirty laundry" on national television. But McCarty was interested in the opportunity for Coors to come out into the public spotlight. Richard knew that granting interviews with Wallace and permission to film the Coors plant came with enormous risk.

Richard was frustrated by growing support for the boycott, and her own strategies to deal with the problem had been unsuccessful. She believed the interview with CBS might only exacerbate an already difficult situation. Her own public relations effort had been an attempt to portray the circumstances as she believed them to be: good management harassed by disgruntled labor organizers. She was convinced that her job was not an effort to cover up Coors's employment practices.

Richard debated how the company should handle the proposal from CBS. Any decisions about approaching "60 Minutes" also would have to be approved by the Coors brothers. Richard felt uncertain about how much control she would ultimately have over the communications strategy. Joe Coors, an ardent conservative and defender of private enterprise, would undoubtedly resist an open-door policy with the network. At the same time,

Source: This case was prepared by Professor Paul A. Argenti, Tuck School of Business at Dartmouth. © 2001 Trustees of Dartmouth College. All rights reserved.

Richard wondered if she should attempt to convince the management of this traditionally closed company to open itself to the scrutiny of a "60 Minutes" investigation or whether the best defense would be a "no-comment" approach. But with no comment from Coors, anything organized labor was willing to say on camera would go uncontested.

HISTORY OF THE ADOLPH COORS COMPANY

The Coors brewery was established in 1880 by Adolph Coors, a Prussian-born immigrant who came to the United States in 1868. Having trained as an apprentice in a Prussian brewery, 22-year-old Adolph Coors became a foreman at the Stenger Brewery in Naperville, Illinois, in late 1869. By 1872, Coors owned his own bottling company in Denver, Colorado. With his knowledge of brewing beer and the financial assistance of Joseph Schueler, Coors established his own brewery in Golden, Colorado. His product was an immediate success. In 1880, Adolph Coors bought out Joseph Schueler and established a tradition of family ownership that was maintained for almost a century.

Famous for its exclusive "Rocky Mountain spring water" system of brewing, the Adolph Coors Company soon became something of a legend in the beer industry. The Coors philosophy was one of total independence. A broad spectrum of Coors subsidiaries combined to create a vertically integrated company in which Coors owned and managed every aspect of production: The Coors Container Manufacturing plant produced aluminum and glass containers for the beer; Coors Transportation Company provided refrigerated trucks to haul the beer to its distribution center as well as vehicles to transport coal to fuel the Golden brewery; Coors Energy Company bought and sold energy and owned the Keenesburg, Colorado, coal mine; the Golden Recycle Company was responsible for ensuring a supply of raw materials for

aluminum can production. By 1980, the recycling plant was capable of producing over 30 million pounds of recycled aluminum a year. Other subsidiaries fully owned by Coors included Coors Food Products Company, Coors Porcelain Company, and the American Center for Occupational Health.

THE COORS MYSTIQUE

A certain mystique surrounding the Golden, Colorado, brewery, and its unique, unpasteurized product won the beer both fame and fortune. Presidents Eisenhower and Ford shuttled Coors to Washington aboard Air Force jets. Actors Paul Newman and Clint Eastwood once made it the exclusive beer on their movie sets. Business magazines lauded Coors as "America's cult beer." As Coors expanded its distribution, the mystique appeared irresistible; Coors moved from 12th to 4th place among all brewers between 1965 and 1969 with virtually no advertising or marketing.

Part of the Coors mystique was attributed to its family heritage. For over a century of brewing, company management had remained in the hands of Adolph Coors's direct descendants. Reign passed first to Adolph Coors Jr., then to his son William Coors. In 1977, Bill Coors turned over the presidency to his younger brother Joseph but continued as chairman and chief executive officer. The company's newest president, Joe Coors, was a well-known backer of right-wing causes such as the John Birch Society; a founder of a conservative think-tank, the Heritage Foundation; and a member of President Ronald Reagan's so-called Kitchen Cabinet. The family name was closely associated with strong conservatism by consumers, labor, and the industry.

The Coors Company was built on a tradition of family and, even after going public in 1975, remained an organization closed to active public relations. Bill Coors recalled that his father, Adolph Coors Jr., was a shy man, and throughout its history the company was reluctant to attract any public attention. In 1960, the sensational kidnapping and murder of brother

Adolph Coors III focused the public eye on the family and the business, but Coors maintained a strict "no-comment" policy.

THE NATURE OF THE BREWING INDUSTRY

From the mid-1960s through the 1970s and into the 1980s, the brewing industry was characterized by a shrinking number of breweries coupled with a growing volume of production and consumption. In 1963, Standard and Poor's Industry Surveys reported 211 operating breweries. Ten years later that number had dropped to 129, and by 1980 there were only 100 breweries in operation. On the other hand, per capita consumption of beer rose from 15 gallons a year in 1963 to 19.8 gallons in 1973. By 1980, per capita consumption had jumped to 24.3 gallons a year.

Until the mid-1970s, beer markets were essentially local and regional, but as the largest breweries expanded, so did their share of the market. Combined, the top five brewers in 1974 accounted for 64 percent of domestic beer production, up from 59 percent in 1973. Previously strong local and regional breweries were either bought by larger producers or ceased operations.

A notable exception, however, was the Adolph Coors Company, which dominated the West. Until 1976, the company's 12.3-million-barrel shipment volume was distributed only in California, Texas, and 10 other western states. Coors's share of the California market alone was well over 50 percent in 1976. Coors dominated its limited distribution area, capturing at least 35 percent of the market wherever it was sold statewide. The Coors Company ranked fifth in market share nationally throughout the 1970s, trailing giants Anheuser-Busch, Joseph Schlitz, Phillip Morris's Miller, and Pabst, all of which had much broader distribution areas.

Competition for market share among the top five brewers was intense during the 1970s and led producers to more aggressive attempts to win consumers. According to compilations by Leading National Advertisers, Inc., advertising expenditures for the first nine months of 1979 were up 37 percent from the previous year for Anheuser-Busch, 18 percent for Miller, 14 percent for both Schlitz and Pabst, and 78 percent for Adolph Coors.

MARKETING AND DISTRIBUTION AT COORS

Industry analysts criticized the Coors Company's sales strategy for stubbornly relying on its product's quality and image rather than marketing. In 1976, the Coors mystique appeared to be losing its appeal to strong competitors—for the first time since Prohibition, Coors could not sell all of its beer. The company finally responded to competition by intensifying its marketing and development operations. Between 1976 and 1981, the company attempted to revive sales by adding eight new states to its distribution. In May 1978, Coors began to market its first new product in 20 years: Coors Light. In 1979, Coors began the first major advertising campaign in its history to defend itself against aggressive competitors such as Philip Morris's Miller Brewing Company and Anheuser-Busch. The company's 1981 annual report pictured Coors's newest product—George Killian's Irish Red Ale—along with a newly expanded package variety designed to "keep pace with consumer demand."

The Coors Company went public in 1975, but investors did not fare well as stock prices declined for the rest of the decade. Coors entered the market at a share price of $31 but by 1978 had fallen to $16—a loss of about 50 percent for the first public stockholders. Net income, according to the company's annual report, was $51,970,000 in 1981, or $1.48 per share. That figure reflected a 20 percent drop from $64,977,000, or $1.86 per share, in 1980.

MANAGEMENT–LABOR RELATIONS AT COORS

During pre-Prohibition years, breweries, including Coors, were entirely unionized. In 1914, the first vertically integrated industrial union in the country established itself at Coors. When the country went dry, Coors remained viable through alternative operations, but the workforce still had to be reduced. Coors offered older workers employment but fired younger employees. A strike of union employees resulted and remained in effect until 1933, when Prohibition was repealed. The company, however, continued to operate without a union until 1937 when Adolph Coors Jr. invited the United Brewery Workers International (UBW) into the Coors Company.

In 1953, the company experienced an abortive strike by the UBW to which a frightened management immediately gave in. In 1955, Coors's organized porcelain workers struck because their wages were less than those of brewery workers. Although the plant continued to operate, all of Coors's unionized workers engaged in a violent strike that lasted almost four months. The union ultimately lost the battle 117 days after the strike, when workers returned to the plant on company terms.

Negotiations over a new union contract in 1957 ended in a stalemate between labor and management, and workers again decided to strike. For another four months, workers were torn between paternalistic and small-town personal ties to management and the demands of the union. Bill Coors, who was then the plant manager, recalled that during the strike, management had wanted to show the union it was not dependent on union workers. Coors hired college students during the summer of 1957 as temporary replacements for the striking brewers. When the students left, the picketers were threatened by management's vow to hire permanent replacements and returned to the plant. The strike was a clear defeat of the union's demands and ultimately left international union leaders with an unresolved bitterness toward Coors. Back in full operation by the fall of 1957, Coors management believed it had won complete control.

By the end of the 1950s, 15 local unions were organized at Coors. Management tolerated the unions but claimed they did not affect wages or employment practices. The Coors family firmly believed that good management removed the need for union protection and that management could win workers' loyalty. In 1960, the plant's organized electricians went on strike but failed to garner the support of other unions, and the plant continued to operate with nonunion electricians hired to replace the strikers. Similar incidents occurred with Coors's other unions. A 1968 strike by building and construction workers ended with Coors breaking up 14 unions. By 1970, Coors's workforce was predominantly nonunion.

A contract dispute between Coors's management and UBW Local 366 erupted in 1976. Workers demanded a 10 percent wage increase and better retirement benefits. After more than a year of negotiations, union officials rejected management's compromise offer, which labor contended would erode workers' rights. In April 1977, over 94 percent of UBW workers voted to strike. Production at the plant continued at 70 percent of normal capacity, however, and management boldly announced plans to replace striking workers. In defense of the union, AFL-CIO officials declared a nationwide boycott of the beer until a new contract settlement was reached. But within five days of initiating the strike, 39 percent of the union members crossed the picket lines to return to work.

In 1978, Coors management called an election for decertification of UBW Local 366. Because more than a year had passed since the strike began, National Labor Relations regulations restricted striking union members from voting. Only workers remaining at the plant, including "scabs" hired across the picket lines, could vote on whether to maintain the UBW

Local. In December of that year, Coors employees voted a resounding 71 percent in favor of decertifying the Local UBW.

Since 1957, the Coors brewery had been a "closed shop," in which workers were required to pay union dues if they were to benefit from union action. But company officials called the 1978 decertification vote a victory for the "open shop," wherein workers could enjoy union benefits without paying dues as members. Union officials, frustrated over the lack of a new contract and the decertification vote, publicly charged Coors with "union busting."

In fact, according to AFL-CIO officials, the UBW was the 20th Coors union decertified since the mid-1960s. Management consistently argued that employees simply rejected union organization because they didn't require it; good management eliminated the need for a union to protect workers. But organized labor maintained that all 20 unions had been "busted" by votes called while members were on strike and scabs were casting the ballots. By the end of the decade, only one union representing a small group of employees remained active at Coors.

NATIONWIDE BOYCOTT

The AFL-CIO was determined not to be defeated by the ousting of the UBW Local from the Golden plant. In defense of the union, AFL-CIO officials declared a nationwide boycott of Coors beer until a new contract settlement could be reached and soon began to claim that their efforts had a significant effect on sales. In fact, 1978 figures reported a 12 percent profit decline for the brewery during fiscal 1977 and predicted that 1978 figures would fall even lower. Corporate officials conceded the boycott was one factor influencing declining sales but refused to admit the drop was consistent or significant.

The defeat of the Coors local brewers' union fueled the boycott fire, but the protest focused on issues beyond the single contract dispute

begun in 1977. The other issues of protest related to Coors's hiring practices. Labor leaders claimed that a mandatory polygraph test administered to all prospective employees asked irrelevant and personal questions and violated workers' rights. In addition, the protesters claimed that Coors discriminated against women and ethnic minorities in hiring and promotion. Finally, boycotters argued that Coors periodically conducted searches of employees and their personal property for suspected drug use and that such search and seizure also violated workers' rights. The boycott galvanized organized labor as well as minority interest groups that protested in defense of blacks, Hispanics, women, and gays.

The boycott's actual effect on sales was the subject of dispute. Coors's sales had begun to fall by July 1977, just three months after the boycott was initiated. Some analysts attributed the drop not to protesting consumers but rather to stepped-up competition from Anheuser-Busch, which had begun to invade Coors's western territories. Despite a decline, Coors remained the number-one seller in 10 of the 14 states in which it was sold. Labor, on the other hand, took credit for a victory at the end of 1977 when Coors's fourth-quarter reports were less than half of the previous year's sales for the same period. Dropping from $17 million in 1976 to $8.4 million in 1977, Coors was faced with a growing challenge. There was no doubt that management took the AFL-CIO protest seriously and began attempts to counter declining sales through more aggressive advertising and public relations.

FEDERAL LAWSUIT

The AFL-CIO boycott gained additional legitimacy from the federal government. In 1975, the federal Equal Employment Opportunity Commission (EEOC) had filed a lawsuit against Coors for discrimination in hiring and promotion against blacks, Mexican Americans, and women. The suit charged Coors with violating the 1964 Civil Rights Act and challenged Coors's hiring tests, which the EEOC said were aimed at revealing an applicant's arrest record, economic status, and physical characteristics. The lawsuit stated that the company used "methods of recruitment which served to perpetuate the company's nonminority male workforce."

In May 1977, one month after the initiation of the AFL-CIO boycott, Coors signed an agreement with the EEOC, vowing that the brewery would not discriminate in hiring. But according to media reports, Coors still refused to admit any past bias toward blacks, Mexican Americans, and women. Coors said it would continue a program begun in 1972 designed to increase the number of women and minorities in all major job classifications. Striking brewery workers refused to sign the agreement, although the Coors's Operating Engineers Union entered into the agreement.

DAVID SICKLER AND THE AFL-CIO

The principal organizer of the AFL-CIO boycott against the Adolph Coors Company was the former president of the company's Local UBW. David Sickler had been employed by Coors for 10 years, acting as a business manager from 1973–1976. Sickler left the plant in 1976 to take a job with the AFL-CIO in Montana. In April 1977, the AFL-CIO decided to put Sickler in charge of coordinating the national boycott against Coors. Sickler moved to Los Angeles, where he also served as director of the Los Angeles organizing committee and the subregional office of the AFL-CIO.

Sickler initially resisted the AFL-CIO's request to put him in charge of organizing the boycott. He believed that his past employment at the company made him too close to the situation to offer a fair position on the issues at stake. But the AFL-CIO felt that Sickler's tenure with Coors made him an ideal choice; according to Sickler, his personal reports of abuse by the company in hiring and employment practices

were shared by numerous Coors employees and were the central issues of the boycott.

Sickler contended that when hired by Coors, he had been subjected to questions on a lie detector test regarding his personal life and sexual preference. In addition, he reported the company's practice of searching individuals or entire departments for suspected drug use. Despite corporate officials' insistence that the accusations were false, Sickler was convinced that Coors employees were generally "unhappy, demoralized."

Coors management was determined to fight back against the boycott and filed a breach of contract suit against the Local 366. The company charged that any boycott was prohibited under contract agreements. Management also made clear to the public its outrage over the boycott, as chairman Bill Coors began to speak out in the national media. In a 1978 interview with *Forbes* magazine, Coors stated about the AFL-CIO: "No lie is too great to tell if it accomplishes their boycott as a monument to immorality and dishonesty." Earlier that year, Bill Coors defended the company against charges of being antiunion. A *New York Times* report on the dispute quoted the CEO as saying: "Our fight is not with Brewery Workers Local 366. Our fight is with organized labor. Three sixty-six is a pawn for the AFL-CIO; that's where they're getting their money."

CORPORATE COMMUNICATION AT COORS

The 1977 boycott forced company officials to reexamine the area of corporate communication. Because labor leaders set out to "destroy the company," Bill Coors, now chairman and chief executive officer of the company, believed management must relate its side of the story. "There was no lie they wouldn't tell," the CEO recalled. "No one knew about Coors, and we had no choice but to tell the story."

In 1978, John McCarty, responsible for fundraising at Pepperdine University, was hired as the vice president for corporate public affairs. McCarty brought to Coors expertise in minority relations and set out to repair the company's damaged reputation among minority groups. McCarty established a staff of corporate communication officers. The division was organized into four branches under McCarty's leadership: corporate communication, community affairs, economics affairs, and legislative affairs.

In response to the boycott and declining sales, McCarty enlisted the expertise of J. Walter Thompson's San Francisco office to help the company improve its corporate image. Coors launched what analysts termed a strong "image-building" campaign in 1979, with messages aimed at ethnic minorities, women, union members, and homosexuals. The theme throughout the late 1970s was clearly a response to labor's accusations against the company: "At Coors, people make the difference."

Another component of the new image campaign, according to media reports, was to condition company managers to project charm and humility in dealing with reporters. Coors executives participated in a training course designed to help them overcome a traditional distrust of the media.

SHIRLEY RICHARD

Shirley Richard was hired along with McCarty in 1978 to direct the company's legislative affairs function but was familiar with the Coors Company long before joining its staff. From 1974–1978, Richard worked on the Coors account as a tax manager for Price Waterhouse. One important issue for the Coors account, Richard recalled, was the deductibility of lobbying expenses and charitable donations. As part of her job, Richard became involved in the political arena, helping Coors set up political action committees. When Richard decided to leave Price Waterhouse in 1978, she asked Coors's vice president of finance for a job and was hired to head the legislative affairs department, a position she held until 1981.

Richard recalled her first year with the company as a time when Coors was "coming out of its shell"; Philip Morris's purchase of Miller Brewing Company meant increased competition for Coors and a demand for more aggressive advertising. In 1975, the company sold its first public stock. The bad publicity from the 1977 strike and its aftermath combined with greater competition led to a serious decline in sales and disappointed shareholders. Clearly, the Coors mystique alone could no longer speak for itself, and an aggressive public relations campaign was unavoidable.

One year before the "60 Minutes" broadcast of the Coors story, Richard became Adolph Coors Company's director of corporate communications. In that position, she managed 25 people, covering corporate advertising, internal communications, distribution communications, training programs, and public relations personnel.

CONFRONTATIONAL JOURNALISM

The challenge of CBS's "60 Minutes" to any company under its investigation was formidable. The 14-year-old program was consistently ranked in Nielsen ratings' top 10 programs throughout the 1970s. Media critics offered various explanations for the success of this unique program, which remarkably combined high quality with high ratings. A *New York Times* critic summarized the sentiment of many within the broadcast profession when he called "60 Minutes," "without question, the most influential news program in the history of the media."

The program had earned its popularity through consistently hard-hitting investigative reporting. Executive producer Don Hewitt proclaimed "60 Minutes" the "public watchdogs." In his book about the program, Hewitt recalled, "I became more and more convinced that a new type of personal journalism was called for. *CBS Reports, NBC White*

Papers, and *ABC Closeups* seemed to me to be the voice of the corporation, and I didn't believe people were any more interested in hearing from a corporation than they were in watching a documentary." Stories revealing insurance executives taking advantage of the poor with overpriced premiums, companies polluting streams and farmlands by irresponsibly dumping, or physicians gleaning profits from unnecessary surgery had all worked to rally public support and faith in CBS as a sort of consumer protection agency.

The program's success in uncovering scandal was due in large part to the aggressive and innovative technique of Mike Wallace. Wallace had been with the program throughout its history and was responsible for shaping much of the "60 Minutes" image. His reporting was always tough, sometimes theatrical, and was commonly referred to within the media as "confrontational journalism."

Allan Maraynes was assigned to produce the Coors segment. His experience with "60 Minutes" was highlighted by some significant clashes with big business. He had produced stories on the Ford Pinto gasoline tank defects, Firestone tires, I. Magnin, and SmithKline. Maraynes was alerted to the Coors controversy when "60 Minutes" researchers in San Francisco told him they suspected bad things were happening at Coors. The research group told Maraynes that the AFL-CIO was calling Coors a "fascist organization," which sounded to the producer like good material for a story.

Maraynes first flew to California to interview David Sickler. "We said we were setting about to do a story explaining that a fascist state exists at Coors," Maraynes recalled about his conversation with Sickler. "If it's true, we'll do it." Maraynes wanted Sickler to give him as much information about the boycott as he had. Maraynes wanted the angle of the story to be a focus on case histories of the people who had experienced Coors's unfair treatment.

OPEN OR CLOSED DOOR?

With the phone call from Maraynes, all of the pressures from David Sickler, the AFL-CIO, and the boycott were suddenly intensified. Shirley Richard had worked hard in the last year to focus public attention away from the boycott, but now her efforts to project a positive corporate image were threatened. Thinking ahead to the next few months of preparation time, she felt enormous pressure in the face of such potentially damaging public exposure.

Shirley Richard was not naive about Mike Wallace or the power of television news to shape a story and the public's opinion. Richard, along with other Coors executives, believed that the company was not at fault, but that did nothing to guarantee that its story would be accurately portrayed in a "60 Minutes" report. Mike Wallace himself had voiced the reason for a potential subject to fear the program's investigative report. In a *New York Times* interview, Wallace stated: "You (the network) have the power to convey any picture you want."

Richard knew that a big corporation's abuse of employees was just the kind of story "60 Minutes" was built on, and she didn't want Coors to be part of enhancing that reputation, especially when she believed organized labor had fabricated the controversy about Coors. Given Mike Wallace's desire to get the story, Shirley Richard guessed the company would automatically be on the defensive.

"60 Minutes" was determined to do the story, with or without cooperation from Coors. Richard wondered, however, whether an interview with Mike Wallace would do the company more harm than good. On the other hand, she considered the possibility that the company could somehow secure the offensive and turn the broadcast into a final clarification of Coors's side of the boycott story.

Richard was clearly challenged by an aggressive news team, and she was uncertain about cooperation from the conservative Coors brothers. Even if she could convince them that an open door was the best policy, would corporate officials be able to effectively present the facts supporting Coors's position? The national broadcast would reach millions of beer drinkers, and Richard knew that the "60 Minutes" report could either make or break the future success of Coors beer.

CASE QUESTIONS

1. What problems should Richard focus on?
2. What kind of research should she do?
3. What would her communication objective be if Coors agreed to the interview? If the brothers did not do the interview?
4. Should Shirley Richard encourage or discourage the Coors brothers to go on "60 Minutes"?
5. What suggestions would you have for improving media relations at Coors?

Source: This case was researched and written by Professor Paul A. Argenti in 1985 and revised in 1994, 1998, 2002, and 2005.

CHAPTER SEVEN

Internal Communications

For years, managers have focused on "customer care." More recently, they have begun to dedicate the same kind of attention to their own employees, recognizing that employees have more to do with the success of a business than virtually any other constituency. According to a study in which human resources consulting firm Towers Watson analyzed three years of employee data for 40 global companies, separating them into high-employee engagement and low-engagement categories over a three-year period, companies with a highly engaged employee population reported a significantly better financial performance (a 5.75 percent difference in operating margins and a 3.44 percent difference in net profit margins) than did low-engagement workplaces.[1] Human capital consultancy firm Aon Hewitt also reports finding a strong correlation between a company's employee engagement rating and its financial performance, noting that organizations with high levels of engagement (65 percent or greater) outperformed the total stock market index in 2010, posting total shareholder returns 22 percent higher than average. It also found that companies with low engagement (45 percent or less) posted total shareholder returns 28 percent lower than the average.[2]

In this chapter, we examine how organizations can strengthen relationships with employees through internal communications. Internal communications in the twenty-first century is more than the memos, e-mails, publications, and broadcasts that comprise it; it's about building a corporate culture based on values and having the potential to drive organizational change. We start by looking at how the changing environment for business has created the need for a stronger internal communication function. Then we explore ways to organize internal communications through planning and staffing and how to implement a strong program using various communication channels. Finally, we discuss management's role in internal communications.

---

[1] Towers Watson, "Turbocharging Employee Engagement—The Power of Recognition from Managers," October 2010, http://www.towerswatson.com/assets/pdf/2979/TowersWatson-Turbocharge-NA-2010-18093.pdf.

[2] Aon Hewitt, "Trends in Global Employee Engagement," 2011, http://www.aon.com/attachments/thought-leadership/Trends_Global_Employee_Engagement_Final.pdf.

Internal Communications and the Changing Environment

As discussed in Chapter 1, the environment for business has changed dramatically over the last 50 years. Today's employee is a different person in terms of values and needs than his or her counterpart in earlier decades. Most of today's employees are well-educated individuals, have higher expectations of what they will get out of their careers than their parents did, and want to understand more about the companies they work for.

The workplace of today is also different—tighter staffing, longer hours, greater workloads, and more emphasis on performance are the norm. In recent years, the increased outsourcing of jobs to foreign countries has filled many employees with feelings of fear, paranoia, and anger. And in light of the recent recession and downfall of established giants such as Lehman Brothers and Bear Stearns, many employees are functioning with a greater degree of cynicism or distrust of corporations and their senior management. All of these factors are causing employees to look more critically at how senior management is communicating with them, what is being communicated, and whether or not they feel engaged in and aligned with the company's direction.

The increasingly complex and highly competitive nature of today's business environment puts greater pressure on employees and also calls for a more concerted effort in the area of internal communications. "There's a tremendous anxiety in the workplace," says Rick Hodson, a longtime practitioner of employee communications. "When internal communications programs shrink, or disappear, rumor and gossip fill the vacuum. If you keep information from employees, they'll keep their ideas and feelings from you."[3]

For example, in 2008, the entire pharmaceutical industry was facing major changes, including outsourcing, increased costs, and changing regulations. Rather than let employees leap to their own conclusions in these stressful times, AstraZeneca, one of the world's leading pharmaceutical companies, worked with leading European communications specialists Synopsis to engage employees by coaching leadership teams and equipping communicators to support managers.[4]

Today's employees are increasingly demanding participation in the conversations at work that drive organizational change. Allowing this participation is vital to keeping employees at all levels of the organization engaged—regardless of job role or responsibility—fostering a more genuine sense of community in companies large or small. In light of this development, communication must be a two-way process. Employees today expect that when their opinions are solicited and they take the time to share feedback, senior management will listen—and act upon it.

At many companies, senior managers simply do not involve lower-level employees in most decisions. This failure tends to make these employees feel alienated and unappreciated, leading them to be unwilling to accept changes within the company. According to a survey of workers by the American Psychological Association (APA), less than half of employees (43 percent) said they receive

---

[3] John Guiniven, "Inside Job: Internal Communications in Tough Times," *Public Relations Tactics*, November 2009, p. 6.

[4] David Norton, "Leadership Communication—The AstraZeneca Way," *Strategic Communication Management*, January 2008.

1. They're smarter than senior managers think they are.
2. They think senior managers are smarter than they actually are.
3. They hate it when you make them feel stupid.
4. They have short attention spans.
5. They have long memories.
6. They're desperate for direction.
7. They want to be able to think on their own.
8. They want the company to succeed.
9. They don't want to leave.
10. They want to believe in the company.

*Adapted from a speech given at the Tuck School of Business, Dartmouth University, by Rod Odham of Bell South's Small Business Services Division, October 1994.

adequate nonmonetary rewards and recognition for their contributions at work and only 57 percent reported being satisfied with their employer's work/life practices. Just 52 percent said that they feel valued on the job, and only two-thirds reported being motivated to do their best at work.[5]

Managers need to recognize that if they provide information to employees and also listen to them, those employees will be excited about their work, connected to the company's vision, and in a position to further the goals of the organization. Mark Grundy, a vice president in Edelman's CSR practice, says, "At the end of the day, it is structure and culture together. . . . If you get individual employees that believe and walk the walk, you will then have a great deal of credibility."[6]

According to the APA study, almost a third of workers (32 percent) indicated that they intend to seek employment elsewhere within the next year. As the war for talent intensifies, strong internal communications will play a pivotal role in companies' employee acquisition and retention, as well as their overall success.[7]

Organizing the Internal Communication Effort

The best way to assess the effectiveness of a company's internal communication efforts is by determining what employees' attitudes are about the firm. This assessment can be done through an internal *communication audit*. Based on the audit results, communications professionals can design the right program for the organization.

For example, Scott & White Healthcare, winner of *PRWeek's* 2011 Employee Communications Campaign of the Year, conducted one-on-one interviews and focus groups with all staffers—nearly 12,000 employees—to learn how to build stronger engagement. This input formed the company's updated core values and a campaign around the tagline "It Matters." "We need to ensure that each employee understands they contribute to the ultimate success of the organization, whether

[5] American Psychological Association, "APA Survey Finds Many U.S. Workers Feel Stressed Out and Undervalued," March 8, 2011, http://www.apa.org/news/press/releases/2011/03/workers-stressed.aspx.

[6] Erica Ianoco and Chris Daniels, "Employee Involvement Vital to Success in Sustainability," *PRWeek*, May 2010.

[7] "APA Survey Finds Many U.S. Workers Feel Stressed Out and Undervalued," March 8, 2011.

A Towers Perrin survey of 25,000 employees across multiple industries worldwide defines effective communication from an employee's perspective as including the following elements:

1. Open and honest exchanges of information.
2. Clear, easy-to-understand materials.
3. Timely distributions.
4. Trusted sources.
5. Two-way feedback systems.
6. Clear demonstrations of senior leadership's interest in employees.
7. Continual improvements in communication.
8. Consistent messaging across sources.*

*Towers Perrin, "Study Offers Insights on Effective Communication from the Perspective of Employees," January 2005, http://www.towersperrin.com.

they're operating on cancer patients or keeping the grounds clean, open, and inviting," said Scott & White.[8] In addition to conducting an overarching communication audit, regular "temperature checks" of employee opinions can be valuable in ensuring that communication channels and approaches continue to meet employees' evolving needs. Communication audits and periodic employee surveys should be publicly endorsed and supported by senior management. Additionally, management needs to be committed to managing and funding the infrastructure needed to respond to audit and survey findings.

Once it knows how employees really feel about the internal communications they receive, management can create a detailed plan to implement or adjust the internal communication infrastructure to meet its needs. Depending on available resources, audit results, and its goals, management could also consider contracting with a third-party communications consultancy. Failure to implement visible changes in a timely manner following a communication audit, or regular employee surveys, can damage employee morale to a level below where it would be if management had never solicited feedback.

Goals for Effective Internal Communications

Now that we have seen how the changing environment affects the internal communication effort and the importance of collecting employee feedback through a communication audit, we need to explore how companies can organize the function so that it supports the overall mission of the firm. Let's begin by first defining some goals for effective internal communications.

Ultimately, effective internal communications should reinforce employees' beliefs that they are important assets to the firm. This reinforcement can happen only if management believes that is true and if the communication effort is handled by professionals.

[8] *PRWeek* April 2011 Awards Supplement, "Employee Communications Campaign of the Year 2011."

Where Should Internal Communications Report?

In the past, internal communications reported to the human resources area, as traditionally this function dealt with all matters related to employees' welfare. Increasingly, the function is falling under the communications umbrella. The International Association of Business Communicators' 2011 Employee Engagement Survey found that human resources (HR) communications in 51 percent of respondents' companies were handled by "a corporate or organizational communications person or group."[9] Often, both areas have some involvement with internal communications. For instance, at Continental Airlines, responsibility for communicating messages from senior management is shared between human resources and corporate communication.[10]

Ideally, both the corporate communication and the human resources departments in large companies have someone in charge of internal communications. In this case, the human resources professional has responsibility for perfunctory communications such as those regarding explanation of benefits and the new-hire experience. The corporate communications professional takes the lead for major announcements that affect employees, such as significant changes to benefits. If the head of the corporate communication department reports to the vice president in charge of that area and the head of the human resources department to his or her respective vice president, each should have a dotted-line relationship with the vice president in the other area. Other companies actually situate the communicators focused on routine human resources issues in the corporate communication area to create continuity between both general and HR-related communication strategy and execution. These approaches also will help ensure that the goals of each department are fully met and that the lines of communication are kept open between these two critical functional areas.

Large, multidivisional companies often have internal communications representatives within each division who report jointly to the chief of staff for divisional management and to a firmwide corporate communication department. Ideally, each division shares best practices for delivering high-level messages to the employees in their respective areas—understanding the particular needs and nuances of their employee base, which, in turn, affects both the content and tone of communications. However, the channels may be different across divisions; for instance, some divisions may have a voicemail culture, whereas others may pay more attention to e-mail. In larger corporations, there might be vast differences in the online connectivity of employees; those working in production plants or call centers might have no e-mail access whatsoever, whereas other office employees are wholly reliant on e-mail access—whether in-office or remote—to get the job done.

In some cases, companies look outside their own organizations for help with internal communications. For example, Bank of America's Card Services communications team hired Tipton Communications to develop its "Global Card Services

[9] International Association of Business Communicators Research Foundation and Buck Consultants Employee Engagement Survey, 2011, http://www.iabc.com/researchfoundation/pdf/2011IABCEmployeeEngagementReport.pdf.

[10] Lin Grensing-Pophal, "Follow Me," *HR Magazine*, February 2000, pp. 36–41.

Highlights" employee newsletter, which received an Honorable Mention in the E2E 2010 Communication Award competition.[11] As the importance of internal communication gains recognition, it is not surprising that public relations (PR) and consulting firms are developing capabilities in the area of internal communications or that companies are increasingly turning to them for assistance.

Regardless of where the internal communications is positioned and whether or not an outside consultant is used, it must work closely in conjunction with external communicators to integrate the messages disseminated to both internal and external audiences. This approach can help ensure that when significant company news breaks, employees will not be the last to hear about it.

When news about a company hits the press or appears on the Internet, employees should already be equipped with the company's own version of the story so they feel they are being kept in the loop by their own team. This strategy also enables companies to maintain better control of their messages, without being at the whim of how the media position them.[12]

In addition to providing employees with timely and strategic updates on company news to help them feel connected and empowered, companies should understand that employees are often members of multiple constituency groups. As David Verbraska, vice president of worldwide policy and public affairs at Pfizer, writes, "employees wear many hats—they're stockholders, recruiters, customers, and members of the community. . . . Management must understand that the internal audience could be even more important to a company than the external for all the right business reasons, and there are consequences to not aligning the areas."[13] Some companies view the label "internal communications" as archaic. The likelihood of memos or other communications leaking to the outside world with a click of the mouse means that internal communicators should always consider the ramifications of their messages being shared with external audiences, including reporters and investors.

Implementing an Effective Internal Communication Program

Once goals for an internal communication program are established and decisions are made about where the function should report, the program is ready for implementation. In smaller organizations, internal communications may be a part of everyone's job, because the ideal method of communicating with employees is one-on-one or in meetings with small groups of employees.

Even in larger organizations, however, this intimacy in the internal communication effort is a good start for building a more formal program. In this section, we will explore some of the key steps in implementing an effective internal communication program, from personal, one-on-one mechanisms to programs that use technology to distribute messages broadly and instantaneously.

---

[11] Business Wire, "Tipton Communications Is Recognized for Excellence in Employee Communications in E2E 2010 Communications Awards," June 3, 2010.

[12] Ibid.

[13] Richard Mitchell, "Closing the Gap: From the Inside Out," *PRWeek* (U.S.), November 22, 2004, p. 17.

It is the job of the internal communications professional to determine which combination of communication channels is the most appropriate for each message, based on factors such as timing requirements and potential employee reactions. Channel selection can mean the difference between success and failure for an initiative, and it can have a significant impact on employee morale.

Communicate Up and Down

Many large companies are perceived as being faceless, unfeeling organizations, an impression that is only reinforced when no upward communication exists from employees to management. When high-level managers isolate themselves physically and psychologically from other employees, effective communication cannot happen. An overdependence on formal companywide e-mails to disseminate news can compound the "facelessness" of management.

Companies should remember to involve individual supervisors when announcing important news. The International Association of Business Communicators found in its Employee Engagement Survey that the biggest factor contributing to an increase in employee engagement within the organization, by 44 percent, was individual supervisors.[14]

It should be the responsibility of the internal communications professionals to provide supervisors with the information, tools, and ongoing support that they need to present news to their direct reports. Additionally, management should strive to create an environment where all employees feel comfortable sharing candid feedback. In an August 2009 newsletter, McKinsey & Company reported that 27 percent of middle managers, a key communications link between employees and executives, "perceive that is has become riskier to their careers to speak up on difficult decisions when their points of view differ from those of senior managers."[15] Effective internal communications can generate a dialogue throughout the company, fostering a sense of participation that can make even the largest companies feel more personal in the hearts and minds of employees.

The best approach to communicating with employees is through informal discussions between employees and supervisors. Employees need to feel secure enough in their positions to ask questions and offer advice without fear of reprisals from top management. At Accenture, the world's largest consulting firm, senior management involvement begins pre-hire, with executives taking part in the final rounds of interviews. Once on board, each new employee is assigned a career counselor, who can mentor employees throughout their stay at the company. Employees can also join community groups based on their discipline. Neil Hardiman, who transferred to the Chicago office from Dublin, said, "As I would walk around the floors of the office, I would knock on the door and just say hello. . . . That led to some great contacts and connections for me within Chicago."[16]

[14] International Association of Business Communicators Research Foundation and Buck Consultants Employee Engagement Survey, 2011, http://www.iabc.com/researchfoundation/pdf/2011IABCEmployeeEngagementReport.pdf.

[15] Guiniven, "Inside Job: Internal Communications in Tough Times," p. 6-6.

[16] Channick, Robert, The Chicago Tribune, "Emphasis on communication with employees brings good words about these companies," April 17, 2011 http://articles.chicagotribune.com/2011-04-17/business/ct-biz-0417-top-workplaces-20110417_1_town-hall-meetings-communication-employees.

Emkay, one of the nation's largest fleet-leasing companies, keeps employees in the know through quarterly town halls and departmental committees. Employees gain a sense of ownership through exposure to the company's most sensitive information, such as potential acquisitions and earnings projections.[17]

Conversations with management promote feelings that employees themselves are serving as catalysts for organizational change. As Peter Senge highlights with a quote from the ancient Chinese visionary Lao Tsu:

> The wicked leader is he who the people despise, the good leader is he who the people revere, the great leader is he who the people say, "We did it ourselves."[18]

Respecting employees as well as listening to and interacting with them form the basis for an effective internal communication program. Emkay CEO Greg Tepas encourages feedback from employees and allows information to flow up to the top ranks. According to Tepas, "You can accomplish more together if everyone is driving towards those common objectives and understands why you're doing it and how you're going to do it."[19]

Make Time for Face-to-Face Meetings

One means of ensuring that employees have access to senior management is to hold regular, in-person meetings with fairly large groups of employees. Such town hall meetings should take place frequently (at least quarterly) and should be used as opportunities for management to share company results and progress on key initiatives and to demonstrate responsiveness to prior employee feedback. Most important, such meetings should provide employees with an opportunity to ask questions of management in an open forum. If size and geography prevent employees from participating in person, video or telephone conferencing should be used to facilitate their inclusion.

Topics for these types of gatherings should be limited; rather than trying to tackle everything that is going on at the company, managers should survey employees beforehand to find out what is most important to them. Then a presentation can be built around one or two critical issues from the employee perspective, plus one or more messages that management wants to share. Too often, management only sets up such meetings when the company has an important announcement, reducing the likelihood of relevant dialogue.

Gatherings can also be creative to mobilize and inspire employees. Coca-Cola South Pacific held an experiential festival-style day called Live Positively (L+) to support the global sustainability platform introduced in 2008 to help Coca-Cola associates integrate the "Live Positively" maxim into their work and lives.[20] To

---

[17] Robert Channick, "Emphasis on Communication with Employees Brings Good Words about These Companies," *The Chicago Tribune*, April 17, 2011, http://articles.chicagotribune.com/2011-04-17/business/ct-biz-0417-top-workplaces-20110417_1_town-hall-meetings-communication-employees.

[18] Peter Senge, "The Leader's New Work: Building Learning Organizations," *Sloan Management Review* 32 (Fall 1990), pp. 7–23.

[19] Media: Asia's Media & Marketing Newspaper, "Employee Communications Coca-Cola," November 19, 2009.

[20] Lacono, Erica and Daniels, Chris,"Employee Involvement vital to sucess in sustainability, PR Week, May, 2009.

encourage environmental sustainability, GE hosts treasure hunts where staffers examine ways in which the company can be more energy efficient.[21]

Certainly, large-scale events are an effective means to reach out to the greatest number of employees at one time, but managers should not overlook the importance of meeting with employees in smaller groups. If they are seeking feedback or opinions about key initiatives, managers may find that employees are more forthcoming when not in a large-group setting. Smaller groups are also more conducive to resolving specific problems. Companies are increasingly realizing the usefulness and value of face-to-face communications. According to an Allman Communication online survey of senior in-house communicators and HR professionals, 80 percent planned to increase their time spent on face-to-face meetings with staff.[22]

Communicate Online

Although meetings are an important way to communicate with employees, the advent of company intranets in the late 1990s provided a new channel through which companies could reach their employees quickly and broadly with important news on events and key management initiatives. Many company intranets also serve as interactive platforms where employees can rally together and share their views on company programs, activities that contribute to building trust.

The Dow Chemical Company's Dow Advanced Materials division required communication around the 2009 acquisition of Rohm and Haas and the integration of several of its businesses. The communications team created an employee education and engagement campaign around the mantra "Empowered to Deliver." The campaign centered around an internal news hub styled like a consumer news site, which served as the core of the company's intranet. The site included multimedia, articles, and on-demand video, along with both business and consumer news content, quick facts and updates, and longer features. The variety and types of news allowed employees to get the information they needed when and how they wanted it.[23]

The intranet channel is also well suited for employee benefits information. A Watson Wyatt survey of 2,000 employees found that employees rank the intranet as one of their preferred ways to receive benefits information.[24]

Internet technology, though extremely powerful, must be used thoughtfully if it is to enhance communication rather than detract from the impact of management's messages. Employees are bombarded by information, especially given the near

[21] Channick, Robert, The Chicago Tribune, "Emphasis on communication with employees brings good words about these companies," April 17, 2011 http://articles.chicagotribune.com/2011-04-17/business/ct-biz-0417-top-workplaces-20110417_1_town-hall-meetings-communication-employees.

[22] Annie Waite, "Surveys Show Impact of Recession on Employee Communication", The Internal Comms Hub, February 17, 2009.

[23] "Case Study: External Approach to Internal Comms Equals Better Employee Engagement," *PRNews*, April 4, 2011, http://www.prnewsonline.com/free/External-Approach-to-Internal-Comms-Equals-Better-Engagement_14741.html.

[24] "How the 'Google Effect' Is Transforming Employee Communications and Driving Employee Engagement," *Medical Benefits*, June 15, 2007, p. 12-12.

ubiquity of e-mail and voicemail. Communications must be relevant and interesting to effectively reach employees. Welmont Health System's intranet had been "one-size-fits-all." The team redesigned it in-house to include "unique entity news with a system presence" that would engage employees.[25]

Companies need to invest considerable amount of thought into ensuring their messages are getting through to employees and that information is easy to find. Portal technology is being used more frequently to help employees more readily locate and manage online information. Portals combine links to key intranet pages, headlines, and applications on a single screen, similar to a Google or Yahoo! home page. In 2008, BlueCross BlueShield of Tennessee used its intranet to encourage employees to embrace a "Culture of Wellness," posting articles and profiles of employees who had made positive lifestyle changes.[26]

A company intranet should be dynamic and engaging, with the home page regularly refreshed, so it becomes an employee's go-to resource for the latest company information. Ideally, it should be integrated into an employee's workday, so that he or she checks it continuously throughout the day. According to a recent International Association of Business Communicators (IABC) Research Foundation and Buck Consultants Employee Engagement Survey, 75 percent of respondents indicated that the intranet was their organization's most frequently used tool for engaging employees.[27]

As younger employees join the workforce, social media is becoming an increasingly essential tool for communicating and building community internally. Sun Microsystems built communities within the organization through wikis, blogs, Facebook fan pages, and six islands on Second Life, the popular virtual world game.[28] As time goes on, companies will increasingly be moving into this arena to communicate with employees where they live. Managers should resist the impulse to move *all* communication online unless they are sure that all employees will use this medium. Surveys can reveal how employees would like to receive different types of information, which helps determine what types of information a company's intranet will be the best channel for. An effective internal communication strategy should focus on both content and channel, recognizing that the use of multiple channels (some traditional and some more innovative) offers the best potential for success.

And although video and online communication channels are often expedient and engaging, they should not be used as a substitute for personal, face-to-face communication between all levels of management and employees.

Create Employee-Oriented Publications

In addition to online communications, another common form of information sharing in many companies is through the print medium. (Print communications are particularly important to prevent employees without e-mail access from feeling marginalized.)

[25] "For Better or for Worse: Employee Engagement, Recession-Style," *PRNews*. July 13, 2009.

[26] "Case Study: Wellness Done," *PRNews*, February 15, 2010.

[27] Amanda Aiello and Natasha Nicholson, "Engaging Intranets," *Communication World*, March 1, 2011.

[28] Paris Barker, "How Social Media Is Transforming Employee Communications at Sun Microsystems," *Global Business & Organizational Excellence*, May/June 2008.

Unfortunately, most internal company publications are unexciting. How can companies make monthly newsletters or magazines more interesting to employees?

Companies need to realize that their publications are competing with the national and local media for their employees' attention. Today's employee is a sophisticated consumer of information more interested in seeing something akin to *USA Today* and *The Huffington Post* than a list of bowling scores or a photo of the "employee of the month." Ideally, the publication should connect employees with goings-on beyond their local surroundings; it should discuss important happenings and accomplishments across the company and give employees a clear sense of the company's overarching direction and strategy.

Creating an employee publication is an ideal job for a former journalist. The most senior communication official and the CEO also should take an interest in company publications to ensure that employees are getting the real story about what is happening to the company and the industry in the most interesting presentation.

Another way to reach employees through company publications is to send the magazines to their homes rather than distributing them at the workplace. Although this distribution is more expensive, it helps make the company a part of the family, something that will be a source of pride for the employee and his or her spouse. International business consultant Paul Levesque tells the following inspiring story about one of his clients:

> A custodial employee working for one of my clients came up with an ingenious way to eliminate a slip hazard for customers on wet or snowy days. A story about it, with a photo of the employee, was featured in the company newsletter. This company routinely mailed copies of its newsletter to the children of any employees highlighted within its pages, with a personalized note that read, "Your daddy's picture appears on page 2." Several weeks later, management held a staff meeting and invited questions about their quality improvement program. The custodian rose to his feet and reported that the day his two children received the newsletter, he'd been greeted with a hero's welcome when he got home. . . . This expression of pride from his own children, he said, was the most personally rewarding experience in his entire 30-year career with the company—and if this was the kind of thing management meant by "quality improvement," he wanted them to know he was ready to do anything he could to help.[29]

Above all, every publication—just as with any other online or print communication—must be honest about anything that might affect employees. The goal is to make employees feel like a part of the team and on the cutting edge of what is happening within the firm and its industry. The tone of publications also should be realistic, as many employees will see through and distrust anything that seems more like propaganda than a genuine communication.

The messages that go into these periodicals will vary by industry and company, but managers must strike the right balance between what employees are interested in and what they really need to hear from top management. Employees should look forward to the next issue of the company publication in the same way they do their university's alumni magazine. In fact, alumni magazines are excellent models in terms of style and tone for company publications.

[29] Paul Levesque, "Is Your Employee Newsletter Doing Its Job?" *Entrepreneur*, February 22, 2008.

Other print materials also are produced from time to time in response to important events that directly affect employees. For example, the health or retirement benefits areas need a special set of publications. If a company is gearing up for a reduction in health benefits, it may start communicating with employees months before the actual changes take place to put these changes in context for employees. In this situation, the corporate communication staff would likely work with human resources to craft a communication strategy for what could be a year-long communication process. Special welcome publications and materials also must be produced for new employees to create a positive and seamless new-hire experience.

Today, many internal publications are available digitally as well, giving employees the opportunity to interact with newsletters through multimedia and video as well as helpful links to additional information. Management also can use memos and letters to communicate to employees about internal changes, such as management succession, new group structures, or important deals or contracts. These written communications should come out frequently enough so that employees do not feel that it is unusual, but not so often that they stop hearing management's messages. Certainly in the case of major events such as a takeover or merger, employees need to be informed at the same time as external constituencies.

The timing gap between internal and external communications about such events must be narrow, however, as it can be damaging to the company if employees communicate sensitive information haphazardly to external constituencies before the company can make an official statement to the media or its client base. Similarly, as discussed earlier in the chapter, if employees hear critical company news from external sources prior to receiving an internal communication, the impact on morale and trust can be damaging.

Communicate Visually

We know that people are increasingly turning to online video, social media, and websites, in addition to television and traditional news outlets, to get their news. Similarly, employees are becoming more visually oriented in their consumption of information, particularly given increased use of company intranets. As a result, many companies have developed ways to communicate with employees through this powerful medium, now including everything from basic webcasts to multimedia presentations allowing for employee interaction.

Most large corporations have elaborate television studios with satellite capabilities staffed by professionals. Such sophisticated systems are the best mechanisms for communicating with employees through visual channels. Even if your company does not have its own studio, outside vendors can provide these services as needed.

These studios are often used to create "video magazines" that can be made available to employees in outlying areas, helping them feel like part of the organization even when company headquarters is 1,000 miles away. Companies are now broadcasting programs on the Internet for employees to view over the company intranet with increased frequency. Employees without e-mail access can convene in spaces such as cafeterias to view webcasts in large groups, creating a communal experience and encouraging inter-employee discussions.

Webcasts can be used interactively with blogs, streaming video, photos, and audio. At Novartis Oncology, employees submitted videos explaining why they work for the company to be used as an internal engagement tool and for recruiting programs externally. "Offices around the world were given Flip cameras. Videos were submitted, viewed, and voted on via a special Web site. Results include 7,000 site visits, 131 video submissions, 1,541 votes cast, and 37.5 comments."[30] As part of a campaign to generate ideas on improving sustainability, Verizon created a new internal website, used webcasts and e-mails to reach employees, and created VZ Green TV, an internal video channel.[31]

Managers should not see expenditures on such communication as frivolous or wasteful but rather as an investment in the firm, a way to make each employee feel more connected, while also "humanizing" senior management. In contrast to the sometimes impersonal nature of e-mail communication, these communications can offer employees a personal touch—literally bringing a company's leaders and vision to life without the time and expense of traveling.[32] If such a production is well done, it can be a tremendous morale booster as well as a visual history of the company that can be used for years to come.

And visual communication does not always have to be high-tech. At Colgate-Palmolive's Mennen plant, for example, ubiquitous white boards revealed details about breakdowns, production goals, sick leaves, birthdays, vacation schedules, and numbers of units coming off each line. A special racecar billboard depicted the productivity of each line relative to the others—a visual measure of success and a source of motivation and pride. Verizon added a high-tech twist to the low-tech bulletin board. The company provides electronic downloads for supervisors to insert in the board. Each month the content is updated to ensure employees receive up-to-date messages.[33]

Focus on Internal Branding

In this chapter, we have discussed the importance of clear, two-way communication about strategy and direction. Internal branding is also important to building morale and creating a workplace where employees are "engaged" with their jobs. Although communicators do inform employees about new advertising campaigns, they seldom recognize the need to "sell" employees on the same ideas they are trying to sell to the public.[34]

Internal branding is especially critical when an organization is undergoing changes such as a merger or a change in leadership. After Pfizer acquired Wyeth, the company launched a program to unite all employees. The CEO Jeff Kinder sent a letter to staff, hosted two global town hall broadcasts, and created an "Our Path

---

[30] "Employee Communications Campaign of the Year 2010," *PRWeek* 2010 Awards Supplement.

[31] Erica Iacono and Chris Daniels "Employee Involvement Vital to Success in Sustainability," *PRWeek*, May 2010.

[32] Julie Flower, "Seeing You Loud and Clear: Will Visual Technology Ever Make a Real Impact on Business Communication?" *Communication World*, December 1, 2002, p. 18.

[33] Kelly Kass, "Verizon Equips Employees with All the Right Tools," http://www.simply-communicate.com/case-studies/company-profile/verizon-equips-employees-all-right-tools.

[34] Colin Mitchell, "Selling the Brand Inside," *Harvard Business Review*, January 2002, pp. 5–11.

Forward: The Next Step" strategic framework to ensure everyone felt involved and on board.[35]

Internal branding campaigns can be launched when results of internal audits reveal that employees are not connecting with a company's vision or when morale is low. When internal and external marketing messages are misaligned, the customer experience will suffer, with adverse effects on the company. For example, one health care company marketed itself as putting the welfare of its customers as its number-one priority, while telling employees that the number-one priority was cutting costs.

Internal branding campaigns are not just for times of leadership change, mergers, crisis, or low employee morale. The launch of a new advertising or rebranding campaign is also an appropriate time to think of internal branding. Shire Pharmaceuticals asked employees what they think makes the company special as part of the creation of a unified corporate brand identity, and devised a brand initiative based on the responses. The campaign engaged 3,700 employees and included films of patient and executive stories, a podcast series, and intranet videos.[36]

Significant anniversaries and milestones can also be opportunities to launch internal branding campaigns. In the midst of the recession in 2009, DHL Express, a global leader in delivery services, celebrated its 40th anniversary by recognizing the contributions of everyone who worked there. The company created a microsite dedicated to its anniversary and launched a photo contest about what working at DHL has meant to employees. This was a great way to motivate staff and raise morale during tough times.[37]

Even when employees *understand* the company's brand promise or key customer deliverable, it is not until they *believe* it that they can really help the company carry it out. Just as external branding campaigns aim to create emotional ties among consumers to your company, internal branding's goal is to do the same with employees. Focusing attention on this important area will generate improved employee morale and, ultimately, better results for the company.

Consider the Company Grapevine

In considering the formal channels of internal communications discussed in this chapter so far, we cannot neglect the importance of their informal counterparts. The company grapevine—an informal communications network including everything from private conversations between two employees to the latest anecdotes shared in the cafeteria—should be considered as much of a communication vehicle as a company's house organ or employee meetings. In fact, given that nearly half of all employees credit the grapevine with bringing them word of major corporate changes,[38] distributing messages faster and in more credible forms than formal

---

[35] "Employee Communications Campaign of the Year 2011," *PRWeek* 2011 Awards Supplement.

[36] "Employee Communications Campaign of the Year 2010," *PRWeek* 2010 Awards Supplement.

[37] Kelly Kass, "How DHL Express Delivered Deep Engagement during a Recession," http://www.simply-communicate.com/case-studies/company-profile/how-dhl-express-delivered-deep-engagement-during-recession.

[38] Jared Sandberg, "Ruthless Rumors and the Managers Who Enable Them," *The Wall Street Journal*, October 29, 2003, p. B1.

channels, it is even more crucial that managers tap into it. But recent surveys show conflicting perceptions of the reliability of this information: 76 percent of one survey's respondents report that news they hear from co-workers is accurate always, usually, or some of the time.[39] However, another survey reveals that many employers downplay the grapevine's importance, with only 17 percent thinking workers rely on it for information.[40] In fact, statistics reveal that over 90 percent of companies do not have a policy for dealing with the grapevine or for managing any other informal communications network.[41] Ultimately, if employees do not receive complete or timely information from their employers, they will have no choice but to rely on one another—as well as external sources—to fill in the gaps. If rumors do start to circulate, supervisors should have a protocol for alerting senior management and the internal communications team. Management should confirm which, if any, parts of the rumors are accurate as quickly as possible, and internal communications should help to shape a response that supervisors can then share with their direct reports and colleagues.

Managers can find out what employees think by simply asking questions. Scott & White Healthcare conducted research including one-on-one interviews and focus groups with employees to inform its updated core values, out of which evolved the "It Matters" campaign.[42] Surveys can help pinpoint what employees are hearing from management and how they are perceiving things. Some companies conduct such all-employee surveys annually, such as the annual employee survey of PECO Energy, the electric and natural gas utility serving southeastern Pennsylvania.[43] Google conducts its annual employee survey, "Googlegeist," to find out what is and isn't working in engaging employees.[44] The stronger the sense of trust, commitment, and engagement between employees and management, the less often employees will resort to the grapevine as the chief means of expressing their voice and hearing those of fellow employees.

Management's Role in Internal Communications

A common thread in the company examples discussed in this chapter is the involvement in internal communications of CEOs and other senior leaders within organizations. This involvement is critical because these individuals are the "culture carriers" and visionaries within a company, and all communications relating to organizational strategy start with them. Increasingly, CEOs and senior managers—in the tradition of J.P. Morgan's desk on the trading floor—are even positioning themselves in the midst of their employees physically, working at standard desks

[39] Schweitzer, Tamara, "Did He Really Say That?!?!" *Inc.com*, August 20, 2007.

[40] Ibid.

[41] Sheri Rosen, "Carry on the Conversation: Helping Employees Make Sense of What Happens at Work," *Communication World*, March 1, 2005, p. 24.

[42] "Employee Communications Campaign of the Year 2011," *PRWeek* 2011 Awards Supplement.

[43] Suzanne D'Eustachio, "Energizing Employee Engagement at PECO," *T + D*, August 2009.

[44] Shannon Klie, "Google's Innovation Imperative," *Canadian HR Reporter*, September 20, 2010.

and in cubes, to boost camaraderie, engage employees more directly, and create a sense of shared culture and responsibility among employees from the bottom to the top of the ladder.[45]

Robert Dilenschneider, founder of corporate strategic counseling and public relations firm The Dilenschneider Group, describes the type of leader the twenty-first-century corporate landscape demands:

> What's needed now is a different kind of CEO: Men and women who shed the trappings of imperial power, work with their boards of directors in new, dynamic relationships and find fresh ways to unleash the creative potential of their people, from middle managers to front-line workers. This will require a big shift in attitude from change-averse managers. They'll need to get off their private jets and fly with everyone else, shed the large personal staffs that coddle and isolate them and spend real time with the workers who are on the factory floors, behind the sales counters or in the office cubicles.[46]

Physical presence and interaction are an important start. Senior managers, however, also need to work closely with internal communications professionals to ensure their messages are received and, most important, understood by all employees. The "understanding" component is crucial but sometimes overlooked. Donald Sheppard, CEO of Sheppard Associates, an independent consulting agency specializing in internal communication strategy, says, "You can have a vision of 'we want to be this'—that's nice, but the person out there in the plant in Michigan or in India needs to understand how that applies to him or her and what he or she needs to do differently. That can't be done at any macro level."[47]

To achieve this "micro"-level understanding of what strategic goals or initiatives mean to individuals, internal communications professionals should work with front-line managers to help make messages relevant to the employees who report directly to them. KPMG LLP in Montvale, New Jersey, introduced a program called "The Power of One"—an annual event during which key messages and information about firm strategy are shared first with the firm's partners, and then subsequently "cascaded" down through the other layers of the firm. FedEx also focuses development efforts on front-line managers, including work in the area of communication.[48] These individuals, after all, have the greatest potential to help relate management's "vision" to employees' individual business units, and, importantly, to their day-to-day activities.

---

[45] Jared Sandberg, "The CEO in the Next Cube—Bosses Who Abandon Offices Win Kudos for Collegiality, but Make Neighbors Nervous," *The Wall Street Journal*, June 22, 2005, p. B1.

[46] Robert L. Dilenschneider, "When CEOs Roamed the Earth," *The Wall Street Journal*, March 15, 2005, p. B2.

[47] Grensing-Pophal, "Follow Me."

[48] Ibid.

Conclusion Over the last several years, "management by walking around" and other management philosophies basically have come to the same conclusion: Managers need to get out from behind their desks, put down their smartphones, and go out and get to know the people who are working for them. No other method works as well, and no "quick fix" will satisfy the basic need for interaction with other employees.

With all the sophisticated technology available to communicate with employees today, such as e-mail, intranets, blogs, social media, and satellite meetings connecting distant offices, the most important factor in internal communications begins with the manager who has a basic responsibility to his or her employees. That responsibility is to listen to what they have to say and to get to know who they really are as individuals. We have come a long way from Upton Sinclair's *The Jungle* to the modern American corporation. Today's employees do want high-tech and sophisticated communications, but they also want personal contact with their managers. Understanding this fact is the cornerstone of an effective internal communication program.

Case 7-1

Westwood publishing

Dan Cassidy, a 2008 graduate of the Tuck School of Business at Dartmouth College, was driving home from work listening to more depressing news on the radio about layoffs at another large media company. He had just left a meeting with his boss, Catherine Callahan (See Exhibit 7.1), the vice president of human resources at Westwood Publishing. "Dan, we are going to have to let some of the old-timers go," she said. "I'm hoping that the CEO will buy my plan for a voluntary severance and early retirement package. We should be able to move out some of the deadwood in this company as well."

Westwood Publishing had never laid off anyone in the 16 years of its existence. As the director of employee relations, Dan would be responsible for telling employees about the new policy within the next couple of days.
As he looked at the beautiful southern California hills surrounding the freeway, many thoughts were going through his head. How should he identify the issues involved for all employees? Should he get the people in corporate communication involved? Who would be the best person to release the information? What about communication with other Westwood constituencies? And what would be the long-term effects of what would be reported in the media as a "major downsizing"?

EXHIBIT 7.1   The People in the Case

Linda Bosworth	CEO of Westwood Publishing
Catherine Callahan	Vice president of human resources
Eric Ridgway	Vice president of corporate communication
Dan Cassidy	Director of employee relations
Craig Stevens	Outside public relations consultant

WESTWOOD PUBLISHING BACKGROUND

Westwood was started by Linda Bosworth, a brilliant UCLA graduate, following her graduation from college in 1995. With only $10,000 in capital borrowed from her father, Bosworth had built the firm up to a multimillion-dollar trade magazine publisher with hundreds of titles and a broad subscriber base. Beginning in the mid-1990s, Westwood began to focus strategically on high-tech trade publications.

As the business grew, Bosworth gradually turned the day-to-day operation of Westwood over to professional managers, preferring young MBAs from top business schools. But the original group of employees, mostly men in their mid-50s, still represented the bulk of senior management at Westwood.

By the turn of the century, analysts had predicted that the publishing industry in general, and Westwood in particular, was ready for consolidation. Many of Westwood's competitors had trimmed their workforces repeatedly after the dot-com bubble burst in early 2000. By this point, half of Westwood's titles were for high-tech and Internet companies. But Bosworth felt that keeping all of her employees happy

Source: This case was written by Professor Paul A. Argenti in 2012 and updated in 2011. It is a fictionalized version of an actual case, but the industry and characters have been disguised. © 2011 Trustees of Dartmouth College. All rights reserved. For permission to reprint, contact the Tuck School of Business at 603-646-3176.

EXHIBIT 7.1   The People in the Case

Linda Bosworth	CEO of Westwood Publishing
Catherine Callahan	Vice president of human resources
Eric Ridgway	Vice president of corporate communication
Dan Cassidy	Director of employee relations
Craig Stevens	Outside public relations consultant

through good times and bad was more important than anything else.

As other business-to-business publishing companies underwent merger and acquisition (M&A) deals—and trade magazine publishers with solid online media divisions continued to sell themselves to media conglomerates at a tidy profit—Westwood resisted making any deals, instead standing its ground, avoiding the messy consolidation of print titles, and having to deal with potential redundancies in departments such as circulation and office support personnel.

Even though one-third of the American newspaper jobs that existed when Bosworth launched her company in 1995 had already disappeared by 2011, in a speech that Bosworth delivered to all of Westwood's employees that same year, she outlined the company's philosophy toward employee turnover: "You, the employees of Westwood, are the most important asset that we have. Despite the difficult times this company now faces, you have my assurance that I will never ask any of you to leave for economic reasons. This is not General Motors!"

CORPORATE COMMUNICATION AT WESTWOOD

The company relied on a small staff of communication professionals to handle its communication efforts. All of the various activities that could be decentralized (e.g., internal communications, investor relations) were housed in the appropriate functional areas. This organization developed naturally as the company grew to become one of the largest independent trade magazine publishers in the United States.

The main outreaches to employees were annual town hall meetings in the major cities where Westwood had offices, where slide-heavy presentations from Bosworth and other top company executives would draw upwards of 300 employees. Ironically, for a publishing company, Westwood's Intranet was updated infrequently, and the company's communications team re-

lied mostly on desk drops of formal memos and newsletters to get messages out to its workers.

Bosworth, as a young owner and CEO, enjoyed much attention from the press as a result of her meteoric rise in the business world. She relied on an outside consultant, Craig Stevens, to handle her own public relations. Stevens also had a tremendous amount of influence over the communications department at the company itself.

The vice president of corporate communication, Eric Ridgway, was actually one of the several employees who would be affected by the current plan to trim the workforce. He had been hired early on as a favor to Bosworth's father. Ridgway had spent 25 years at the *Los Angeles Times* before signing on at Westwood, and although he had a media background, he did not know much about the trade magazine business or the industries that made up Westwood's primary subscriber base. The problems associated with Ridgway made the communications effort more difficult for both Dan Cassidy and the outside counsel advising him through the process.

THE VOLUNTARY SEVERANCE AND EARLY RETIREMENT PROGRAM

Although the CEO was very much against the two programs that were about to be implemented, she had been convinced by both Callahan, the head of human resources, and her board of directors that something had to be done immediately, or the company itself would be at risk.

The way the programs would work, several senior managers would be told about the generous voluntary severance or early retirement packages and asked to avail themselves of the appropriate plan. Thus, a director who had received less than excellent performance appraisals for two consecutive years would be a prime candidate for voluntary severance, whereas a vice president approaching 60 would be offered the retirement package. Although both of these programs were "voluntary," the supervisors responsible for identifying candidates were urged to get the weaker people to agree as soon as possible.

COMMUNICATING ABOUT THE PLANS

Cassidy reported to work the following day and was asked to attend a meeting with his supervisor, Catherine Callahan, Bosworth, and Craig Stevens. "Well, Dan, how are you going to pull this one off?" joked Bosworth. Cassidy responded, "Quite honestly, Linda, given your position on this issue, my feeling is that you need to get involved with the announcement tomorrow."

As the discussion progressed, however, it was obvious to Dan Cassidy that he was the one that his boss and the head of the company wanted to take the heat. After two hours, Bosworth looked Dan squarely in the eye and said: "This was not my idea in the first place, but I know we have no choice but to adopt the voluntary severance packages and early retirement plans for Westwood Publishing. Unfortunately, I need to leave for a conference in New York the day after tomorrow. You and Catherine are going to have to take responsibility this time."

Dan looked over at Catherine. She was gazing at a drawing on Bosworth's wall. It was a picture of someone about to lose his head by guillotine during the French revolution. Somehow the picture seemed very appropriate to their situation.

CASE QUESTIONS

1. Create a strategy for communicating change at Westwood Publishing that you could give to Bosworth.
2. How do changes in the workforce affect how Cassidy ought to think about communicating the new policy?
3. What advice would you give Cassidy about how communications to employees are structured at Westwood?

CHAPTER EIGHT

Investor Relations

As companies strive to maximize shareholder value, they must continually communicate their progress toward that goal to the investing public. Accordingly, investor relations is an essential subfunction of a company's corporate communication program. While explaining financial results and giving guidance on future earnings are critical investor relations activities, companies today need to go "beyond the numbers"—as Collins and Porras explain in their book *Built to Last*:

> Visionary companies pursue a cluster of objectives, of which making money is only one—and not necessarily the primary one. Yes, they seek profits, but they're equally guided by a core ideology—core values and sense of purpose beyond just making money. Yet, paradoxically, the visionary companies make more money than the more purely profit-driven comparison companies.[1]

Investor relations professionals therefore need to link communications to a company's strategy and "vision" as frequently as possible. Increasingly, the investor relations (IR) function is getting involved in activities traditionally handled by public relations (PR) and media relations professionals and communicating with many of the same constituencies. In addition to a solid understanding of finance, then, IR professionals also need strong communication skills.

In this chapter, we begin our examination of this important subfunction with an overview of investor relations and a brief look at its evolution over the years. We then turn to the goals of investor relations and provide a framework for IR. After discussing important investor constituency groups and how IR reaches them, we look at how the function fits into an organization and conclude with a discussion of investor relations in the changing business environment.

Investor Relations Overview

The National Investor Relations Institute (NIRI) defines investor relations as "a strategic management responsibility that integrates finance, communication, marketing and securities law compliance to enable the most effective two-way communication between a company, the financial community, and other constituencies, which ultimately contributes to a company's securities achieving fair valuation."[2] The chief financial officer of one corporation explained the task of the

---

[1] James C. Collins and Jerry I. Porras, *Built to Last* (New York: Harper Business, 1994), p. 8.
[2] NIRI corporate website, http://www.niri.org.

IR professional as follows: "You're competing for the investment dollar. Your company's story must appeal to the investment world more than the next guy's, or you can't expect to win the coveted shelf space for which everyone is fighting."[3]

As these descriptions illustrate, investor relations is both a financial discipline and a corporate communication function. Changes in the business and regulatory environment over the past decade have affected the way corporations decide how, to whom, and to what extent they convey financial and operating results.

Investors want understandable explanations of financial performance as well as nonfinancial information about companies. According to a report from Ernst & Young's Center for Business Innovation, investors give nonfinancial measures, on average, one-third of the weight when deciding to buy or sell a stock.[4] Examples of nonfinancials include the credibility of management, the company's ability to attract top talent, and the quality and execution of corporate strategy. A survey by McKinsey & Co. found that three-quarters of institutional investors from the United States, Europe, Latin America, and Asia said that board practices are as important as financial results when considering investing in a company.[5]

To ensure that a company presents itself clearly and favorably on all these fronts, then, IR professionals must have both financial acumen and solid communication skills. Access to senior management is also necessary so that the IR function is connected to the company's strategy and vision. An IR department organized in this way is positioned to instill confidence in investors in both good times and bad.

The Evolution of Investor Relations

In the early part of the twentieth century, corporate secrecy was a great concern for companies. Disclosure of any kind was seen as potentially harmful to the interests of the corporation. This perception changed in the 1930s with the passage of two federal securities acts that required public companies to file periodic disclosures with the U.S. Securities and Exchange Commission (SEC). Despite the new reporting responsibilities brought about by the enactment of the Securities Act of 1933 and the Securities Exchange Act of 1934, corporations were interested only in mandatory disclosure, which required little in the way of an investor relations function.

Investor relations did not begin to resemble the discipline we know today until the 1950s. A decade later, the National Investor Relations Institute (NIRI) officially recognized the IR function. NIRI was established as a professional association of corporate officers and investor relations consultants who were responsible for communicating with corporate management, the investing public, and the financial community. Around the same time, the Chicago-based Financial Relations Board (FRB), now a unit of The Interpublic Group, became the first public relations firm dedicated to helping its clients develop relationships with investors.

By the 1970s, FRB had pioneered the distribution of investment profiles that laid out a company's long-term financial goals and strategies. Prior to this innovation,

---

[3] Brett Nelson, "So What's Your Story?" *Forbes*, October 30, 2000, p. 274.

[4] David A. Light, "Performance Measurement: Investors' Balanced Scorecards," *Harvard Business Review*, November–December 1998, pp. 17–20.

[5] Paul Coombes and Mark Watson, "Three Surveys on Corporate Governance," *McKinsey Quarterly*, December 2000.

information reached potential investors through presentations by company representatives to local stockbroker clubs or analysts' societies.

Further regulatory changes altered the landscape for IR in the 1970s. With the enactment of the Employee Retirement Income Security Act (ERISA) in 1974, pension fund managers were legally held responsible for acting in the best interests of their beneficiaries. This new responsibility made pension fund managers more demanding of their portfolio companies. For instance, they sought more detailed explanations of company results, particularly when companies underperformed.

In the 1980s, state and local laws enabled pension funds to increase the equity allocation in their portfolios. That share rose to 36 percent in 1989 from 22 percent in 1982, making institutional investors an even more important constituency for the IR departments of corporations. At the same time, inflation caused many individual investors to flee the stock market, and by the end of the 1980s, institutional investors represented 85 percent of all public trading volume.

The first conference calls were held for hundreds of institutional investors at a time in the 1980s. Soon thereafter, quarterly conference calls were standard practice at many companies. A decade later, the Internet provided yet another channel for communicating company financials to large numbers of investors. Organizations began to create investor relations areas within their corporate websites to post information such as news releases, annual reports, 10-Ks (SEC required annual filing) and 10-Qs (SEC required quarterly filing), and stock charts. Today, the Internet is one of the most important channels for IR. The Internet is unparalleled in terms of cost, speed, and reach, and continues to grow in importance. Companies are using interactive online charts, downloadable data, webcasts, and even apps to communicate their information in new ways. In 2011, Microsoft won the IR Magazine Award for the best investor relations website.[6] Microsoft's website included a visual interactive annual report, videos from recent events, blog posts, interactive and detailed financial data that could be downloaded to Excel, and a constantly updated stream of news. [7]

Even with mass communications such as conference calls and webcasts, however, IR professionals still arranged for periodic private meetings between large institutional investors or sell-side analysts with the chief financial officer (CFO) or the chief executive officer (CEO). These meetings allowed the analysts to ask specific questions and get management's feedback on their own earnings models and projections.

These practices changed with the enactment of legislation designed to put individual investors on a level playing field with large institutions. The 1990s saw a resurgence of individual investor participation in the stock market and, at the same time, deepening concerns that these individual players were not afforded the same access to company information as their institutional counterparts. This theory was supported when two studies showed that volatility and trading

[6] "IR Magazine US Awards 2011," *Inside Investor Relations*, March 24, 2011, http://www.insideinvestorrelations.com/events/2011/ir-magazine-us-awards-2011/.

[7] Microsoft corporate website, http://www.microsoft.com/investor/default.aspx.

increased immediately after quarterly conference calls (which were only open to institutional investors).[8]

In response, in late 2000, the SEC passed Regulation Fair Disclosure, commonly referred to as "Reg. FD," prohibiting companies from disclosing "material non-public information" to the investment community (e.g., institutional investors, analysts) that has not already been disclosed to the general public. One of the immediate effects of this legislation was the opening up of conference calls to *all* investors, inviting individuals into a previously closed forum that allowed them to hear about company results and strategy directly from senior management. Generally, it set a more formal and coordinated tone for guidance—companies could no longer give "selective guidance"; that is, they could no longer provide some investors with information on earnings projections before others. Some of the other implications of Reg. FD for IR will be discussed later in this chapter.

Over the last 50 years, IR has gained the respect and attention of senior management, who increasingly acknowledge it as a vital corporate communication function. The majority of the largest publicly held corporations in the United States and a growing number of small and mid-sized companies are now members of NIRI; by 2011, NIRI had more than 3,500 members representing 2,000 publicly held companies and $5.4 trillion in stock market capitalization in 32 chapters in the United States.[9]

Beginning in mid-2007, investor relations became an even more prominent force in business, both for the United States and, as globalization continued to break down borders, the global economy. A global recession, kick-started in part by the subprime mortgage crisis, prompted markets in North America, Europe, and Asia to sink at alarming rates. The turbulence that began in the United States quickly spread around the world, throwing investors and financial analysts into a frenzy. Equity markets both in the United States and abroad tumbled; in January 2008, world equity markets suffered a $5.2 trillion loss, according to Standard & Poor's.[10]

What became known as the Great Recession lasted 17 months, from October 2007 through March 2009, with the S&P 500's total loss reaching 56.4 percent. In February 2007, the market had fallen to its lowest level since 1997.[11] The years following the recession did not prove to be the rebound investors were hoping for, with the S&P 500 finishing 2011 exactly flat with a 0.00 percent change.[12] This brief snapshot has a number of implications for business executives in the current environment: Investor relations is all the more important as local markets shift and merge into a global economy and as increased globalization enables one country's recession to initiate a domino effect around the world. Given that investors today demand more communication, more transparency, and more access to companies

[8] Steve Davidson, "Understanding the SEC's New Regulation FD," *Community Banker*, March 2001, pp. 40–42.

[9] National Investor Relations Institute website, http://www.niri.org/about/mem_profile.cfm.

[10] "World Equity Markets Lost $5.2 Trillion in January," February 8, 2008, http://money.cnn.com/2008/02/08/news/economy/world_markets/

[11] "11 Historic Bear Markets," MSNBC, http://www.msnbc.msn.com/id/37740147/ns/business-stocks_and_economy/t/historic-bear-markets/#.TxXfh6US2Hc.

[12] "S&P Perfectly Flat for 2011, Dow up 5.5% for year," *USA Today*, January 3, 2012.

than they have in the past, corporations competing for their investment dollars need to create IR programs that deliver on these requirements. In the next section, we will explore how organizations can accomplish this.

A Framework for Managing Investor Relations

How do companies attract and retain investors? When you consider that monthly turnover velocity (defined as share turnover in $/market capitalization) was 130 percent on the New York Stock Exchange in 2010, you begin to appreciate the challenges facing investor relations officers (IROs).[13] The following section addresses the key objectives of investor relations and also provides a framework for the implementation of a successful IR program.

The Objectives of Investor Relations

Although the structure of an investor relations program will vary from one organization to the next based on the size of the company, the complexity of its businesses, and the composition of its shareholder base, the main goal of any IR program is the same: to position the company to compete effectively for investors' capital. To achieve this goal, companies need to focus on the following objectives:

1. *Explain the company's vision, strategy, and potential to investors and intermediaries such as analysts and the media.* One of the most critical duties of an IR professional is to get messages about company results and potential future results across as understandably as possible to the investing public. We further examine the various investor constituencies later in this section.

2. *Ensure that expectations of the company's stock price are appropriate for its earnings prospects, the industry outlook, and the economy.* IROs need to understand investor concerns and expectations for their organizations and relay this information to management to develop so that there is a high-level understanding of what the market anticipates from the company. If management does not see the company as being able to meet market expectations, it needs to work with IR to craft a communication plan to explain why and to manage expectations appropriately. Conversely, if management feels that the company's potential is not reflected in its stock price (that the stock is undervalued), an IR strategy should be developed to help investors see that potential and, accordingly, drive the stock to appropriate levels.

3. *Reduce stock price volatility.* Corporate investor relations activity could account for as much as a 25 percent variance in a company's stock price, according to a recent survey that polled 243 buy-side investment professionals from mutual, pension, and insurance firms; additionally, 82 percent of the survey's respondents believe that good investor relations affects a company's valuation, that "superb" IR is associated with creating a stock price premium of 10 percent, and

---

[13] World Federation of Exchanges website, http://www.world-exchanges.org/statistics/annual/2010.

FIGURE 8.1
Investor
Relations
Framework

Source: Adapted from
Markus Will and
Anna-Lisa Wolters,
"Interdependencies
of Financial
Communications
and Corporate
Reputation,"*Proceedings
of the 5th International
Conference on Corporate
Reputation, Identity, and
Competitiveness,* Paris,
France, May 17–19, 2001.
p. 14.

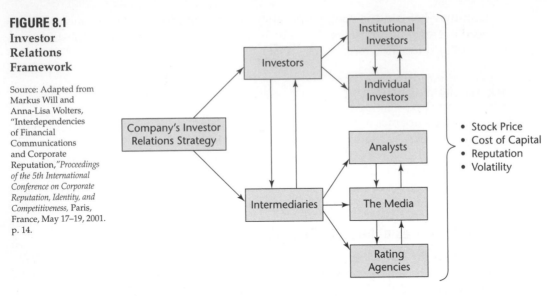

that "poor" IR is associated with a median stock price discount of 15 percent.[14] This capability can be accomplished through the related goal of optimizing the company's shareholder structure to include primarily long-term owners of the stock. Companies with stable share prices typically enjoy a lower cost of capital and thus can issue new equity more economically. In addition to the more strategic goal of stabilizing share price over the long term, IROs often have to respond to market news or events that have the potential to affect stock price negatively in the short term. We will cover some examples of this later in this chapter.

Now that we understand what investor relations is designed to accomplish, let's look at how it achieves these objectives. Figure 8.1 depicts how the IR function communicates both directly and indirectly with investors. The indirect communication occurs through intermediaries such as analysts, the media, and rating agencies. Communication with these constituencies influences stock price, volatility, and, in turn, the firm's cost of capital and reputation.

Types of Investors

A company's IR strategy should address both retail investors (individual shareholders) and institutional investors (pension funds, mutual funds, insurance companies, endowment funds, and banks). These constituencies, however, place different demands on the IR department and require the use of different communication channels. For example, individuals often require substantially less detailed information due to their relative lack of sophistication but require more hand holding with respect to routine matters such as stock split transactions. In addition, compared with individuals, institutions provide companies with access to larger,

[14] Holmes Report, "Investment Professionals Believe Communications Adds—or Subtracts—Value," July 2007.

fairly concentrated pools of capital, affording them greater efficiencies in message delivery and market impact (defined as the combination of trading volume and price movement).

Institutional Investors

At the end of 2010, institutional investors owned $11.5 trillion of the $23.29 trillion U.S. equity market.[15] The proportion of U.S. equities held by institutional investors has been growing, creating concern among market participants on the impact this may have on market volatility and information asymmetry between individual investors and more sophisticated institutional investors. The latest research suggests, however, that it is not overall ownership by institutions that matters, but the concentration of ownership in particular stocks.[16] The proportion of the top 1,000 U.S. companies held by institutional investors was at an all-time high of 73 percent, according to a 2010 Conference Board Report.[17] There is continuing debate about the influence of institutional investors on the volatility of individual stocks and the market as a whole, but the importance of this subset of investors is clear.

IR departments can identify and target multiple categories of institutional investors. For instance, institutions can be broken down into groups based on portfolio turnover (high, moderate, and low) as well as investment styles (e.g., growth, value, income, index). By grouping investors into smaller constituencies with similar characteristics, IROs can efficiently communicate their message to appropriate target audiences. For example, explaining a company's vision and outlook to index investors will yield little benefit because index fund managers do not have the discretion to change portfolio holdings away from index weightings.

IR professionals (or their agencies) can use databases to gather information on institutional stock holdings, turnover rates, and basic portfolio characteristics to identify institutions whose portfolio characteristics closely coincide with their company's price/earnings (P/E) ratio, yield, market capitalization, and industry classification. A company with a low price/book ratio, for instance, might focus on marketing itself to mutual fund managers who specialize in "value" investments. A small company will similarly target small-cap managers and possibly start raising awareness among mid-cap managers if it is approaching a larger capitalization. This kind of research will prevent the company from spending too much time communicating with uninterested investors.

Having identified those institutions whose investing criteria match its characteristics, the company should develop a plan to interest them in investing for the long term. IROs can then reach those institutions in a variety of ways, including day-to-day phone contact and one-on-one meetings with analysts. For meetings with representatives of large, influential institutions the company would like to have a relationship with, the CEO and/or CFO are often involved.

[15] The 2012 Statistical Abstract, U.S. Census Bureau, Report 1201—Equities, Corporate Bonds, and Treasury Securities—Holdings and Net Purchases, by Type of Investor, http://www.census.gov/compendia/statab/cats/banking_finance_insurance/stocks_and_bonds_equity_ownership.html.

[16] "Institutional Ownership Nears All-Time Highs. Good or Bad for Alpha-Seekers?" AllAboutAlpha.com, February 2, 2011.

[17] "The 2010 Institutional Investment Report: Trends in Asset Allocation and Portfolio Composition," The Conference Board, 2010.

More formal gatherings are another way to access large groups of institutional investors. For example, CEOs often address analyst or brokerage societies, industry conferences, and conferences geared toward particular kinds of organizations (such as small-cap, high-tech firms). Companies also host their own meetings in major financial centers such as New York and Boston and invite institutional investors who either own or might want to buy the company's stock.

Individual Investors

Individual investors in the United States owned $8.5 trillion in equities at the end of 2010.[18] Like institutions, individual investors are not a monolithic constituency group. They may own stock directly or through mutual funds, company stock plans, or 401(k) plans. They may actively trade securities to generate trading profits on an intraday basis, apply "buy-and-hold" strategies to save for retirement, or anything in between.

Compared with institutions, individual investors have smaller account sizes and generate lower trading volume. In addition, as mentioned previously, they tend to require different types of information than institutional investors.

We talked in Chapter 2 about the blurring lines between a company's constituency groups. As an example, individual investors also can be employees of the company whose stock they are investing in, through a 401(k) program, bonus compensation in the form of company stock, or options. Employees read about the financial performance of their own companies in the media and expect to see information that is consistent with what they are hearing internally. Companies thus should be prepared to respond to employees' concerns about depictions of their organization appearing in the press that are inconsistent with management's own messages to them.

Reaching individuals is more difficult than connecting with institutions, as they are more numerous and harder to identify. The channels companies use to communicate with individual investors include direct mail to affinity groups (e.g., current shareholders, employees, customers, suppliers), the brokerage community to promote their stocks with individuals, and visibility generated through the media and advertising (see Chapter 4 for more on financial advertising).

In recent years, the Internet also has proved to be a powerful channel for providing investors with real-time information about companies. In 2010, an Investment Company Institute Study found that, 89 percent of U.S. households that owned mutual funds had Internet access.[19]

The Internet is certainly used by institutional investors as well as individuals—portfolio managers and analysts can now use it to quickly and easily obtain baseline information about a company's financials and see up-to-date press releases—but for individuals who do not also have relationships with company IROs or CFOs, it has provided previously unparalleled access to company information. The Internet has begun to mimic the in-person experience once only accessible to large institutional shareholders via technology such as on-demand videos, webcasts, and presentations being available in real time.

[18] The 2012 Statistical Abstract, U.S. Census Bureau, Report 1201.

[19] *2011 Investment Company Factbook,* Chapter 11, "Characteristics of Mutual Fund Owners," http://www.icifactbook.org/fb_ch6.html.

Intermediaries

Investors often learn about corporations through sources other than the company itself. In particular, the media and the analyst community are key conduits. Companies provide information to them through conference calls highlighting quarterly achievements, press conferences announcing annual financial results, and face-to-face meetings to discuss company developments and strategy. Reporters and analysts often present management with probing and difficult questions and report the company's responses to the investing public. Accordingly, management should present honest answers and messages that are consistent with what the organization communicates to investors directly.

The Media

We learned in Chapter 6 that the business world increasingly attracts print, television, and online media coverage. Business network news hosts regularly discuss earnings announcements on their programs and often invite equity research analysts to appear and comment on developments within companies they follow.

Media coverage of business can have a dramatic effect on a company's stock price. As an example, in early 2002, shares of Krispy Kreme Doughnuts fell nearly two points after a *Forbes* article pointed to an "off-balance-sheet trick" in the company's financial statements. Unfortunately, the reporter had read the wrong line of the balance sheet to come to this conclusion. After Krispy Kreme's chief operating officer (COO) drafted a letter to the editor of *Forbes* and spoke with wire services so that they could issue articles the next day pointing out the error, the stock price returned to prior levels.[20]

For another, more global example, consider South African beer producer SABMiller PLC. The company had long maintained a stellar reputation among investors for its ability to manage acquired brewers in emerging markets. However, a year after taking over Miller Brewing Company in 2002 for $5.6 billion, its stock began to plummet. In an effort to understand the reasons behind this sudden decline, executives enlisted the help of Echo Research to analyze financial media coverage of the company and pinpoint a potential catalyst. Based on Echo's findings, which compared coverage and analyst commentary against SAB's stock movement, the executives gained insight into which journalists and analysts had the most influence over the strength of the stock. Furthermore, the research identified the biggest factor contributing to the stock's demise: Miller's ongoing and consistently poor performance. With this information in its arsenal, SAB executives retooled their communications strategy to restore investor confidence. Since then, the stock has been on the rise, which shows that media coverage and investor relations are key drivers of financial performance.[21]

Certainly, having a strong media relations function coordinated with the IR department will be beneficial to a firm's investor relations effort by maximizing access to media outlets and ensuring consistency in the messages each group sends to the media.

[20] Robin Londner, "Investment Insiders Grow Skeptical of Financial Data," *PRWeek*, February 11, 2002, p. 3.

[21] "What Price Reputation?" *BusinessWeek*, July 9, 2007.

Additionally, for low-visibility companies looking to attract investors, obtaining the right kind of media coverage can be a critical component of an IR strategy. In response to the rising influence of the financial media, some IR and PR consulting firms offer "financial media relations" programs to help companies target media strategically.

As we will discuss in the next section, the media also play an important role in bringing the views of prominent analysts to the investing public as well, giving a voice to this other very influential intermediary.

Sell-Side Analysts

IR functions target the financial community through "buy-side" and "sell-side" analysts. Buy-side analysts typically work for money management firms (mutual funds or pension funds, for example) and research companies for their own institutions' investment portfolios. They sometimes use sell-side research in their analysis, but many perform proprietary analysis, including company visits and their own review of company financials. As such, for the purposes of our investor relations framework, buy-side analysts belong in the institutional investor constituency group and are not intermediaries.

Sell-side analysts, however, cover stocks within certain industries and generate detailed research reports that offer "buy," "sell," or "hold" recommendations. This research is then provided to clients of investment banks such as JP Morgan Chase or retail brokerages such as Charles Schwab. Thus, sell-side analysts are intermediaries between a company and existing and potential investors. According to research from UCLA Accounting Professor Michael Brennan, strong IR can increase interest in the company from both investors and sell-side analysts. Brennan's research shows that sell-side analysts are able to positively affect the trading of stocks by improving market liquidity, increasing trading volume, and tempering marketing reaction to news affecting the company.[22]

In the late 1990s and with the crash of the Internet bubble in 2000, sell-side analysts came under fire for continuing to issue "buy" recommendations on severely underperforming stocks. The media raised awareness of the inherent conflicts of interest in the job of a sell-side analyst working for an investment bank. Traditionally, companies covered by a firm's research team were also important banking clients who could take their business elsewhere or cut off the analysts' access to information if offended by an unfavorable rating.

As the Internet economy was thriving and stock prices seemed to be on an unstoppable upward trajectory, many of these sell-side analysts enjoyed near-celebrity status. Merrill Lynch entertainment analyst Jessica Reif Cohen could, in her own words, "instantly add—or subtract—billions in market value."[23] As business coverage received increasing attention in the media, these analysts became household names.

---

[22] Michael Brennan and Claudia Tamarowski, "Investor Relations, Liquidity and Stock Price," *Journal of Applied Corporate Finance* 12, no. 4 (2000).

[23] Nina Munk, "In the Final Analysis," *Vanity Fair*, August 2001, p. 100.

This kind of visibility meant that analyst recommendations carried tremendous weight. According to Zacks Investment Research, between 1985 and 2000, stocks that attracted coverage by three or more analysts fared 37 percent better over the ensuing six months than stocks that did not receive the same coverage.[24] However, when that period is viewed from 1996 to 2003, buy recommendations by independent securities firms (those that have no investment banking business) outperformed the buy recommendations issued by analysts at investment banks by an average of 8 percent annually. In the period following the NASDAQ market peak, buy recommendations underperformed by 17 percent annually.[25]

Even when the Internet bubble burst in early 2000, many analysts maintained sky-high valuations on companies whose stocks were simultaneously plummeting to maintain their firm's investment banking relationships with the companies they covered. Investors, who had come to view these analysts as trusted advisors, felt betrayed and misled. Media coverage of these "star analysts" was just as prevalent as it had been in the dot-com heyday, but its angle on the analysts was decidedly changed. A *Vanity Fair* article characterized the group as "superstar analysts who were no longer objective observers of the market: they were insiders with inherent conflicts of interest."[26] Mary Meeker, once dubbed "Queen of the Net," appeared on the cover of *Fortune* magazine in a feature article entitled "Can We Ever Trust Wall St. Again?"[27]

The burst of the bubble ushered in an era of analyst regulation that changed the landscape for communicating with the sell-side analyst community.

In October 2000, the SEC-proposed Regulation Fair Disclosure (Reg. FD) took effect. Previously, corporations had communicated with analysts through many of the same channels they used for institutions. One-on-one meetings or lunches with the CEO or CFO were common. On a day-to-day basis, IR professionals spent a great deal of time on the phone with analysts, going over specific inquiries or providing feedback on their models. Reg. FD sought to eliminate this standard practice of company executives sharing nonpublic information with security analysts by requiring parallel public disclosure of this information. It had been alleged that executives would reveal material financial and operational information to analysts of investment banks with which their companies had business relationships. And as a result of Reg. FD, corporations were no longer free to give specific feedback on analysts' earnings models beyond corrections to factual data—much to the dismay of the analyst community.[28] This correction had long been standard practice and a key mechanism used by analysts in formulating their own estimates for companies. Many companies responded to the new rule by providing their own models to analysts instead of providing specific feedback on models the analysts created.

In April 2003, the Securities and Exchange Commission, the New York Stock Exchange (NYSE), the National Association of Securities Dealers (NASD), and the

[24] Brett Nelson, "So What's Your Story?" *Forbes*, October 30, 2000, p. 274.

[25] UCLA Anderson School of Management, http://www.anderson.ucla.edu/x5046.xml.

[26] Robin Londner, "Street Cleaning," *PRWeek*, July 23, 2001, p. 17.

[27] Peter Elkind, "Can We Ever Trust Wall St. Again?" *Fortune*, May 14, 2001, p. 69.

[28] Tommye M. Barnett, "To Speak or Not to Speak," *Oil and Gas Investor*, September 2001, pp. 73–75.

New York attorney general announced the $1.4 billion Global Analyst Research Settlement with 10 of the largest U.S. investment banks. The Global Settlement was the result of a long investigation by the New York district attorney that found evidence of investment banks inappropriately influencing the work of research analysts. The settlement sought to eliminate the inherent conflicts of interest in the job of a sell-side analyst working for an investment bank. The settlement imposed $1.4 billion in fines and penalties on 10 of the largest U.S. investment banks, mandated structural changes to ensure research and coverage decisions were independent, and prohibited improper interactions between a firm's investment banking and research functions. In particular, analyst compensation could no longer be based directly or indirectly on investment banking revenues, and research analysts were prohibited from participating in investment banking sales efforts, such as pitches and roadshows.[29]

Several changes have occurred in the IR environment post–Reg. FD and the Global Settlement. The strict regulations surrounding research on stocks have made it more difficult and costly for investment firms to maintain research coverage of as many stocks as they used to. Major broker-dealers are concentrating their research on large-capitalization stocks. Forty percent of all NASDAQ-listed companies have two or fewer analysts covering them.[30] As a result of regulation, there are now fewer analysts and therefore fewer channels for public companies to communicate with the investor community.

The settlement also has meant that analyst coverage has become less optimistic. In 2000, at the height of the boom, 95 percent of stocks in the S&P500 had no "sell" ratings, and no stock had more than one sell rating.[31] Since the settlement, however, according to research from Washington University in St. Louis, analysts have become more cautious in issuing forecasts and recommendations.[32] Among investment firms that had both research and investment banking practices, "strong buy" recommendations were made on stocks 37 percent of the time in the period prior to the Global Settlement, versus 21 percent of the time following the settlement; "buy" recommendations were made 40.6 percent of the time before versus 32.2 percent of the time after; and "hold" recommendations, which have traditionally been seen as bad news by the market, were made 19.9 percent before and 43.3 percent after by affiliated analysts.

One certainty is that the relationships between analysts and the companies they cover can be fraught with tension if not handled strategically. Consider the story of Tad LaFountain, a long-time analyst at Wells Fargo Securities, who announced in July 2005 that he was dropping coverage of semiconductor giant Altera Corporation because company management would not take his calls or provide adequate information to analyze the business. According to LaFountain, who had a "sell" rating on the stock, the company objected to his negative opinion. One of

---

[29] SEC Fact Sheet on Global Analyst Research Settlement, http://www.sec.gov/news/speech/factsheet.htm.

[30] NASDAQ.com stock screening service, http://www.nasdaq.com/screening/companies-by-industry
.aspx?exchange=NASDAQ.

[31] William H. Donaldson, "Speech by SEC Chairman: CFA Institute Annual Conference," Philadelphia, PA, May 8, 2005.

[32] Stephen Taub, "Spitzer Pact Cut Analyst Bias: Study," CFO.com, November 10, 2004.

31 analysts covering Altera, LaFountain says he was told by Altera's VP of investor relations Scott Wylie and CFO Nathan Sarkisian that "it was not in the shareholder's interest to facilitate" his analysis.[33]

Media coverage of this decision was fast and negative, with many seeing the move as an attempt to manipulate opinion. A few days later, Altera was forced to apologize, saying, "In retrospect, our decision to disengage was in error, and (we) apologize to Mr. LaFountain, our investors and the investment community."[34]

Clearly, analyst reports contain much more than a simple buy or hold recommendation, and despite the recent crisis of confidence in the objectivity of these ratings, other information about companies contained in these reports is often used by institutional investors to help them with their investment decisions. Analysts remain an important intermediary for a company's IR strategy. IROs also should be prepared to communicate strategically with and handle downgrades from analysts with a communication plan.

Rating Agencies

In the United States, examples of rating agencies include McGraw-Hill's Standard & Poor's (S&P), Moody's Investors Service, and Fitch Ratings. These agencies analyze companies in much the same way that buy-side and sell-side analysts do, but with a specific focus on their creditworthiness. The ratings that these agencies assign to a company reflect their assessment of the company's ability to meet its debt obligations. This rating, in turn, determines the company's cost of debt capital (the interest rates at which it borrows).

These agencies make their ratings available to the public through their ratings information desks and published reports. The highest ratings are AAA (S&P, Fitch) and Aaa (Moody's), and the lowest are D (S&P) and C (Moody's), representing companies that are in default of existing loan agreements. Companies rated BBB/Baa or above are considered "investment grade," and those below are considered not investment grade, or "high yield." The more perjorative term *junk bonds* also refers to below-investment-grade bonds. The lower the rating, the higher the agency's assessment of the company's potential to default on its loans, thus making it more expensive for the company to raise capital by issuing debt.

Debt ratings affect more than a firm's cost of capital. Senator Joseph Lieberman, chair of the Senate Committee on Governmental Affairs, put it this way:

> The credit raters hold the key to capital and liquidity, the lifeblood of corporate America and of our capitalist economy. The rating affects a company's ability to borrow money; it affects whether a pension fund or money market fund can invest in a company's bonds; and it affects stock price. The difference between a good rating and a poor rating can mean the difference between success and failure, prosperity and bad fortune.[35]

---

[33] Gretchen Morgenson, "An Analyst Receives a Time Out from Altera," *The New York Times*, July 27, 2005.

[34] Gretchen Morgenson, "With Apology to an Analyst, Altera Seeks to Repair a Rift," *The New York Times*, July 29, 2005.

[35] Statement by Chair Joseph Lieberman, "Rating the Raters: Enron and the Credit Rating Agencies," U.S. Senate Committee on Governmental Affairs website, March 20, 2002, http://www.senate.gov/~gov_affairs/03202002lieberman.htm.

A recent example of the ripple effects that debt ratings can have on a company is that of the many bond insurance companies that fell victim to the credit crisis, started in large part by 2007's subprime mortgage collapse. Take FGIC Corp.: On January 30, 2008, Fitch Ratings slashed the bond insurer's rating from AAA to AA after its capital shortfall ballooned to $1.3 billion. Ambac Financial Group Inc. was also downgraded after its aborted attempt to raise the $1 billion necessary to satisfy Fitch's requirements. Ambac's stock value alone dropped from a 2007 high of $96.10 (on May 18, 2007) to a 52-week low of $4.50 on January 17, 2008. In February 2008, Berkshire Hathaway Chairman Warren Buffett stepped up to offer backup insurance on up to $800 billion in municipal bonds to help the troubled industry, but in November 2010, Ambac filed for bankruptcy protection under Chapter 11. In January 2012, the stock was trading at around $0.08 per share.[36]

Credit rating analysts are similar to equity research analysts when it comes to their relationship with a company, with the obvious exception that they will focus a great deal more on the company's debt structure. Additionally, many buy-side and sell-side analysts rely on the research and ratings of credit analysts as a component of their own assessment of the company overall, especially for firms within capital-intensive industries characterized by heavy debt loads.

Credit rating analysts at the major rating agencies came under fire after the 2008 debt crisis for many of the same reasons equity analysts were under fire in 2001. The rating agencies had lost credibility with the investing public due to perceived conflicts of interest and a lack of willingness to downgrade both corporate and sovereign debt until significantly after the crisis hit.[37] The rating agencies were also being blamed and held directly responsible for some of what happened in the credit crisis. As a result, they have been faced with lawsuits, investigations, and much public and media criticism. By early 2012, the Department of Justice was stepping up its investigations on Standard & Poor's in particular, focused on its mortgage bond ratings during the financial crisis.[38]Although regulation of these agencies is still in progress, one would expect the agencies to undergo a transformation similar to what equity sell-side research did in the early 2000s.

Developing an Investor Relations Program

Now that we understand who the key investor constituencies are, let's look at how IR functions are structured to communicate with them: in-house, delegated to an agency, or some combination of the two. This section also will take a closer look at some of the activities that make IR such an important function within a company.

How (and Where) Does IR Fit into the Organization?

A company's IR function can be structured in a number of ways, from fully in-house to fully outsourced. In-house IR teams are typically small: According to

---

[36] http://ir.ambac.com.

[37] "Rating Agencies under Fire, But Big Reform Unlikely," CNBC, May 13, 2010.

[38] "Justice Department Ramps Up Probe into S&P's Ratings During Financial Crisis," Reuters, January 17, 2012.

NIRI, the average size of a corporate IR department is between one and two people. At smaller organizations, the CFO might handle IR responsibilities directly and use an agency to perform some of the more routine report-writing tasks.[39]

When companies do turn to agencies for assistance, they can choose from agencies that specialize in IR work, such as Kekst & Company, Abernathy MacGregor, and the Financial Relations Board, or full-service PR firms that have strong IR specialty groups, such as Fleishman-Hillard or Burson-Marsteller. Agencies can help with projects and activities across the spectrum of IR, from report-writing and arranging analyst conferences to higher-end services such as bankruptcy and litigation communications, mergers and acquisitions, and initial public offerings. In the last decade agencies also have focused on fully understanding Reg. FD so that they can help companies with their disclosure policies, Reg. FD has a goal of creating an even playing field between institutional and individual investors; however, it also adds a burden to the company to identify and disclose all material information to all constituents at the same time.[40]

The division of responsibilities between what is done in-house versus what is handled by the agency depends on several factors, including the size of the firm and its IR objectives. However it is arranged, the individuals responsible for a company's IR efforts should have access to senior management, including the CEO and CFO. This situation appears to be the case for in-house IR professionals—two-thirds of corporate NIRI members report to the CFO.[41]

Given the increasing overlap between IR and areas such as media relations, in some organizations, IR and corporate communication are linked or part of the same group. The advent of Regulation Fair Disclosure also made many companies consider the merits of combining these areas—or at least ensuring that they are closely coordinated—to avoid inadvertent selective disclosure.[42] If companies applaud the idea of a closer partnership between IR and corporate communication, however, it is not yet broadly reflected in corporate structure. According to a recent survey, corporate communication and marketing departments still report independently of IR at over 70 percent of companies.[43]

Using IR to Add Value

As mentioned previously, the investor relations function assumes a marketing role with respect to a company's equity and debt, which involves much more than producing and distributing annual and quarterly reports, responding to shareholder inquiries, and sending information to securities analysts. IR plays both proactive and reactive roles within an organization.

Proactively, IR targets investors to market the company's shares to and provides regular informational updates and explanations of performance to the

[39] "Investor Relations: Corporate," *PRWeek*, September 24, 2001, p. 19.

[40] Ibid.

[41] "Understanding IR," *PRWeek*, September 24, 2001, p. 17.

[42] Robin Londner, "IR-PR Link Not Seen in Chain of Command," *PRWeek*, March 4, 2002, p. 3.

[43] "Business Wire Announces Survey Results on the Consolidation of Communications in IR and PR," *Business Wire*, March 21, 2002, Online Lexis-Nexis Academic, April 2002.

The SEC's reporting requirements create the need for a number of documents to be produced periodically, such as the annual report, form 10-K, and form 10-Q, for example. Companies can file these reports electronically with the SEC, and investors can download them from the SEC's online database, EDGAR, or the company's own website in addition to, or instead of, receiving hard copies.

Among all these documents, the annual report is the most time-consuming, expensive, and high-profile endeavor. An annual report is a company's equivalent to a "coffee table piece" and is now being used by companies as much as an image vehicle as a reporting tool. Annual reports have played a role in shaping corporate reputation and public perception for decades. Investors can obtain the financial information contained in a printed annual report faster online, yet there is still great demand for the printed piece.

Executives surveyed by Roper Starch ranked the printed annual report as the single most important document their company produces. One executive said that the annual report "should be the face of the firm."[1] An annual report gives a company the opportunity not only to share and explain results for the prior year but also to communicate the company's vision.

Annual reports typically have themes that are carried through the piece in graphics and text. Thompson Reuter's highly regarded 2010 annual report, had the theme "The Knowledge Effect." The theme emphasized Thompson Reuter's role in improving the signal-to-noise ratio for professionals around the globe and the belief that the right information in the right hands leads to amazing things. The annual report was a dynamic website with downloads, photographs, data, and client stories embedded throughout. The annual report won awards for its online report, written text, chairman's letter, and photography.[2]

Thompson Reuters reflects another prominent theme in annual reports today: the rise of sustainability reporting. Its report was addressed to "Our Shareholders and Other Stakeholders," emphasizing the importance of multiple stakeholder groups, and included not only the typical sections on strategy, vision, and financials, but also ones on allocating capital and talent globally and the greater good. In their 2010 annual reports and 10-Ks, 52 percent of *Fortune* 100 companies included statements of corporate social responsibility.[3] According to *Forbes*, "a company that does it right quickly realizes that effective and transparent communication is key to maximizing investments, as well as transforming the company and its brand."

Today, annual reports are used as reporting vehicles, brand builders, recruiting pieces, marketing brochures, corporate image books, and strategic positioning tools.[4] Even as more companies post their annual reports online, in new, innovative ways, it doesn't appear that the hardcopy version will go away.

---

[1] "Are Annual Reports Still Relevant?" *@ ISSUE, The Journal of Business and Design* 6, no. 2 (2001), pp. 26–31.

[2] Thompson Reuters corporate website, http://thomsonreuters.com/about/awards_recognition.

[3] CSR in Annual Reports: 7 Conflicting Trends, Forbes CSR blog, July 20, 2011, http://www.forbes.com/sites/csr/2011/07/20/csr-in-annual-reports-7-conflicting-trends/.

[4] Ibid.

---

marketplace. Proactive communications can go beyond traditional analyst calls and include activities such as "field trips" for analysts and portfolio managers. Plant tours and meetings or lunches with key company executives can provide investors and potential investors with a true feel for the company and its management.

IROs also craft communication strategies in response to certain internal or external events. Internal events such as mergers, acquisitions, or the sale of a part of the business allow time to confer with the CEO and CFO, develop a communication strategy around the event, and script answers to anticipated questions and concerns. External events, however, such as an unanticipated crisis (see Chapter 10) require much more rapid damage control.

Charles C. Conaway, former president of the drugstore chain CVS Corp., explained: "Unless you have a very targeted investor-relations program that communicates your message, you're going to get in trouble."[44] CVS underwent a restructuring in 1995 that resulted in a complete turnover of its shareholders in the course of one year. To resolve this instability, CVS bolstered its investor relations program and began actively recruiting longer-term institutional investors to suit its new growth profile.[45] In 2004, CVS won the Interactive Investor Relations Award from the Web Marketing Association.

Companies with extensive IR resources can conduct research to identify their most influential shareholders and seek to understand what motivates them, allowing management to predict more accurately the effect on share price of various events or announcements. Research on the changing stock prices of large U.S. and European public companies over a two-year period showed that a company's share price is significantly influenced by a maximum of 100 current and potential shareholders.[46] By identifying these investors and creating profiles on each of them that detail how they make decisions and what motivates them, companies can better perform scenario analysis on the potential effect on the stock price of certain announcements. If necessary, management can modify plans to bring them in line with the desires of key shareholders and minimize negative effects on stock price.[47]

Management also must be careful, however, not to become beholden to investor demands in the short term. The bull markets of the 1980s and 1990s were a major cause for the short-term orientation of the investment community. As Darrell K. Rigby, Bain & Company director, has commented: "I've seen so many senior executives saying and doing things to deliver short-term news lately that it's a little frightening. . . . Their time horizons are shortening. They're thinking more about retiring rich at 45 or 50 and less about the institution they will leave behind."[48] Perhaps a certain corporate strategy will not deliver the earnings that investors and analysts are expecting in the short term, but if the strategy is one that the company believes is right for the long term, management must clearly explain the reasons for this to the investment community. Indeed, the former CEO of Merck says "CEOs are focusing way too much on maximizing shareholder value and on short-term results."[49]

NIRI's former head, Lou Thompson, maintained that the 19 percent drop in Hewlett-Packard (HP) shares and nearly 10 percent decline in Compaq shares that occurred the day the HP–Compaq merger was announced in 2004 could have been mitigated by stronger IR efforts. If Compaq and HP had identified market

---

[44] John A. Byrne, "Investor Relations: When Capital Gets Antsy," *BusinessWeek*, September 13, 1999, p. 72.

[45] Ibid.

[46] Kevin P. Coyne and Jonathan W. Witter, "What Makes Your Stock Price Go Up and Down," *McKinsey Quarterly*, no. 2 (2002), p. 28.

[47] Thomas F. Garbett, *How to Build a Corporation's Identity and Project Its Image* (Lexington, MA: Lexington Books, 1988), p. 99.

[48] Byrne, "Investor Relations."

[49] Raymond V. Gilmartin, "CEOs Need a New Set of Beliefs," Harvard Business Review Blog Network, September 26, 2011.

skepticism over the merger, Thompson argued, the companies could have addressed these concerns before investors "voted with their feet" and took a toll on both companies' stock prices.[50]

When a crisis hits, or a company undergoes some structural change that the market reacts to negatively, investors have already lost money, as the stock price usually adjusts downward nearly instantaneously. Either shareholders can join in the selling, or they can continue to hold the company's stock, hoping that it will recover. To ensure that shareholders do not sell, companies must be prepared with swift, honest communications to investors when the stock price starts spiraling downward.

Management must identify the problem (or perceived problem), what caused it, and, importantly, what it is doing to address it. In these types of "damage control" situations, channel choice matters: A webcast or conference call with the CEO or CFO will carry much more weight than a press release posted to the company website.

Similarly, when a company is not performing as well as it should, IR professionals should proactively communicate to analysts and investors what management is doing about the situation. Such candor is definitely in the company's best interests. As Thomas Garbett says:

> Information reduces risk. The stock market, as a process, arrives at a stock price based upon all known elements relating to the company. Some of the unknown factors add to the price, others subtract. Areas about the company that are unknown usually contribute to the minus side of the price equation.[51]

Investor Relations and the Changing Environment

In this chapter, we have discussed the evolution of the IR function over the years and some of the external developments that have shaped it. Over the past decade, technological advances, decreasing investor confidence, and the changing business environment have been significant influences on the field of investor relations.

As mentioned previously, many companies are creating investor relations areas on their corporate Web sites that make stock quotes and charts, news releases, webcasts, and company financial statements available to anyone with Internet access. Investors find this kind of instantaneous access to information reassuring, particularly during periods of market volatility and uncertainty. Earnings webcasts also are quite popular. These events enable participants to witness firsthand how companies' top executives handle themselves and can bring an otherwise two-dimensional upper management to life for current and potential investors.

Web-based IR has become increasingly prevalent and is supported by external vendors and agencies that can help a company create effective sites. Jeffrey Parker

---

[50] "Understanding IR," *PRWeek*, September 24, 2004, p. 17.

[51] Thomas F. Garbett, *How to Build a Corporation's Identity and Project Its Image* (Lexington, MA Lexington Books, 1988), p. 99. CCBN company Web site, http://www.ccbn.com/about/faqs.html (accessed April 11, 2002).

(who founded First Call) and Robert Adler established Corporate Communications Broadcast Network (CCBN) in 1997, recognizing that "the concept of 'Internet time' has created pressure on corporations to do everything better, cheaper, sooner and faster."[52] CCBN (acquired by Thomson in 2004) now builds and manages the IR portions of the websites of over 2,500 publicly traded companies. Shareholder .com also emerged in the 1990s to provide an array of online IR services— including website hosting, webcasts, and integrated e-mail broadcasts—and now works with over 750 companies, including Johnson & Johnson, Kellogg's, and Microsoft.[53]

The Internet enables greater transparency by providing nearly real-time information about companies to a wide audience, and this transparency is especially valued in the current business environment. Indeed, see Chapter 9 for more information on the Sarbanes-Oxley legislation and the effect it has had on the need for transparency in business.

Considering that of the 20 largest bankruptcies in history, all but 3 occurred between 2000–2009, investors witnessed previously unimaginable corporate events over a short period of time. Financial Services giant Lehman Brothers was by far the largest single bankruptcy in history, totaling $691 million in September 2008 at the peak of the credit crisis.[54]

Even America's most admired corporations have felt the ripple effects of investor insecurity. General Electric, for example, saw its shares plummet to $35 in February 2002, even though the company had not—unlike an increasing number of companies in today's media spotlight—been accused of any sort of misdeed. However, GE's financing subsector, GE Capital, came under fire for not offering substantial earnings information to the public, forcing the company to make reporting changes to increase transparency.[55]

Cendant, formerly one of the foremost providers of travel and real estate services in the world, was created by the merger of HFS Inc. and CUC International in December 1997. In April 1998, Cendant was hit hard when it was discovered that CUC's financial statements had been overstated by hundreds of millions of dollars in both revenues and profits. Following this discovery, the market value of Cendant dropped more than 40 percent, threatening the credibility of both the company and CEO Henry Silverman. It also brought a barrage of questions from numerous constituencies. How could the company and its CEO not have conducted adequate "due diligence" to uncover CUC's fraudulent reporting before the transaction was completed? Silverman realized that, to regain credibility, complete honesty and financial transparency were the only viable course of action. He established the mantra, "Tell the truth. Tell it all. Tell it now," insisting that all the

---

[52] Business/technology editors, "Shareholder.com Clients Showcase Strength at *IR Magazine* U.S. Awards," *Business Wire*, April 3, 2002, Online Lexis-Nexis Academic, April 2002.

[53] shareholder.com corporate website.

[54] "The Biggest Bankruptcy Filings of the Decade," http://www.totalbankruptcy.com/blog/the-biggest-business-bankruptcy-filings-of-the-decade/.

[55] Paul A. Argenti, Robert A. Howell, and Karen A. Beck, "The Strategic Communication Imperative," *MIT Sloan Management Review* 46, no. 3 (Spring 2005), pp. 83–89.

accounting irregularities be acknowledged as soon as they were known. Silverman and the company's head of corporate communication and investor relations, Sam Levenson, continued to tell the Cendant story as frequently and as clearly as possible to restore investor confidence in the company. "I can never be far away from investor relations or public relations. At the end of the day, I'm accountable," says Silverman. "You can never over-communicate. There is no such thing."[56]

As seen from these examples, investor relations is even more important to companies against this backdrop of uncertainty and mistrust. Clear, full disclosure of business results will put companies in a strong position in the competition for investor capital.

Conclusion

Many activities fall under the IR function, from planning and running annual meetings and putting together reports for SEC filings to targeting and marketing the company's shares to investors. The way all these should be approached is no different from any other communication activity: Companies need to follow a communication strategy that includes a clear understanding of the company's objectives and a thorough analysis of all of its constituencies so that appropriate messages can be crafted and delivered.

Unfortunately, efforts to quantify IR's direct effect on stock price and/or a company's cost of capital have yielded little in the way of results. Today's equity markets are influenced by many factors beyond companies' control, and thus, although it is still used as a broad indicator, stock price does not single-handedly signal an IR success or failure. Anecdotal evidence, however, does provide a basis for the simple conclusion that IR is a required communication function in today's marketplace.

No company can afford to deal with the current investment community without developing an effective investor relations function, whether it is fully in-house, fully contracted to an outside agency, or a combination of the two. The price paid for overlooking this advice is far greater than the investment made in the personnel that staff this important function.

[56] Ibid.

Case 8-1

Steelcase, Inc.

Perry Grueber sat at his desk at Steelcase, Inc., on a bright day in July 2000, thinking about the work that lay ahead for him. Grueber had just joined Steelcase, a maker of office furnishings and workspace solutions, as director of investor relations. Steelcase was dedicated to improving IR at the company and had promised Grueber the resources he required to make the department an effective tool for communicating to key constituencies.

When Grueber accepted his job in May, Steelcase was trading at $11.56 per share, just above its all-time low of $10.38 per share and down 70 percent from a high of $37.94. Steelcase's operating performance was mostly to blame for the declining share price; however, the company's communications with its investors also had played a role. The company had high turnover in its institutional shareholder base and, since the time of its IPO, had not actively marketed itself to sell-side analysts. These analysts, in return, expressed little interest in the company. At the same time, insider sales were increasing, sending more shares into the market amid soft demand. Grueber needed a new IR strategy to help Steelcase turn its situation around. As he settled into his new office at the company's headquarters in Grand Rapids, Michigan, he began to assess the challenge that lay before him objectively.

HISTORY OF STEELCASE, INC.

Steelcase was founded in 1912 by Peter Wege, Henry Idema, and 12 other investors under the name Metal Office Furniture. Wege hoped to capitalize on the benefits of metal furniture

over its more flammable wooden counterparts. This original vision found success early on, as government architects began to specify metal in their designs and turned quickly to Metal Office Furniture to fill their demand. Early company successes included the development of the metal wastebasket and the later invention of the suspension cabinet, which became the foundation for all modern filing cabinets.

Company sales in 1913, the first full year of operations, were $76,000. As revenues began to increase, Metal Office Furniture hired a media consultant, who created the trademark Steelcase name in 1921. World War II and the resulting war material contracts benefited the company, and the boom years of the 1950s and 1960s catapulted Steelcase further forward in terms of revenues and profits. By the late 1960s, Steelcase had become the largest manufacturer in the office furniture industry. It retained that status through the year 2001, when the company reported revenues of $4.1 billion.

Steelcase's founder, Peter Wege Sr., died in 1947. Wege's partner Henry Idema died four years later in 1951, and control of the company fell to Henry Idema's son Walter. The Idemas began a tradition of family stewardship over the company that continued when Walter Idema's son-in-law Robert Pew II assumed leadership in 1966. Pew became executive chairman in 1974 and retained that title until his retirement in 1999, although James P. Hackett became president and CEO in 1994. By 2000, Steelcase as a company retained the imprint of the vision and the direction it had received through the founding families' descendents.

IDENTITY, VISION, AND REPUTATION

Steelcase built its image from the set of values held by founders Peter Wege and Henry Idema,

Source: This case was prepared by Thomas Darling under the supervision of Professor Paul A. Argenti at the Tuck School of Business at Dartmouth. Information was gathered from public and corporate sources, including interviews with Perry Grueber at Steelcase in May 2002.
© 2008 Trustees of Dartmouth College. All rights reserved. For permission to reprint, contact the Tuck School of Business at 603-646-3176.

clearly articulated in its organizational goals: "Steelcase aspires to transform the ways people work . . . to help them work more effectively than they ever thought they could."[1] Every employee read the company's core values statement:

"At Steelcase, We:
Act with integrity
Tell the Truth
Keep commitments
Treat people with dignity and respect
Promote positive relationships
Protect the environment
Excel"[2]

Steelcase had a history of "putting people before profits"[3] and dealing fairly with its employees. As Grueber explained, "There is very much a family atmosphere . . . I've never seen a better benefits package and it is not just executive benefits; it's all the way down the line." Tenure with the firm averaged nearly 18 years. The values embodied in Steelcase's treatment of its employees applied to other constituencies as well, including dealers, vendors, and the communities in which Steelcase operated.

As Grueber described it, Steelcase prided itself in "communicat[ing] values through actions. It's not just the corporate line." For example, shortly after becoming CEO, James Hackett voiced his concern that Steelcase's offices did not communicate the company's goal of transforming the way people work to be more effective. Outdated headquarters designs from the 1960s and 1970s isolated executives in their offices. When company management wanted to conduct brainstorming or other creative sessions, they often "fled headquarters."[4]

James Hackett challenged senior management to trade their traditional offices for a new office environment one floor below. This office overcame the existing separation and used a quarter less space. The offer contained an escape clause, allowing management to move back to the traditional offices after a trial period. But the redesigned offices proved to be an unmitigated success, increasing workplace effectiveness and becoming the prototype for a new line of systems furniture called Pathways.[5]

Internally, all members of Steelcase acted in a way that reinforced the company's message of open communication at every level. As Grueber explained, "Executives all have an open-door policy. If you came to visit our offices, you would see that our senior executives reside in an open-plan environment. They don't have enclosed offices and so, we go to great lengths to live our vision."

Externally Steelcase's strong values helped create a dealership network that was the envy of the industry and demonstrated the extent to which Steelcase's values shaped its business. Steelcase relied heavily on its dealers to support its "made to order" business model and made a point of treating them with respect, as primary purchasers of their products and as fellow businesspeople whose own businesses would prosper as Steelcase's had prospered.

THE INITIAL PUBLIC OFFERING

During its time as a private company, Steelcase had developed a much-admired reputation stemming from its well-articulated identity, vision, strategy, and culture. Although Steelcase intended to continue its focus as defined by the original families into the future, the company also believed that it had reached a point where it would benefit from a changeover to public ownership. This change would provide increased liquidity to the company's founding families and give them the ability to diversify their holdings. As the list of Steelcase heirs grew, liquidity became more important to these private owners. Many of the family

[1] Steelcase, Inc., website, "Our Company: Overview," http://www.steelcase.com/servlet/OurCompany.

[2] Ibid.

[3] Conversation with Perry Grueber.

[4] Marc Spiegler, "Changing the Game," *Metropolis Feature*, July 1998, http://www.metropolismag.com/html/content/0798/jl98game.htm.

[5] Ibid.

members wanted to diversify their long-term holdings and allow for distributions to charities and other philanthropic activities. During its 90 years as a private company, Steelcase had grown to become a member of the *Fortune* 500 and the largest manufacturer of office solutions in the United States. By 1998, the private ownership structure for this organization was simply too inflexible.

The economic environment at the end of the 1990s was prime for Steelcase's initial public offering. Data from the Business and Industrial Furniture Manufacturer's Association (BIFMA) forecast double-digit increases in office furniture shipments throughout the first three quarters of 1997.[6] Steelcase was the leader in this growing furniture shipment industry, which was already worth $10 billion in 1996. Furthermore, the U.S. economy overall was still growing at an impressive pace (though the Asian crisis had sparked some doubt in late 1997), white-collar job growth remained strong, and companies were flush with cash from the booming stock market.

Steelcase came to market on February 18, 1998, with a 9.4-million-share offering priced at $28 per share; the proceeds went entirely to family stakeholders. The offering proved very popular with money managers and was oversubscribed to such a degree that the number of shares was increased from 9.4 million to 12.5 million and the IPO price quickly exceeded the originally projected range of $23–$26 per share. On the first day of trading, Steelcase shares rose from the opening offer to close at $33.63, up approximately 20 percent.

After the IPO, 156 million total shares were outstanding, with 12.5 million in public hands, and the balance owned by the founding family. Employees received a gift of 10 shares each and options allowing them to purchase shares at below-market rates. One-third of the IPO shares went to employees. Institutions were the largest

purchasers of the 12.5 million shares sold to the public.

STEELCASE AS A PUBLIC COMPANY (IPO TO JUNE 2000)

Steelcase hit an all-time closing high of $37.94 per share on March 13, 1998, less than one month after the IPO. Almost everything that followed with respect to the company's stock price, however, was disappointing. Uncertainty caused by the 1997 Asian crisis and the 1998 Russian default significantly disturbed many companies' capital expenditures. In addition, as the year 2000 approached and "Y2K" fears loomed, corporate spending was focused almost solely on technology and information systems. Although traditional indicators of furniture system demand remained strong, those indicators did not translate into end demand for Steelcase's products.

In 1999, just as the company's profitability started to weaken, Steelcase purchased the remaining 50 percent of Strafor, a previous joint venture interest in Europe and Africa with annual sales of $500 million. Because the two companies concentrated on different aspects of the furniture business, the addition of Strafor's business to the balance sheet had a material effect on several of Steelcase's financial ratios. All of Steelcase's products were "made to order." This business model had allowed it to carry only a small amount of inventory, and Steelcase's dealers typically paid for purchases in less than 30 days. Strafor did not have the same inventory constraints. Also, many of Strafor's customers were accustomed to paying closer to 90 days after receiving an order. With the Strafor acquisition, inventory at Steelcase rose and inventory turnover fell. At the same time, however, a new customer base increased Steelcase's collection risk. The softness in the balance sheet reinforced investor concerns over deteriorating earnings performance.

Steelcase performance in 1999–2000 was mediocre. Sales slumped or, at best, remained flat.

[6] Mahua Dutta, "Steelcase Builds IPO," *IPO Reporter*, February 16, 1998.

Cost control initiatives and a cut in bonuses brought earnings up in 1999, but gains were erased by a significant fall in earnings reported in 2000.[7] Steelcase had expected a certain amount of business in 1999 that never materialized, throwing off the company's cost structure and causing the gross margins to drop.

Sagging sales turned into a flood of orders in early 2000, coming in from companies that had delayed renovation projects until after Y2K. Investors expected Steelcase to bounce back quickly. Unfortunately, the company had underestimated the costs associated with serving a rush of new orders. According to Grueber, "As the surge in business came in 2000, when our system should have been there to meet the needs without any difficulty, the customer service requirements were so rigorous in terms of delivering product, getting it there on time, and the pricing environment so tough that we had further erosion of our gross margins and operating margins." Profitability and operating margins continued to slump.

At the time of the IPO, Steelcase had no debt on its balance sheet. It had positive earnings of $1.40 per share (including shares issued through the IPO) and an overall strong demand for its product. Two years later, Steelcase faced increasing volatility in end demand and a weaker balance sheet. The company was unsure of what strategy to communicate to investors. "The market just didn't understand what was happening," said Grueber, "and we were not in a position to articulate a great strategy."

THE INVESTOR RELATIONS EFFORT (1998–2000)

STRUCTURE

Because of its large size and market-leading position, Steelcase had the potential to be a credible, attractive investment for multiple types of institutional investors. But institutions didn't flock to Steelcase shares. The company was large enough to be included on several indices; however, its percentage weighting was often adjusted to reflect its small float (the number of shares owned by the public, not including insiders). Many institutional investors chose to steer clear because of its relative illiquidity. SEC filings showed only 28 institutional holders of Steelcase in 2000, representing between 5 and 8 percent of the shares available to the public. With the exception of several small index players, turnover among institutions was well over 50 percent, meaning that the institutional shareholder base changed every two years.

When Perry Grueber arrived at the company to take over as director of investor relations, he replaced Gary Malburg, who was both the vice president of finance and treasurer and head of IR. Malburg had been responsible for communicating with investors, answering questions, and assisting with the financial statements. However, because of significant and growing responsibilities in the Treasury department, only about a quarter of his time was available for investor relations activities. The company's corporate communications director, Allan Smith—who reported to the vice president of global marketing and communications, Georgia Everse—also assisted the IR effort, crafting and disseminating press releases and creating the company's annual report. The staff in these two divisions had few formal channels for interaction. Steelcase's internal structure lacked a clear conduit for IR staff to respond to the concerns of its shareholders.

Although IR had not been a priority at Steelcase, the company had not remained inactive in its attempts to communicate with existing shareholders following its IPO. It engaged the services of Genesis, Inc., a highly respected investor relations consulting company. According to Deborah Kelly, a partner at Genesis, "The good news was that Steelcase was widely respected by its core constituencies as the dominant force in the office furniture

---

[7] Note that Steelcase operates on a fiscal year ending in February, and all references to financial statements are for the year ending in February.

industry and as being guided by people with strategic vision and a solid grasp of trends."[8] Nonetheless, she continued, "there was frustration among investment analysts regarding performance."

Genesis's main responsibility was advising Steelcase in the creation of the company's annual report. Genesis also helped plan Steelcase's first "analyst day" in November 1999, an event hosted by the company for buy- and sell-side analysts and portfolio managers. Leading up to this "analyst day," Steelcase hired another outside consultant to perform a perception study of investors' opinions about company communications. The report produced from the study, according to Kelly, revealed that "investors were looking for a more proactive IR program that could help them better understand strategic objectives and they wanted to have greater access to management, so they could get more than just the phone answered." Analyst day helped open up the decision-making process to many analysts and portfolio managers, but this important first step lacked the vital follow-up that additional proactive communications might have provided. As he entered the company, Grueber had the opportunity to launch a renewed and sustained effort to implement strong IR strategies at Steelcase.

GUIDANCE AND REPORTING

In some ways, Steelcase resembled a public company even before its IPO in 1998; it had a board of directors, audited financial statements, and a large shareholder base. Once Steelcase became a public company, however, the previous shareholder makeup led to an "inner circle" mentality that proved difficult to change. Management was not used to the additional requests for information and sometimes assumed a defensive posture toward inquisitive analysts or investors. "Once we had come public," said Grueber, "we were providing

the required elements but not a great deal of insight into decision making at the company or the strategic direction of the company."

Steelcase's reluctance to share publicly its inner decision-making processes extended a company approach to communications that had developed during its decades of heavy reliance on the controlling families' leadership. The families chose board members as their representatives, who then hired and supervised the management team. The company under this system earned a strong level of trust both in its direction and in the quality of the reported information. Very little information was ever questioned or requested by non–board members. Along with this trust, though, came a highly conservative outlook from company leadership with regard to the amount of information shared and prospective statements regarding business performance.

In each earnings release, Steelcase typically disclosed very specific guidance for the upcoming year or quarter. Grueber noted, "The company, due to its conservative nature, has been very cautious about selective disclosure throughout its public life. The way they communicated to the Street was through a press release." Due to advice from internal counsel and a desire to prevent selective disclosure, management never "walked the Street up or down" with its estimates.[9] Another major factor in Steelcase's conservative approach to its disclosures after it became public was the lack of incentives to develop a strong quarterly forecasting discipline during its years as a private company. In addition, the company did not strive to build and maintain relationships with its analysts, so when it came time to disseminate information, it didn't have a receptive ear through the sell-side analysts.

[8] This and all quotes are from interviews with Deborah Kelly in May 2002.

[9] "Walking the street" is a practice that includes providing material information to analysts during conversations, making excessive statements concerning future earnings prospects, or blatantly encouraging analysts to raise a lower earnings estimate. Some of these tactics have since been prohibited through the Regulations for Fair Disclosure, enacted in 2000.

Steelcase's inexperience with releasing company information to the analyst community cost it credibility in the years immediately following its IPO. Press releases assumed relatively high importance at Steelcase in a company environment that both lacked a strong channel for adjusting guidance and reflected the company's inherently conservative attitude toward providing information to outside parties. Unfortunately, if the information published in a current press release was inconsistent with earlier guidance, Steelcase could do very little to minimize the surprise the information caused investors. Several pre-announcements in 1998 and 1999 damaged Steelcase's reputation with investors and increased the perceived risks associated with owning the stock. As Genesis's Deborah Kelly explained, "They kept missing quarters and it was an unusually large number. For a company that has just gone public, usually you want to have 4–5 quarters in the bag. . . . Not here."

NEXT STEPS FOR STEELCASE

Overall, Steelcase put a tremendous amount of effort into its IPO and into readying itself for the rigor of being a public company. Unfortunately, assumptions about public company communications that Steelcase made based upon past experiences as a private company often led to disappointment for investors. In addition, the equity markets entered into an extremely turbulent period after the IPO, which caused significant shocks to equity values and corporate capital spending and also created a harsh environment for a newly public company to develop its investor relations acumen. Deborah Kelly summed up Steelcase's situation as follows:

> I think they put a lot of effort into getting a grasp for what being a public company meant from a communications perspective. They are such good people. They are a terrific management team in terms of doing the right thing, integrity, and caring about what happens. The shock was that they had spent so many years communicating with owners that I don't think they realized there might be a difference when you go public. It was kind of a shock that you had to do things a little differently and have a different sensitivity with this group.

As part of Steelcase's effort to readdress its corporate communication to investors, the company had hired Grueber, and now it was up to him to outline his goals and strategy for the IR department.

CASE QUESTIONS

1. As part of creating the full-time IR position, Steelcase had to decide where to place Grueber in the company hierarchy. Given the issues facing Steelcase when Grueber arrived, what are the strengths and weaknesses of placing Grueber under the CFO versus the corporate communication department?

2. What resources should Grueber ask for? How should he organize the function (reporting lines, internal staff versus agencies, etc.)?

3. What investor constituencies should Steelcase try to interest in the company's stock? What channels should Grueber use to attract them? What message would Steelcase deliver to them?

4. What mistakes did Steelcase make in its past IR efforts?

5. What are the biggest challenges facing Steelcase in mid-2000 and beyond? How would you position the IR function to handle those challenges?

6. What can you learn about Steelcase today through an online search of its website?

CHAPTER NINE

Government Relations

Government and business in the United States tend to have an adversarial relationship, as business attempts to minimize government involvement in the private sector and Washington attempts to manage the needs of all citizens by exerting its power over the corporate realm.

Government influences business activities primarily through regulation. Originally, government regulation managed market competition. The first government regulations applied to industries such as railroads and telecommunications, in which high barriers to entry facilitated the emergence of monopolies that could hurt the consumer. In these cases, regulation replaced Adam Smith's "invisible hand" to protect citizens from high prices, bad service, and discrimination.

Governmental regulation of monopolies has not prevented large corporations, however, from wielding impressive political and social power. The predominance of global corporate giants such as ExxonMobil, Coca-Cola, and Microsoft transcends voting districts and political borders. Some of the most politically active organizations in the United States are, in fact, domestic or multinational corporations and trade associations.[1] Political largesse on the part of big business has made many Americans cynical about the integrity of the political process in Washington and its ability to govern the corporate world properly. The recent activities under the "Occupy Wall Street" umbrella are a good example of this.

On the list of trustworthy professions, many voters rank politics below business. However, 90 percent of incumbents still return to Congress each election, indicating that if this distrust does exist, Americans are not translating their dislike of politicians in general into action against particular office holders by voting them out.[2] At the same time, as Chapter 1 has shown, action against particular corporations that the public perceives as corrupt is increasingly prevalent. Anticorporate campaigns range from boycotts and demonstrations to support of legislation to restrict corporate influence on Capitol Hill. In the fall of 2008, the near collapse of the financial banking system and the following global credit crisis led Congress to pass a bill in July 2010 that increased government's role in the financial markets. For the first time in nearly a century, the government cracked down on the U.S. banking sector, an industry that historically had been ignored by Congress. When Congress grilled U.S. bank chief executives on how Troubled Asset Relief Program (TARP) money was used, the government appeared to take the protestors' side.

---

[1] Wendy L. Hansen and Neil J. Mitchell, "Disaggregating and Explaining Corporate Political Activity: Domestic and Foreign Corporations in National Politics," *American Political Science Review*, December 1, 2000, p. 891.

[2] Douglas G. Pinkham, "How'd We Get to Be the Bad Guys?" *Public Relations Quarterly*, July 22, 2001, p. 12.

In this chapter, we first examine the nature of the relationship between government and business. Then, we discuss the importance of government relations departments within companies and how the function itself has developed over the past few decades. After seeing how businesses today manage government affairs, we highlight some of the political activities that companies use to advance their agendas in Washington.

Government Begins to Manage Business: The Rise of Regulation

Government regulation began over 100 years ago with state regulation of the railroad companies. By the mid-nineteenth century, trains had triumphed over rival forms of land transportation. Railroad systems opened travel opportunities to people all over the United States and drove the growth of industry, shipping goods quickly across long distances. However, the railroads also presented the country with enormous problems. Although proponents of a laissez-faire approach to the markets maintained that competition would regulate business, it failed to regulate the railroads and corruption ensued.

The federal government's regulation of business began in 1887, with the Act to Regulate Commerce and the establishment of the Interstate Commerce Commission (ICC). Then, in 1890, another critical piece of legislation was passed: the Sherman Antitrust Act. This act established a legal framework to prevent trusts from restricting trade and reducing competition and remains the main source of antitrust law in the United States. From the ICC and the Sherman Antitrust laws to the hundreds of regulations currently in place, covering topics that range from the environment to pornography to food quality, the government is actively engaged in business affairs. Each year the federal government passes laws, and even creates new agencies, to correct what it perceives as market externalities produced by private business.

Some examples of past bills that affected business include the Cigarette Labeling and Advertising Act (1965), which requires all cigarette packages to carry warnings about the hazards of smoking; the Clear Air Act Amendments (1970), which outlined procedures for monitoring air quality; and the Employee Retirement Income Security Act of 1974 (ERISA), which set new federal standards for employee pension programs.

One of the most significant acts to affect business was the Sarbanes-Oxley Act of 2002, officially titled the Public Company Accounting Reform and Investor Protection Act. It came in the wake of a series of corporate financial scandals, including those affecting Enron, Arthur Andersen, and WorldCom.

The act was designed to review dated legislative audit requirements and, by doing so, protect investors by improving the accuracy and reliability of corporate disclosures. The act covers issues such as establishing a public company accounting oversight board, auditor independence, corporate responsibility, and enhanced financial disclosure. It also eliminated some of the most egregious practices in the accounting world, such as using auditing as a loss leader to encourage companies to buy their higher-profit consulting services. It also mandated that companies test their internal financial controls to help ensure that fraud doesn't happen.[3]

Historically, business has resisted new regulations, especially laws that mandate costly additions to existing procedures. One example of these regulations is the "best available technology" clauses of many environmental laws, which demand that polluting companies maximize investment in "clean" equipment when they update their facilities. American industry has complained that regulations hurt American businesses and their efforts to compete with foreign rivals. Regulatory bills, they argue, add costs to American companies not incurred by foreign competitors. These costs could drive up the price of American products, making them comparatively less attractive than foreign substitutes. Corporate America has especially been hit with the costs associated with Sarbanes-Oxley compliance. The amount and extent of government involvement in the market has fluctuated with changes in White House administrations, but in spite of these fluctuations, and the arguments against regulation, business will always have to deal with a baseline level of government regulation.

The Reach of the Regulatory Agencies

Through the years, regulatory agencies have evolved into sophisticated organizations. Franklin D. Roosevelt's New Deal gave government incredible power to regulate business. The Securities and Exchange Commission (SEC) was created to stabilize financial markets and the National Labor Relations Board to remedy labor problems. The Federal Communications Commission (FCC) regulated radio, television, and telephones, and the Civil Aeronautics Board (CAB) regulated the airlines. The safety rulemaking powers would later be transferred to a new agency, the Federal Aviation Administration (FAA), and then undergo yet another restructuring as part of Homeland Security after the terrorist attacks of September 11, 2001. More recently, Sarbanes-Oxley (SOX) increased the power of the Public Company Accounting Oversight Board (PCAOB), a congressionally created private-sector, nonprofit corporation. The PCAOB has sweeping powers over the nation's external auditors with respect to their auditing of publicly held companies. This description is only a small selection of the regulatory agencies that have emerged over the last century.

Government is involved in virtually all stages of business development. Many enterprises cannot begin operations until they receive a license from a regulatory agency such as the Interstate Commerce Commission (ICC), the FCC, or the Food and Drug Administration (FDA). Once an enterprise has its license to operate, the same government agencies must then inspect and approve its products. The Consumer Product Safety Commission (CPSC) helps set safety standards for consumer products, and most products must pass its "tests" before they ever reach the market.

Beyond approving which products become available to the public, the government also can influence the prices of goods and services. Agricultural goods,

[3] American Institute of Certified Public Accountants website, http://www.aicpa.org.

forest products, and metals are examples. Using congressionally approved formulas, federal agencies set floor prices, volume-based subsidies, and quota systems that shape the prices in these markets. The government also heavily influences the prices set by transportation, communications, and utility companies—industries that provide the basic infrastructure for society.

Since the days of the Sherman Antitrust Act, the U.S. government has continued its efforts to prevent monopolies and other anticompetitive business practices. One example is the Federal Trade Commission's (FTC) rejection of a merger between Staples and Office Depot. The FTC argued that, if they merged, each superstore would lose its largest competitor. Without the check that direct competition between Staples and Office Depot had placed on prices, the merged office supply store would gain considerable control over what it charged its customers. The FTC viewed the joining of Staples and Office Depot as more of a threat to consumers than a benefit and so prevented the merger.[4] A more recent example of the power of the FTC is AT&T's abandoned takeover of T-Mobile following stiff opposition from the Justice Department and the FCC, which reported that a merger of the two companies "would substantially lessen competition and its accompanying innovation, investment, and consumer price and service benefits."[5] AT&T deserted the acquisition and instead paid T-Mobile a $4 billion breakup fee.[6]

In the next section, we will look at how business has responded to government regulation and ways in which companies work with lawmakers to ensure their own voice is heard when drafting business-specific legislation.

How Business "Manages" Government: The Rise of Government Relations

In light of the government's heavy involvement in commercial affairs, business eventually realized that instead of fighting regulation, a more effective approach would be advocating its own positions to key political decision makers. Companies began to protect their interests with well-crafted lobbying and negotiating tactics, particularly when they were facing substantial opposition from consumer and community groups that politicians were eager to appease.

One of Wall Street's most largest investment banks, Goldman Sachs, is a good example of a politically active organization in a sometimes controversial industry. The bank was the second largest corporate donor from 1989 to 2012, distributing $36,215,437 in political donations.[7] During the 2008 election cycle, a period of great economic and regulatory uncertainty for securities firms facing the fallout of the economic crisis, Goldman Sachs made $6,025,681 in political donations, a 36 percent increase from 2000. Goldman Sachs was not the only securities firm concerned about economic policy and regulation of the financial sector. In 2008,

---

[4] John M. Broder, "FTC Rejects Deal to Join Two Giants of Office Supplies," *The New York Times*, April 5, 1997, p. 7.

[5] "FCC Report Slams AT&T's Takeover of T-Mobile," CNET, November 29, 2011, http://news.cnet.com/8301-1035_3-57333499-94/fcc-report-slams-at-ts-takeover-of-t-mobile/.

[6] "AT&T Admits Defeat over T-Mobile Takeover, Will Pay $4 Million Break Up Fee," http://arstechnica.com/tech-policy/news/2011/12/att-admits-defeat-on-t-mobile-takeover-will-pay-4-billion-breakup-fee.ars.

[7] Data from www.opensecrets.org, http://www.opensecrets.org/industries/totals.php?cycle=2012&ind=F07.

securities and investment firms poured dollars into Washington, contributing $167,106,035 to campaigns, more than doubling donations from the previous two years. As an industry group, securities and investment firms ranked third in 2008 campaign donations, up from a fifth-place ranking in 2006.[8]

In 2011, the top five groups lobbying the American government were all private-sector industries, with single-issue and legal lobbyists falling significantly below the spending of the private sector. Topping the list was the health care industry, totaling $500 million in spending on lobbying in 2011 alone.[9] GE was the single corporation that spent the most over the last several decades as well as in 2011, with a lobbying budget of over $26 million in 2011.

In the interaction between business and Capitol Hill, powerful lobbies and trade unions are prevalent—subjecting the government to a multitude of pressures. As Alfred D. Chandler Jr., an economic historian, wrote, "the visible hand of management [has] replaced what Adam Smith referred to as the invisible hand of market forces. . . . [As business has] acquired functions hitherto carried out by the market, it [has become] the most influential group of economic decision makers."[10]

Businesses use a number of tactics to further their own agendas in Washington. In this section, we look at the rise of the government relations function within companies.

The Government Relations Function Takes Shape

In the late 1960s and early 1970s, government regulations placed on certain industries significantly raised the cost of doing business. Thus, "It became apparent to American business leaders that in order to win in Washington, they would have to adapt the rules to their advantage, and that meant playing Washington's game."[11] "Playing the game" became the job of a company's government relations, or government affairs, department. This function concentrated specifically on the positive and negative effects of policy and policy changes, as well as monitoring shifts in ideology and agendas on Capitol Hill and accurately identifying emerging trends. By being knowledgeable about government and getting involved in the development of regulatory policy, business could better protect itself from damaging regulations while taking advantage of any positive opportunities that governmental regulation created.

Since the 1980s, government relations departments have improved their effectiveness by studying the methods of other companies, hiring consultants, organizing popular support, learning to use the media properly, making alliances, creating political action committees, and establishing connections with influential Washington insiders. By applying business and marketing techniques to politics and combining traditional organizational tools with advanced technology

---

[8] http://www.opensecrets.org/industries/totals.php?cycle=2012&ind=F07.

[9] http://www.opensecrets.org/lobby/top.php?indexType=c.

[10] Walter Adams and James W. Brock, *The Bigness Complex: Industry, Labor, and Government in the American Economy* (New York: Pantheon Books, 1986).

[11] Sar A. Levitan and Martha R. Cooper, *Business Lobbies: The Public Good and the Bottom Line* (Baltimore, MD: Johns Hopkins University Press, 1984), pp. 4–5.

(i.e., computerized association memberships, the Internet, electronic and paper newsletters), business has increased its influence over Washington's policymakers. Over 50 percent of *Fortune* 500 corporations had representatives in Washington or retained counsel there.[12]

Many companies, such as Bridgestone/Firestone and Walmart, have learned the costs associated with *not* having a Washington presence. When the National Highway Traffic Safety Administration forced Bridgestone/Firestone to recall millions of tires, the company had no Washington office in place and had lost most of its outside consultants. The company needed to seek out new representation in the midst of a highly publicized crisis. Bridgestone/Firestone learned from this mistake and now has a dedicated Washington office and several consultants.[13]

In the 1990s when China entered the World Trade Organization, Walmart executives discovered a problem: U.S negotiators had agreed to a 30-store limit on foreign retailers operating in China—a major roadblock for Walmart's expansion plans. So, in 1998, Walmart hired its first lobbyist in an attempt to build a Washington presence for a company that had traditionally shunned political involvement but found itself facing legal challenges from unions, workers' lawyers, and federal investigators. Today, Walmart's political action committee (PAC) is one of the largest donors to federal parties and candidates, giving $3.2 million in the 2010 election cycle.[14]

The Foundation for Public Affairs (FPA) conducted a survey in 2000 to define the responsibilities of public affairs executives. Two-thirds of the 223 executives polled reported that they provide senior management with political and social trend forecasts as one of their duties, with over half reporting directly to the company CEO, president, or chairperson.[15] Sixty percent of respondents reported a direct correlation between this trend monitoring and the company's overarching strategy. The duties for government affairs executives revealed in the FPA statistics point to a greater interaction between this department and public affairs, two functions that were once separate within companies.[16] A more recent survey by the FPA found that six professionals and two administrative staff people comprise the median corporate public affairs department, and the median annual budget is $2 million to $3.5 million.[17]

Along with a strong internal team for government relations, a number of businesses today outsource certain functions to external firms in a "divide-and-conquer" strategy. In responses to a recent survey, half of the public affairs executives reported an increase in outsourcing from 1997 to 2000.[18] Not surprising, then,

[12] Hansen and Mitchell, "Disaggregating and Explaining Corporate Political Activity," p. 891.

[13] Shawn Zeller, "Lobbying: Saying So Long to D.C. Outposts," *National Journal*, December 1, 2001, http://www .nationaljournal.com.

[14] http://www.opensecrets.org/pacs/lookup2.php?strID=C00093054.

[15] "Survey Shows Public Affairs Emerging as Top Management Function," *Public Relations Quarterly*, July 22, 2000, p. 30.

[16] Ibid.

[17] "CEOs More Politically Involved, but Many Shrug Off Value of Crisis Planning," *Executive Update Magazine*, February 2003.

[18] "Survey Shows Public Affairs Emerging as Top Management Function," p. 30.

is the fact that the number of registered lobbyists in Washington has more than doubled since 2000 to more than 34,750, and the amount that lobbyists charge their new clients has increased by as much as 100 percent. A 2011 survey by the FPA found that even in the recent recession, 80 percent of companies maintained their public affairs budgets despite cost-cutting pressures.[19]

The external lobbying consultants in Washington to whom companies often turn for advice and guidance on political activities can command rates of $15,000 to $25,000 per month. Hewlett-Packard Co., the California computer maker, nearly doubled its budget for contract lobbyists to $734,000 in 2004 and added the elite lobbying firm of Quinn Gillespie & Associates LLC. Its goal was to pass Republican-backed legislation that would allow it to bring back to the United States at a dramatically lowered tax rate as much as $14.5 billion in profit from foreign subsidiaries. The extra lobbying paid off. The legislation was approved, and Hewlett-Packard saved millions of dollars in taxes. "We're trying to take advantage of the fact that Republicans control [*sic*] the House, the Senate and the White House," said John D. Hassell, director of government affairs at Hewlett-Packard. "There is an opportunity here for the business community to make its case and be successful."[20]

These steep costs make relying entirely on outside counsel to oversee all government affairs activities unrealistic for most companies. For example, after closing its Washington office in 2000, Lucent Technologies soon discovered that depending on external consultants can be just as expensive as operating a small-scale Washington office.[21] Microsoft has successfully built a strong in-house government relations function. After the Justice Department filed an antitrust suit against the company in 1998, Microsoft initiated a government relations overhaul of unprecedented magnitude. The end result was a team of 15 savvy government affairs staffers in Washington—a presence three times larger than the average corporation's lobbying presence in D.C.—as well as lobbying representatives in every major state nationwide.[22] Microsoft also implemented a number of less conventional strategies, including constructing a website tailored to generate nationwide support for the company from individuals. The company has spent more than $61 million on lobbying since 1998, according to the Center for Public Integrity, a political watchdog group. The majority of Microsoft's 20 registered lobbying companies are law firms dealing with legal and tax issues.[23]

As with all other corporate communication functions, companies must measure the impact of their government affairs program to gauge whether it is properly tailored to the existing political environment. Today, businesses use a range of methods to track and evaluate their efforts. A 2011 survey by the FPA revealed that 68 percent of companies use objectives set and achieved, 70 percent use costs

[19] Foundation for Public Affairs, "State of Public Affairs 2011–2012," 2011.

[20] Ibid.

[21] Zeller, "Lobbying."

[22] Jeffrey H. Birnbaum, "How Microsoft Conquered Washington," *Fortune*, April 29, 2002, pp. 95–96.

[23] Alicia Mundy, "Consultants for Microsoft Aren't Such Odd Couples," *Seattle Times*, May 4, 2004.

reduced/avoided, 68 percent use internal customer satisfaction, and 59 percent use legislative wins and losses as measuring sticks for their performance in government affairs. A results-focused approach will help ensure that a government affairs program stays strategically on track.

The Ways and Means of Managing Washington

An internal staff of government relations professionals and senior leaders who are engaged in the issues that affect their companies are two important components of any business's strategy to stay tapped into Washington. In this section, we look at some of the specific activities that companies use to advance their positions with lawmakers.

Coalition Building

The 1970s saw a great "political resurgence of business." Many of the methods used by government relations departments today became established or perfected during this period. In particular, coalition building emerged as a popular form of political influence. Many businesses previously acted to defend only their individual interests when faced with legislative problems, without considering the ways in which their own concerns might coincide with those of other groups or organizations. When a particular company was in trouble, it often battled Washington alone, even when the same issues applied to many other corporations within its industry.

The times of each business standing alone in Washington ended when legislation that affected most, if not all, businesses became more common than the earlier regulations that had affected one or a small collection of industries. Laws concerning consumer safety and labor and wage reform led the wave of these broader regulations. Companies soon learned the benefits of working together. When one company was affected by new regulations, it would find other firms in a similar position to form ad hoc committees. In these committees, the companies forged alliances of support on the business level, which then translated into channels for expressing their views in a greater number of congressional districts and states.

Although loosely formed ad hoc coalitions are still common, companies also often join established industry associations that pool financial and organizational resources for representing their positions in Washington. The Consumer Electronics Association (CEA), for example, advocates that industry's collective viewpoint on issues that include government regulation of broadband, consumer home recording rights, and copyright protection. The National Cattlemen's Beef Association (NCBA) is a similar organization presenting the unified views of thousands of ranchers and beef producers with respect to public policy affecting the cattle industry.

By joining forces through either ad hoc coalitions or more formalized industry associations, companies can assert greater power and have a better chance of affecting legislative outcomes than they would have acting alone.

CEO Involvement in Government Relations

Large and small companies alike strengthen their government relations programs through actively involving senior management in political activities. As they have recognized the importance of gaining a seat at the policy discussion table for their companies, increasing numbers of CEOs are stepping into the policy debate. This trend does not surprise most executives. As John de Butts, former chairman of AT&T, remarked, "So vital . . . is the relationship of government and business that to my mind the chief executive officer who is content to delegate responsibility for that relationship to his public affairs expert may be neglecting one of the most crucial aspects of his own responsibility."[24]

Indeed, 98 percent of the chief executives of the 115 major companies surveyed in 2002 took part in some form of political involvement activity, including extensive government relations work for trade or business associations. Other activities drawing CEO participation were correspondence to federal legislators or regulators, endorsement of the company's political action committee, direct lobbying of federal legislators, and attendance at candidate fundraisers.[25]

Frederick W. Smith, founder, chair, and chief executive officer of FedEx, exemplifies the benefits of CEO involvement in government relations. Since FedEx's inception in 1973, Smith has advanced his company's interests by using ingenuity, networks of personal alliances, and strategic charitable contributions. Examples of his creative political outreach range from FedEx's maintenance of a small corporate jet fleet ready to fly members of Congress across the country at a moment's notice to Smith's preservation of his longstanding relationship with former Yale fraternity brother George W. Bush.[26] As Wendell Moore, chief of staff to Tennessee Governor Don Sundquist, explained: "The fact that Smith knows his members of Congress on a first-name basis is a significant reason that the company has been so successful."[27]

FedEx continues to reap the rewards of its CEO's notable presence on Capitol Hill. For example, during the Clinton administration, Smith's rapport with Clinton undoubtedly led to his place as part of the official delegation on a trade mission to China. In 2002, the U.S. Postal Service announced a seven-year partnership with FedEx worth up to $7 billion, in which FedEx planes provided the postal service an air-delivery network in exchange for having its branded drop boxes installed at 10,000 post offices nationwide, a major accomplishment for Smith's well-positioned company.[28]

Lobbying on an Individual Basis

When business leaders realized the importance of having a say in the activities on Capitol Hill, they turned to lobbying groups to help them successfully advance their viewpoints with congressional decision makers. (Lobbying is

[24] James W. Singer, "Business and Government: A New 'Quasi-Public' Role," *National Journal*, April 15, 1978, p. 596.

[25] "CEOs More Politically Involved."

[26] Michael Steel, "FedEx Flies High," *National Journal*, February 24, 2001, http://nationaljournal.com.

[27] Ibid.

[28] Ibid.

any activity aimed at promoting or securing the passage of specific legislation through coordinated communications with key lawmakers.) In recent decades, as government intervention has grown, so has the number of organizations in Washington that present the views of business to Congress, the White House, and the regulatory agencies.

Using individuals in lobbying (which typically consists of activities such as letter writing, inserting editorials or op-ed pieces in print news media, and office visits to lawmakers) can have a significant impact in Congress. The U.S. Chamber of Commerce, for one, has conducted very effective and sophisticated grassroots campaigns to increase its influence. With state and local chapters that contain thousands of members, the Chamber of Commerce has a wide base from which to work. By 1980, it had established 2,700 "Congressional Action Committees" that consisted of executives who were personally acquainted with their senators and representatives. These executives received information about events in Washington through bulletins from the Chamber's Washington office and remained in touch with their representatives so that they might contact them when called upon to do so. In 2011, they were the largest single lobbying group in the United States, with annual spending of over $800 million.[29]

This method of lobbying through far-reaching constituencies has produced good results: "within a week [the Chamber of Commerce] . . . can carry out research on the impact of a bill on each legislator's district and through its local branches mobilize a 'grassroots campaign' on the issue in time to affect the outcome of the vote."[30] Today, the Chamber of Commerce—once poorly regarded in Washington— has a grassroots network of 50,000 business activists, an expansive membership of 3 million businesses, 830 business associations, and 102 American Chambers of Commerce abroad.[31]

Returning to our Microsoft example, this company's lobbying efforts, which included its "grassroots" website campaign, clearly have paid off as well. In 2001, Microsoft's lobby against copyright violators resulted in a government crackdown on software piracy. Later that year, in the wake of the September 11 terrorist attacks, Microsoft led the charge in persuading the Bush administration to allot over $70 million to improve "cybersecurity" in America.[32]

Success stories like Microsoft and the U.S. Chamber of Commerce have prompted many major corporations to establish campaigns that target individuals. In addition to achieving desired legislative outcomes, "a prudently managed grassroots program can be a team-building exercise. Providing information about legislation that will affect current and future company activities will be of interest to many employees at all ranks. . . . [B]uilding a grassroots program with employees makes them part of the team."[33] Using individuals in lobbying is one of the most popular methods for companies and their employees to get involved in politics. Blogs have

[29] http://www.opensecrets.org/lobby/top.php?indexType=s.

[30] Graham Wilson, *Interest Groups in the United States* (New York: Oxford University Press, 1981).

[31] American Chamber of Commerce website, http://www.uschamber.com.

[32] Birnbaum, "How Microsoft Conquered Washington."

[33] Gerry Keim, "Corporate Grassroots Program in the 1980's," *California Management Review* 28, no. 1 (Fall 1985), p. 117.

become an important part of grassroots campaigns, as they can be targeted to niche groups of constituents very easily.

Political Action Committees

Another popular method of getting involved in government is the formation of political action committees (PACSs). The idea for this movement came from organized labor, which created official committees responsible for raising and dispersing money to support political campaigns. In 1980, 1,200 companies had their own PACs; today there are 4,700.[34] Approximately 58 percent of *Fortune* 500 companies currently have a PAC.[35] Industry leaders such as Walmart, UPS, and AT&T have some of the largest and most active PACs, giving between $1 and $1.8 million to candidates in 2003.[36] Nineteen percent of Walmart's 60,000 domestic managers contribute to its PAC, mostly through payroll deductions that average $8.60 a month.[37]

To target their funding efficiently, PAC administrators need to have access to information about each political candidate and the races they support. The Business-Industry Political Action Committee (BIPAC) was formed to meet these information needs. Although this group does contribute directly to candidates, its most important role is to research candidates and identify close races. During each election year, BIPAC holds monthly information briefings for PAC managers, along with providing daily updates on congressional races through a BIPAC telephone service. By using their national organization, individual PACs remain well informed and are able to direct their funds intelligently.

PACs have arisen out of an increased political awareness in corporate managers. They provide a simple framework for getting employees involved in political issues that could determine their employers' well-being into the future. Employee involvement is key, as federal election law prevents direct corporate contributions to party committees and candidates. According to one executive, "PACs are one of the most effective vehicles to generate individual participation in the political process to come along in a long time." Another executive further commented, "Our first goal is to involve our people in the political process. Only about five percent of our time is devoted to fund raising and the distribution of funds; ninety-five percent is devoted to political education. Our philosophy is to encourage long-term understanding and continuing involvement in the political process."[38]

The Public Affairs Council estimates that PAC contributions currently account for between 1 and 10 percent of political donations.[39] The amount of money that businesses contribute to politicians in any given election cycle is staggering. According to the Center for Responsive Politics, business interests contribute far

[34] Federal Election Commission website, http://www.fec.gov.

[35] Tim Reason, "Campaign Contributions at the Office," *CFO Magazine*, July 12, 2004.

[36] Wesley Bizzell, "Office Politics," *Corporate Counsel*, March 2004.

[37] Cummings, "Joining the PAC," p. A1.

[38] Edward Handler and John R. Mulkern, *Business in Politics* (Lexington, MA: Lexington Books, 1982).

[39] Jeffrey H. Birnbaum et al., "The Influence Market: Capitol Clout: A Buyer's Guide: Access in Washington Comes at a Price," *Fortune*, October 26, 1998, p. 177ff.

more money to candidates and political parties than do labor unions or ideological groups. In the 2010 election cycle, total lobbying spending was over \$3.5 billion, with the vast majority coming from corporations.[40] The flow of money from business interests to political campaigns has alarmed some sectors of the American public, who fear that businesses' ability to back their agenda with large sums of money gives them an unfair advantage in having their voice heard in Congress. A recent *BusinessWeek*–Harris Poll revealed that three-quarters of Americans think large companies are too influential in Washington. The same poll concluded that 84 percent of the public believes campaign contributions made by big business have too much influence on American politics.[41] Clearly, not everyone views business political spending as a positive trend.

Conclusion

Corporate America's relationship with various levels of government extends far beyond licenses, safety standards, and product prices. Today, the influence of private business on public affairs, and vice versa, has become so established that we often assume changes in one arena will lead to changes in the other. Democratic reform in Latin America and Eastern Europe came hand in hand with market reform, and upon China's acceptance into the World Trade Organization (WTO), President George W. Bush declared, "I believe a whiff of freedom in the marketplace will cause there to be more demand for democracy."[42]

In America, defining the roles of government and business with regard to each other is an ongoing process. In the summer of 2002, a crisis of confidence in business ethics and corporate governance compelled President Bush to address Wall Street leaders, saying, "We must usher in a new era of integrity in corporate America."[43] Congress moved rapidly to negotiate a raft of bills and proposals that would regulate not only how businesses are run but also how they report their activities to the public. Several years later, government and business are still struggling to find a balance, with business claiming that the cost of complying with Sarbanes-Oxley is too high.

On the technology front, the Internet has significantly shaped companies' approaches to government affairs. Companies may rely heavily on web monitoring, issues-focused websites, and online networks of grassroots lobbyists to track important legislation and broaden the reach of their own coalitions.[44] At the same time, as we learned in Chapters 1 and 4, the speed at which information flows over the Internet means news of corporate wrongdoings or legal violations has the potential to reach many constituencies before senior management can prepare for the crisis at hand.[45]

[40] Center for Responsive Politics website, http://www.crp.org.

[41] http://www.opensecrets.org/lobby/index.php.

[42] Lawrence F. Kaplan, "Why Trade Won't Bring Democracy to China," *New Republic*, July 9, 2001, p. 23.

[43] Randall Mikkelsen, "Update 4: Bush Seeks 'New Era of Corporate Integrity,'" Reuters, July 9, 2002, http://www.Forbes.com.

[44] Pinkham, "How'd We Get to Be the Bad Guys?" p. 12.

[45] Douglas G. Pinkham, "Corporate Public Affairs: Running Faster, Jumping Higher," *Public Relations Quarterly*, no. 5 (Summer 1998), pp. 33–37.

With the complexities of globalization and the ever-increasing speed of information flows, businesses must devote attention and resources to actively manage their relationships with the government and its lawmakers. Successful companies recognize the importance of staying abreast of what happens on Capitol Hill. "It is essential we have a very strong presence," said Robert L. Garner, president of the American Ambulance Association. "It's pricey, but it's the cost of doing business in the federal environment."[46] Critical to this effort to keep connected with Washington is the government relations function, which, whether entirely internal or partially outsourced, must be an integrated function of a company's overall communication strategy.

---

[46] Jeffrey H. Birnbaum, "Lobbying Firms Hire More, Pay More, Charge More to Influence Government," *Washington Post*, June 22, 2005.

Disney's America Theme Park: *The Third Battle of Bull Run*

When you wish upon a star, makes no difference who you are. Anything your heart desires will come to you. If your heart is in your dreams, no request is too extreme . . .

—Jiminy Cricket

On September 22, 1994, Michael Eisner, CEO of the Walt Disney Company, one of the most powerful and well-known media conglomerates in the world, stared out the window of his Burbank office, contemplating the current situation surrounding the Disney's America theme park. Ever since November 8, 1993, when *The Wall Street Journal* first broke the news that Disney was planning to build a theme park near Washington, D.C., ongoing national debate over the location and concept of the $650 million park had caused tremendous frustration. Eisner thought back over the events of the past year. How could his great idea have run into such formidable resistance?

THE CONTROVERSY COMES TO A HEAD

Eisner's secretary had clipped several newspaper articles covering two parades that took place on September 17. In Washington, D.C., several hundred Disney opponents from over 50 anti-Disney organizations had marched past the White House and rallied on the National Mall in protest of the park. On the same day in the streets of Haymarket, Virginia, near the proposed park site, Mickey Mouse and 101 local

Source: This case was written by Elizabeth A. Powell, Assistant Professor of Business Administration, and Sarah Stover, MBA 1997. It was written as a basis for class discussion rather than to illustrate effective or ineffective handling of an administrative situation. Copyright © 2001 by the University of Virginia Darden School Foundation, Charlottesville, VA. All rights reserved. To order copies, send an e-mail to dardencases@virginia.edu.

children dressed as Dalmatians had appeared in a parade that was filled with pro-Disney sentiment. Eisner was particularly struck by the contrast between the two pictures: one showing an anti-Disney display from the National Mall protest and another of Mickey and Minnie Mouse being driven through the streets of Haymarket during the exuberant community parade.

Despite the controversy depicted in the press, on September 21, Prince William County, Virginia, planning commissioners had recommended local zoning approval for Disney's America, and regional transportation officials had authorized $130 million in local roads to serve it. It appeared very likely that the project would win final zoning approval in October. At the state level, Virginia's Governor George Allen continued his strong support of the park's development.

Over the past three weeks, however, Eisner had been ruminating over a phone call he received in late August from John Cooke, president of the Disney Channel since 1985. Although Cooke had no responsibility for Disney's America, he had more experience in the Washington, D.C., political scene than any other of Disney's highest-ranking managers and was one of Eisner's most trusted executives. Cooke was not encouraging about the park's prospects. Quite familiar with many of the park's opponents, he believed they would not give up the fight under any circumstances. Given the anti-Disney coalition's considerable financial resources, the nationally publicized anti-Disney campaign could go on indefinitely, inflicting immeasurable damage on Disney's fun, family image. Cooke advised Eisner to think very seriously about ending the project.

Since the mid-1980s, Eisner's business strategy was to revitalize Disney by broadening its brand into new ventures. Although promising at first, now the wisdom of some of the ventures seemed less certain. The worst example, EuroDisney, the new Disney park located outside Paris, continued to flounder. The numbers for fiscal year 1994, due in just a couple of days on September 30, didn't look promising. Estimates said net income would be down to $300 million from $800 million the year before, mostly because EuroDisney lost $515 million from operations and $372 million from a related accounting charge.[1]

The good news was that due to cost cutting, EuroDisney's losses were actually less than in the previous year, while the bad news was that attendance was also down. Prince al-Waleed bin Talal bin Adulaziz of Saudi Arabia had agreed to buy 24 percent of the park and build a convention center there, thus relieving some of the financial pressure, but it seemed that the negative press coverage of that park's troubles would never end.

The Disney's America problem was particularly bothersome, however. Eisner realized that the controversy surrounding the park, coupled with the many other highly publicized problems of 1994, was damaging Disney's image. Due to publicity about its highly visible corporate problems, Disney's image as a business threatened to tarnish its reputation for family-friendly fun and fantasy.

Personally, Eisner was particularly fond of the Disney's America concept. He had helped develop the original idea and had personally championed it within the Disney organization. He recalled the early meetings during which several Disney executives, including himself, had brainstormed an American history concept. He and the other executives had strongly believed that Disney had the unique capability of designing an American history theme park that would draw on the company's technical expertise and offer guests an entertaining, educational, and emotional journey through time. They envisioned guests, adults and children alike, embracing a park dedicated to telling the story of U.S. history. Eisner had hoped the park would be part of the personal legacy he would leave behind at Disney. As he told a *Washington Post* reporter, "This is the one idea I've heard that is, in corporate locker room talk, what's known as a no-brainer."[2]

THE DISNEY'S AMERICA CONCEPT AND LOCATION

The idea of building an American history theme park originated in 1991 when Eisner and other Disney executives attended a meeting at Colonial Williamsburg in southeastern Virginia. The executives were impressed by the restored pre-Revolutionary capital. Disney had already been thinking about locations for theme parks that were on a somewhat smaller scale than the company's massive ones. Visiting Williamsburg helped Disney make the connection to a new park based on historical themes.

Disney's attention soon shifted focus to Washington, D.C. As the third-largest tourist market in the United States and the center of American government, the nation's capital seemed a natural location for an American history park. The abundance of historical sites in the area broadened its appeal as a center of American history. Disney's other parks were located on the fringes of developed urban centers (Anaheim, Orlando, Tokyo, and Paris). The parks gained advantages from their proximity to urban centers, but due to their peripheral locations, Disney was able to acquire lower-priced land and ensure a safe environment for visitors, far from inner-city congestion and crime.

[1] The Walt Disney Company Annual Report, 1995. See also Kim Masters, *The Keys to the Kingdom* (New York: William Morrow, 2000), p. 299.

[2] William M. Powers, "Michael in Eisnerland: Disney's Chairman's Sense of Wonder, Will to Win Drive for Virginia Theme Park Plan," *Washington Post*, January 23, 1994, p. H1.

Disney needed a location with easy access to an airport and an exit off an interstate highway. Executives hoped to find land that had already been zoned for development as well as local and state politicians who would be open to economic growth. In Prince William County, located in the heart of Virginia's Piedmont region, Disney found all these things. Dulles International Airport was located just east of Prince William County. U.S. Interstate 66 (I-66), the main traffic artery connecting Washington, D.C., with its western suburbs, could transport tourists straight from Washington's monuments and museums into Prince William County, a distance of about 35 miles.

The political and economic context also made Prince William County attractive to Disney. Virginia had long been a pro-growth state, and its governors were constantly under pressure to bring in new business. Democratic Governor Doug Wilder would leave office in November 1993, having lost some notable campaigns to bring growth to Virginia's economy. Polls showed that he would likely be replaced by Republican George Allen, the son of a former Washington Redskins American football coach and a graduate of the University of Virginia. If elected, Allen would be under instant pressure to create state economic growth. Most Prince William County officials were also "pro-growth," though not well prepared for it. The county's growing population of middle-class residents (up 62 percent since 1980) paid the highest taxes in the state of Virginia due to a dearth of economic development within the county. The Virginia legislature set an ambitious goal in 1990 to attract 14,000 jobs and $1 billion in nonresidential growth to the county to fund more and better schools and county administrative services, in addition to reducing residential taxes paid by each family.

In the spring of 1993, Peter Rummell, president of Disney Design and Development, which included the famous Imagineering group, as well as the real estate division, identified 3,000 acres in Prince William County near the small town of Haymarket (population 483). The largest property was a 2,300-acre plot of land, the Waverly Tract, owned by a real estate subsidiary of the Exxon Corporation. Waverly was already zoned for mixed-use development of homes and office buildings, yet due to a weak real estate market, Exxon had never broken ground on the undeveloped farmland. For a modest holding price, Exxon was willing to option the property. Using a scheme that had worked years before in Orlando, the Disney real estate group bought or put options on Waverly and the remaining 3,000 acres without revealing the company's corporate identity in any of the transactions.

THE VIRGINIA PIEDMONT

The northeast corner of Virginia comprises the Piedmont region. The region contains countless significant sites related to U.S. history, including, for example, the preserved homes of four of the first five U.S. presidents: Washington, Jefferson, Madison, and Monroe. According to Pulitzer Prize–winning historian David McCullough, "This is the ground of our Founding Fathers. These are the landscapes—small towns, churches, fields, mountains, creeks, and rivers—that speak volumes."[3] Thomas Jefferson loved the agrarian life he found on the farms east of the Blue Ridge Mountains. In his letters, he exulted over the region's "delicious spring," "soft genial temperatures," and good soil.[4]

The region is also home to more than two-dozen Civil War battlefields. The U.S. Civil War was fought largely over the issue of slavery, pitting northern states against the southern states that had seceded from the union. Just a

[3] Richard L. Worsnop, "Historic Preservation," *The CQ Researcher,* October 7, 1994, p. 867.

[4] Rudy Abramson, "Land Where Our Fathers Died," *Washingtonian Magazine,* October 1996, p. 62.

few miles from the Waverly tract is Manassas National Battlefield Park, land that is protected and preserved by the U.S. National Park Service, commemorating two major Civil War battles. The first battle in 1861 was the Civil War's first major land engagement. The second, in 1862, marked the beginning of Confederate General Robert E. Lee's first invasion of the North. On what would become some of the bloodiest soil in U.S. history, Lee reflected at Bull Run in 1861, "The views are so magnificent, the valleys so beautiful, the scenery so peaceful. What a glorious world the Almighty has given us. How thankless and ungrateful we are, and how we labor to mar His gifts."[5]

Although largely rural and predominantly middle class, the region was also notable as home to some of America's most wealthy and influential citizens. The largest estates suggested the presence of privilege at every turn: perfect fences built from stone or wood, carefully manicured pastures, large barns requiring lots of hired help, a few private landing strips, and long private lanes lined with boxwood or dogwood that led to magnificent private homes. Exhibit 9.1 provides a map showing the locations of some of the area's most wealthy homeowners.

The Piedmont also had a history of successfully fighting local development projects. In the late 1970s, the Marriott Corporation had proposed building a large amusement park, and, in the late 1980s, a development group had planned to develop a major shopping mall in the area. Both projects were defeated by local opposition.

DISNEY'S PLANS REVEALED

To keep its land acquisition secret, Disney had done little to work with local government and communities, but by late October of 1993, Eisner learned that Disney's plans had begun to leak.[6] It would only be a matter

[5] "Making a Stand," *Conde Nast Traveler*, September 1994, p. 148.

[6] Michael D. Eisner, *Work in Progress* (New York: Random House, 1998), p. 323.

EXHIBIT 9.1 **The Third Battle of Bull Run: The Disney's America Theme Park. Proximity of Disney's America Site to Several Wealthy Residents. Design: Bob Mansfield/Forbes**

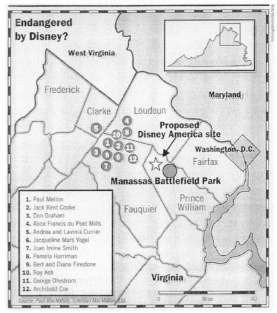

Source: Lisa Gubernick, "The Third Battle of Bull Run," *Forbes 400*, October 17, 1994, p. 72. Reprinted by permission of *Forbes* Magazine, copyright 2002, Forbes, Inc.

of days before the news would hit the media, so in the meantime Disney had to act quickly. Behind the scenes, Eisner contacted outgoing Governor Wilder and Governor-elect Allen. Both gave their immediate support and agreed to attend a public announcement scheduled for November 11. The company hired a local real estate law firm, and also retained the services of Jody Powell, former press secretary to President Jimmy Carter, who ran the Powell Tate public relations firm.

On November 8, 1993, a brief item appeared in *The Wall Street Journal* stating that Disney was planning to build a theme park somewhere in Virginia. That same day, Disney officials confirmed the story but provided no additional details. The next day, a *Washington Post* reporter identified Prince William County as the targeted area. Disney spokespeople confirmed the location and added some details. Disney officials briefed reporters and legislators, stating that they had investigated possible obstacles to the project, including environmental and historic preservation concerns, and believed there would be no serious problems.[7] They also stated that they had studied traffic patterns on I-66 and believed additional theme park traffic would not exacerbate rush-hour congestion.[8] Because 65 percent of Prince William County residents commuted to jobs in other northern Virginia counties, traffic congestion was a primary concern. On November 10, the *Post* ran the first full news story, headlined "Disney Plans Theme Park Here; Haymarket, VA: Project to Include Mall, Feature American History."

As local discussion increased, Disney held an upbeat news conference on November 11 and also issued a press release. Rummell, flanked by the governors and local officials, revealed an architectural model of the theme park and the surrounding development plans. The park logo featured a bold close-up of a stylized bald eagle rendered in navy blue, draped in red and white striped bunting, and with the words, "Disney Is America" emblazoned in gold across the eagle's chest.

Disney's America was presented as a "totally new concept . . . to celebrate those unique American qualities that have been our country's strengths and that have made this nation the beacon of hope to people everywhere." Disney would draw upon its entertainment experience in multimedia and theme park attractions. Disney officials emphasized the park's focus on the Civil War. Guests would enter the park through a detailed Civil War–era village and then ride a steam train to explore nine areas, each devoted to an episode from American history. One of these included a Civil War fort,

complete with battle reenactments. Other exhibits included "We the People," depicting the immigrant experience at Ellis Island, and "Enterprise," a factory town featuring a high-speed thrill attraction called "The Industrial Revolution."

Disney officials predominantly sold the park on its economic benefits to the local area, stating that the park would directly generate about 3,000 permanent jobs[9] along with 16,000 jobs indirectly.[10] Around the park, the company would develop resort hotels, an RV park, a 27-hole public golf course, a commercial complex with retail and office space, and 2,300 homes.[11] Disney projected $169 million in tax revenues for the first 10 years after the park opened in 1998 and nearly $2 billion over its first 30 years.[12] In addition, Disney would donate land for schools and a library and reserve up to 40 percent green space as a buffer around the core recreational area.[13]

In part, the announcement came off better in print than at the conference. In the press release, Bob Weis, senior vice president of Walt Disney Imagineering, was quoted as saying, "Beyond the rides and attractions for which Disney is famous, the park will be a venue for people of all ages, especially the young, to debate and discuss the future of our nation and to learn more about its past by living it." In the conference, however, Weis said of the attractions, "We want to make you a Civil War soldier. We want to make you feel what it was like to be a slave, or what it was like to escape through the Underground Railroad." Weis's intended meaning was to refer to the new technology of

---

[7] Kirsten Downey and Kent Jenkins Jr., "Disney Plans Theme Park Here; Haymarket, VA.: Project to Include Mall, Feature American History," *Washington Post*, November 10, 1993, p. 59.

[8] Ibid.

[9] Spencer S. Hsu, "Disney Project Runs into Concern about Traffic Pollution," *Washington Post*, November 12, 1993, p. A18.

[10] Lisa Gubernick, "The Third Battle of Bull Run," *Forbes* 400, October 17, 1994, p. 68.

[11] Ibid.

[12] Ibid.

[13] Park Net, National Park Service, "More Battles: The Horse and the Mouse, Battling for Manassas," http://www.cr.nps.gov/history/online_books/mana/adhi11b.htm (last modified August 8, 2001).

virtual reality that would be used, but critics quickly jumped on the statement. *Washington Post* columnist Courtland Milloy contrasted the description to "authentic history" that would have to portray atrocities like slave whippings and rape.[14] Author William Styron wrote that he believed the comment suggested that slavery was somehow a subject for fun or that the escape route used for slaves was similar to a subway system.[15]

PIEDMONT OPPOSITION

Almost immediately after Disney confirmed its plans to build a park in Prince William County, anti-Disney forces began organizing their opposition. To many who were alarmed, the plans seemed already so well-developed that they gave the impression of a *fait accompli*. Just days after Disney's formal announcement, a meeting was held at the home of Charles S. Whitehouse, a retired foreign service officer who had owned property in the Virginia hunt country since the early 1960s. The dozen guests included William D. Rogers, former undersecretary of state under Henry Kissinger and now a senior partner in a powerful Washington law firm; Joel McCleary, a former aide in the Carter White House and former treasurer of the Democratic National Committee; and Lavinia Currier, great-granddaughter of Pittsburgh financier Andrew Mellon. William Backer, a former New York advertising executive who had created slogans for Coca Cola ("Coke— It's the Real Thing") and Miller Beer ("If you've got the time, we've got the beer"), also attended. The group worried that the proposed development would undermine the upper Piedmont's "traditional character and visual order."

The phrase came from the charter of the Piedmont Environmental Council (PEC), a rural-preservation group co-chaired by Whitehouse. Many of Whitehouse's guests

had donated considerable amounts of time and money to the organization, which fought development and bought land and easements to preserve the area. The PEC was originally founded in 1927 by a group of prominent landowners. Over the years, it fought successfully against uranium mining and plans for a "western bypass" highway.

The group discussed the options for stopping Disney's encroachment upon the hunt country. There was the possibility of derailing the project during the Virginia legislature's next session. Rogers discussed some of the legal options. Backer suggested a negative publicity campaign, possibly at a national level, to force an image-conscious company like Disney to retreat. He argued for a subtle approach rather than a straight-on NIMBY (Not In My Backyard) campaign. He didn't want the opposition campaign to be viewed simply as a group of wealthy landowners who wanted to prevent a theme park from disturbing their fox hunting. As the meeting ended, Backer agreed to come up with a slogan.

A few days later, Backer presented his "Disney, Take a Second Look" slogan to the group. Backer's angle was to convince Disney that it should reassess the idea of building a theme park amid the beauty of the Piedmont. Although the campaign addressed Disney directly, it would remind anyone who saw it of the Piedmont's unspoiled and now-threatened natural beauty. Within a few days, the slogan was running in radio ads and incorporated on a letterhead. The logo accompanying the slogan showed a balloon, with a barn and farmhouse inside, drifting away in the breeze.

This initial meeting was followed by dozens of others in the coming weeks and months. One week after the gathering at the Whitehouse home, over 500 people attended a meeting at the Grace Episcopal Church in The Plains, Virginia, about seven miles west of the Disney site. The meeting's attendants represented a wide range of economic backgrounds, but they were united in their preference for the rural life they enjoyed.

[14] Ibid.
[15] Chris Fordney, "Embattled Ground," *National Parks*, November/December 1994, p. 28.

Megan Gallagher, Whitehouse's co-chairman of the Piedmont Environmental Council, led the meeting, reminding the group of other local protest movements that had stopped big projects. Other speeches rounded out the audience's concerns: The park's estimated 9 million annual visitors would spark low-density ancillary development like that around Anaheim and Orlando. The pristine countryside would be overcome by cheap hotels, restaurants, and strip malls. Already problematic traffic congestion would be exacerbated. The park would create low-wage jobs and not provide the tax base that the Disney plan promised. During the meeting, the Piedmont Environmental Council, which had already committed $100,000 of its $700,000 annual budget to the project, emerged as the leading opposition group.

A few days after the meeting at Grace Episcopal Church, another meeting organized by the Prince Charitable Trusts of Chicago, another land preservation group, was held at a local restaurant. This meeting brought together several regional and national environmental groups concerned about the Disney's America project. Eventually, the Prince Charitable Trusts would give over $400,000 to 14 different anti-Disney groups that conducted studies, gave press conferences, and attacked Disney from every possible environmental angle. The largest grant went to the National Growth Management League, which mounted a local advertising campaign under the name "Citizens Against Gridlock." The campaign depicted I-66, already one of northern Virginia's busiest and congested highways, as "Disney's parking lot."

In early December, the PEC held a news conference in a Washington hotel to increase the reach of its "Second Look" campaign. It retained a prominent Washington law firm as well as a public relations firm. The group began recruiting and organizing dozens of volunteers. It sent out a fundraising letter seeking $500,000 in contributions. It also commissioned experts to assess the park's impact on the environment, urban sprawl, traffic, employment, and property taxes.

DISNEY'S CAMPAIGN

Soon after the public announcement, Disney undertook concerted efforts to win over state and local government, as well as constituencies within proximity of the proposed site. Virginia's new governor, George Allen, immediately promoted the Disney project. He believed Disney's worldwide reputation would make Virginia an international tourist destination, bringing millions of travelers to the state. In numerous press releases, Allen endorsed Disney's belief that the project would create 19,000 jobs and bring millions of new tax dollars to state and municipal coffers.

By early January, Disney asked the state to bear some of the costs of the new park. Disney requested $137 million in state highway improvements and $21 million to train workers, move equipment from Orlando, pay for advertising, and put up highway signs directing tourists to the park. These funds would have to be guaranteed by the end of the current state legislative session to allow Disney to move ahead with development in early 1995. Now focusing on Virginia's state capital, Disney retained a well-connected Richmond law firm to handle its lobbying efforts. It also hired a Richmond event-planning firm to organize two large receptions. Lobbying expenses alone reached almost $450,000, including $32,000 for receptions and $230 for the Mickey Mouse ties given to state legislators.

Allen supported Disney's request and argued the highway improvements Disney planned would ease traffic problems that already existed in northern Virginia, in addition to accommodating the extra traffic generated by the park. The state's support of the project, he said, would send a message that Virginia was "open for business."

Disney officials met with African-American legislators and promised to ensure that minorities got a good shot at contracts and jobs. They invited a dozen officials from area museums and historic sites, including Monticello and Colonial Williamsburg, to a meeting in Orlando to discuss their plans for portraying history in the northern Virginia park. In Richmond, Disney lobbyists portrayed opponents as wealthy landowners who simply did not want Disney in their backyards.

Disney sought and received strong support from the Prince William business community, especially realtors, contractors, hotels, restaurants, and utility companies. The Disney staff also poured tremendous effort into harnessing the support of local citizens, spending hours preaching the message of neighborliness to local groups. Groups formed to support the park, including the Welcome Disney Committee, Friends of the Mouse, Youth for Disney, and Patriots for Disney. Members of these groups attended state legislative hearings, gave testimony, and handed out bumper stickers and buttons. Disney sent newsletters to 100,000 local households and retained a second Washington public relations firm to handle grassroots support.

Disney also entered negotiations with the National Park Service at the Manassas park, which was three miles away. The company agreed to limit the height of its structures to 140 feet so they would not be visible from the park and to develop a special transit bus system that would transport 20 percent of Disney guests and 10 percent of employees. The company promised to promote historic preservation and the Manassas National Battlefield Park within Disney's America and immediately donated money to an allied nonprofit group.

THE PEC'S CAMPAIGN

The PEC mounted a strong effort in Richmond as well, including hiring two full-time lobbyists.[16] Its campaign was based on the premise that the park was a bad business deal for Virginia.

The PEC claimed that the park would generate fewer jobs than Disney and Governor Allen had promised—6,300 rather than 19,000—and that the jobs would not pay well. They accused Allen of exaggerating the tax benefits. They emphasized the traffic and air pollution that would be caused by the park. They also suggested 32 other sites in the Washington area that would be more suited to the project than the current site. Finally, they suggested that the state should not have to fund any of the park's development. The PEC spent over $2 million in its campaign against Disney, including lobbying and public relations.[17]

THE VOTE

On March 12, 1994, when the Virginia state legislature voted on a $163 million tax package for Disney, the results clearly favored Disney. This wasn't even a close call for the state government officials. Disney won 35 to 5 in the Virginia Senate and 73 to 25 in the House of Delegates. Things were looking up for the Disney's America project, although a new bumper sticker appeared in the Piedmont that said, "Gov. Allen Slipped Virginia a Mickey."

THE HISTORIANS AND JOURNALISTS TAKE OVER

Disney officials were elated after their victory in Richmond. It seemed likely that construction could begin in early 1995 after all. Meanwhile, Disney's opponents were not ready to give up the fight. Public debate on local issues such as traffic congestion and pollution had failed to keep Disney out of the Piedmont, and the anti-Disney crowd realized they needed to change the theme of their campaign. They needed a grander, more significant argument—something that would gain national attention.

---

[16] Ibid.

[17] Ibid.

The kernel of that argument had appeared in December 1993, in an editorial written to the *Washington Post* by Richard Moe, president of the National Trust for Historic Preservation and former chief of staff for Vice President Walter Mondale.[18] In his article, Moe suggested that Disney's development would engulf "some of the most beautiful and historic countryside in America."[19] He predicted that the park would reduce attendance at authentic northern Virginia historic landmarks, including the Manassas battlefield. Moe also questioned Disney's ability to seriously portray American history when the success of its other theme parks was based on simply showing visitors a good time.

Moe's article was followed by a similar piece published in mid-February 1994 by the *Los Angeles Times*, the newspaper serving Disney's southern California headquarters. This editorial was written by Pulitzer Prize–winning journalist Nick Kotz, whose Virginia farm happened to be located three miles from the Disney site. Like Moe, Kotz based his article on the premise that Disney's park would desecrate land that should be considered a national treasure. He suggested that Disney would cheapen and trivialize its historic value. After the article was published, Kotz met with Moe over breakfast at the Mayflower Hotel in Washington to discuss anti-Disney strategy. This meeting was one of many that would follow among a growing network of the nation's most elite journalists and historians who were becoming increasingly concerned over Disney's plans.

As the network grew, several prominent historians joined the fight against Disney and formed a group that became known as Protect Historic America. An early recruit was David McCullough, the author of several best-selling books, including a Pulitzer Prize–winning biography of Harry Truman. McCullough was very well known, particularly of late for his narration of the highly acclaimed Ken Burns "Civil War" series on PBS. Another prominent member was James McPherson, a Princeton University professor and author of the Pulitzer Prize–winning *Battle Cry of Freedom*. Exhibit 9.2 provides a list of many prominent authors and historians who joined Protect Historic America in its early stages.

By May, Protect Historic America was prepared to launch a national campaign in partnership with Moe's National Trust for Historic Preservation. Using funds donated by Piedmont residents, the group placed a full-page ad in the May 2 edition of the *Washington Post*, asking Eisner to reconsider the Haymarket site. The ad included a tear-away response form at the bottom and generated over 5,000 responses. Nine days later, on May 11, the group held a news conference at the National Press Club that featured McCullough and Moe, among others. The prominent journalists in the group virtually assured that the conference would receive national news coverage.

During the well-publicized news conference, the speakers argued that Disney threatened the Piedmont countryside, including historic towns and battlefields. The region's rich heritage made it valuable to all Americans. David McCullough stated, "We have so little left that's authentic and real. To replace what we have with plastic, contrived history is almost sacrilege."[20] James McPherson, in a written statement presented to reporters, said, "A historical theme park in Northern Virginia, three miles from the Manassas National Battlefield, threatens to destroy the very historical landscape it purports to interpret."[21]

The press conference, along with personal correspondence from McCullough and McPherson, convinced over 200 historians and writers to endorse the fight against Disney.

[18] Ibid.

[19] Richard Moe, "Downside to Disney's America," *Washington Post*, December 21, 1993, p. A23.

[20] Larry Van Dyne, "Hit the Road, Mick," *Washingtonian Magazine*, January 1995, p. 59.

[21] Paul Bradley, "Prominent Historians Join Disney Foes," *Richmond Times-Dispatch*, May 12, 1994, p. B1.

EXHIBIT 9.2   The Third Battle of Bull Run: The Disney's America Theme Park. Partial List of Historians and Authors in the Anti-Disney Campaign

James David Barber	Professor of political science, Duke University
Frances Berry	Professor of American social thought, history, and law, University of Pennsylvania
William R. Ferris	Director, Center for the Study of Southern Culture, and professor of anthropology, University of Mississippi
Barbara J. Fields	Professor of history, Columbia University
Shelby Foote	Author of four-volume *Civil War*, which was made into a popular PBS miniseries
George Forgie	Associate professor of history, University of Texas at Austin
John Hope Franklin	Former president, American Historical Association
Ernest B. Furgurson	Journalist and historian
Gary Gallagher	Chairman, History Department, Pennsylvania State University
John Rolfe Gardiner	Piedmont Virginian and author of novels set in the Piedmont region
Doris Kearns Goodwin	Professor of government, Harvard University
Ludwell H. Johnson III	Professor emeritus of history, College of William and Mary
Richard M. Ketchum	Editorial director, American Heritage Books
Nick Kotz	Journalist, author of four books on American history and politics
Glenn LaFantasie	Deputy historian and general editor of the *Foreign Relations of the United States* series, U.S. State Department
David Levering Lewis	Professor of history, Rutgers University
David McCullough	Author, Pulitzer Prize winner, *Truman*
James McPherson	Professor of history, Princeton University, and Pulitzer Prize winner, **Battle Cry of Freedom**
Holt Merchant	Professor of history, Washington and Lee University
Richard Moe	President, National Trust for Historic Preservation
W. Brown Morton III	Chairman, Department of Historic Preservation, Mary Washington College
Neil Irvin Painter	Professor of American history, Princeton University
Merrill D. Peterson	Professor emeritus and former chairman, History Department, University of Virginia
James L. Robertson Jr.	Professor of history, Virginia Polytechnic and State University
George F. Scheer	Author specializing in colonial and Revolutionary War history
Arthur Schlesinger Jr.	Author of 16 books on American history
William Styron	Author, Pulitzer Prize winner, *The Confessions of Nat Turner*
Dorothy Twohig	Associate professor of history, University of Virginia
Tom Wicker	Former Washington bureau chief of *The New York Times*
Roger Wilkins	Former advisor to President Johnson and professor of history, George Mason University
C. Vann Woodward	Professor emeritus of American history, Yale University

Source: Paul Bradley, "Prominent Historians Join Disney Foes," *Richmond Times-Dispatch*, May 12, 1994, p. B1.

Several historians wrote articles in national publications, attacking the Disney project. C. Vann Woodward, the noted Southern historian, wrote an article for the *New Republic* in which he stated:

What troubles us most is the desecration of a particular region . . . historians don't own history, but it isn't Disney's America either. Nor is it Virginia's. Every state . . . in the country sent sons to fight here for what they believed, right or wrong. They helped make it a national heritage, not a theme park.[22]

The historians and journalists attempted to limit their arguments to the importance of preserving the Piedmont land and its historic heritage. Concerned that they would be regarded as cultural elitists, they tried to avoid

[22] C. Vann Woodward, "A Mickey Mouse Idea," *New Republic*, June 20, 1994, p. 16.

the argument that Disney should not attempt to portray history in a theme park. There were several notable deviations from this strategy, however. McCullough once referred to Disney's plans as "McHistory."[23] Shelby Foote, a Civil War historian, made it clear that he believed Disney would sentimentalize history as it had done to the animal kingdom.[24] Commentator George Will asked facetiously, "Is the idea to see your sister sold down the river, then get cotton candy?"[25] Around this same time period, a lively online discussion took place on the H-Civwar listserve, whose members included academics and historians who were Civil War buffs.[26]

DISNEY'S RESPONSE

Following the May 11 press conference, a Disney spokesperson reiterated the park's intended effect: "Disney's America will bring America's history to life, celebrate America's diversity, provide a road map to other attractions throughout the region, and encourage Americans to go further into their history."[27] Governor Allen also defended Disney after the press conference, saying, "I majored in history. I love history, and I think it's one of the best selling points for tourism. As much as I respect Shelby Foote and enjoyed the *Civil War* series, we shouldn't set ourselves up as censors." He added, "Hopefully, it will get people interested and want to go see the real thing."[28]

At Disney the situation seemed reminiscent of a 1991 controversy over an exhibit on Abraham Lincoln, which was criticized for its cursory treatment of slavery. Disney responded by redesigning the exhibit with the help of Eric Foner, a history professor from Columbia University who had made the complaint. To avoid costs associated with designing and redesigning an entire park based on varying interpretations of history, Disney had already begun to seek advice as early as mid-December 1993.[29] The company turned again to Foner, as well as other historians. As Weis put it, "We all share a common interest to make sure that our treatment of history is sensitive, honest, and balanced."[30] Disney invited a group of historians to Orlando to help them envision what Disney had in mind for Disney's America. Though at first skeptical, some came away thinking that Disney's America might work.[31] James Oliver Horton, a professor of African-American history and American studies at George Washington University, who also designed exhibits for the Smithsonian, took the view that Disney's technological expertise might indeed help audiences learn more about history.[32]

Even with this much foresight, the strength of Protect Historic America's objections caught Eisner off guard. In April, the U.S. Transportation Department had decided to assess the environmental impact of Disney's proposed development. In light of impending federal involvement and PHA's national campaign, Eisner decided to personally visit Washington in mid-June to meet with reporters and editors from the *Washington Post*, Interior Secretary Bruce Babbitt, U.S. House Speaker Thomas Foley, and 30 other legislators. Eisner defended the company's intentions and expressed his frustration openly to the press, saying:

> I'm shocked because I thought we were doing good. I expected to be taken around on people's shoulders.... If this was any other city in the country, the (Federal government) wouldn't even be

[23] Sarah Skolnik, "The Mouse Trapped: Horton Gives a Hoot; Professor James Oliver Horton Retained by Walt Disney Company as a Consultant," *Regardie's Magazine*, September 1994, p. 44.

[24] Bradley, "Prominent Historians Join Disney Foes," p. B1.

[25] Van Dyne, "Hit the Road, Mick," p. 122.

[26] Archived by Avon Edward Foote, "Disney Documents Plus," last modified March 26, 2002, http://www.chotank.com/disvasav2.html.

[27] Bradley, "Prominent Historians Join Disney Foes," p. 1.

[28] Ibid.

[29] Eisner, *Work in Progress*, p. 326.

[30] Skolnik, "The Mouse Trapped," p. 44.

[31] Eisner, *Work in Progress*, pp. 329–31.

[32] Park Net, National Park Service, "More Battles."

interested. . . . (I was unaware) so many wealthy people (lived west of Washington). . . . Disney's America will offer an alternative approach to history that may have more effect on people than conventional history. . . . It's private land that is in the middle of a historic area, but it's not in the middle of a battlefield. . . . We have a right to do it. . . . If people think we will back off, they are mistaken.[33]

Then Eisner threw in at least one other comment that came back to haunt him, "I sat through many history classes where I read some of their stuff, and I didn't learn anything."[34] A few days later, PHA responded with a full-page advertisement in *The New York Times*. The ad headlined "The Man Who Would Destroy American History" reiterated the quote and commented, "Unfortunately, he means it."[35] In an attempt to generate some positive publicity for Disney, Eisner's visit coincided with the Washington movie premier of *The Lion King*. The plan backfired when the event attracted over 100 protesters from the PEC and other organizations, including a couple dressed as lions carrying a sign reading "Michael Eisner, the Lyin' King."

CONGRESSIONAL HEARING

Protect Historic America's leaders next met with several U.S. senators and congressional representatives. As a result, Arkansas Senator Dale Bumpers, chairman of the subcommittee with jurisdiction over national parks, agreed to hold a hearing on the issue, which was held on June 21. McCullough, McPherson, and Moe represented the historians' point of view, while Governor Allen and several Disney executives presented their side of the issue. The historians presented a legal brief prepared *pro bono* for the hearing by a Washington law firm, stating that the Interior Department had a responsibility to investigate the project, given its proximity

to the Manassas battlefield and Shenandoah National Park. They added that Virginia's historic landmarks were threatened, and because the landmarks were national treasures, the federal government had a responsibility to protect them. Allen and the Disney officials countered, arguing that the park was a local land use issue that should be handled within the state of Virginia.

Most of the congresspeople sympathized with Disney, believing that Congress and the federal government should stay out of the situation, and Bumpers said he would take no further action. Although it had no legislative impact, the hearing spurred thousands of newspaper stories, cartoons, and editorials nationwide, greatly increasing national awareness of the issue. Protect Historic America's clipping service pulled over 10,000 items covering the hearing. At this point, national television and radio shows began covering the issue in depth, and political cartoonists were having a field day.

THE DEBATE CONTINUES

Despite outraged or lampooning overtones in the press, a few columnists supported Disney in the debate. For example, columnist Charles Krauthammer wrote,

Those who fear that a children's entertainment will destroy real history have little faith in history. Disney's America is an amusement for kids who bring their parents along for the ride. The issue of urban sprawl is serious. The suggestion of cultural desecration is not. As the kids would say, "Lighten up, guys."[36]

In another instance, William Safire of *The New York Times* called the opposition group "a little band of well-credentialed historians, litigating greens, liberal columnists, and self-protected landowners."[37]

---

[33] William F. Powers, "Eisner Says He Won't Back Down," *Washington Post*, June 14, 1994, p. A1.

[34] Ibid.

[35] Van Dyne, "Hit the Road, Mick," p. 123.

[36] Charles Krauthammer, "Who's Afraid of Virginia's Mouse," *Time*, June 6, 1994, p. 76.

[37] Quoted in Van Dyne, "Hit the Road, Mick," p. 123.

Overwhelmingly, however, press opinion sided with the historians, and the criticism became increasingly vicious over time. Pat Buchanan suggested that Eisner should "take his billions and go back to Hollywood . . . where they are impressed by . . . swagger."

Eisner remained steadfast as he continued in his attempt to build public support for the project. On July 12, *USA Today* printed Eisner's retort to the historians' arguments to build national support for the project. Eisner wrote, "When we began developing plans for a northern Virginia park to celebrate America's heritage, we expected to encounter hurdles. . . . But we did not expect that our creative reputation and talent for educating while entertaining would be attacked with such invectiveness." He continued, "We see Disney's America as a place where people can celebrate America, her people, struggles, victories, courage, setbacks, diversity, heroism, dynamism, pluralism, inventiveness, playfulness, compassion, righteousness, tolerance. . . . [O]ur goal is to instill visitors with a desire to see and learn more."[38]

On September 12, Protect Historic America and the National Trust for Historic Preservation invited journalists and politicians to a program at Ford's Theater in Washington celebrating Virginia as the "Cradle of Democracy." Foote, Styron, and several other authors gave readings from their work, and each guest received a binder of anti-Disney news clippings.[39] Don Henley, co-founder of the rock group the Eagles, also read a brief passage and donated $100,000. Several years earlier, Henley had been involved in the fight to save nineteenth-century American essayist Henry David Thoreau's Walden Pond in Massachusetts. Then, just five days after the Ford's Theater event, the anti-Disney national mall demonstration and the pro-Disney parade in Haymarket took place concurrently.

[38] Michael D. Eisner, *USA Today*, July 12, 1994, p. 10A.

[39] See Protect Historic America, "Reaching the People: News Media Coverage of the Controversy over the Siting of 'Disney's America,'" Washington, D.C., May 11, 1995.

THE DECISION

Eisner watched the beautiful California sunset and pondered the situation. Could he come up with an argument that would sway public opinion in Disney's favor? Would the public tire of the issue, or would the debate continually resurface? How many lawsuits would Disney have to become involved in, and what would be the cost of litigation? What would the historians do once the park opened? Would Disney continually be engaged in a costly process of redesigning exhibits that were objectionable to various factions of historians? Could the park's theme be changed or repackaged?

If Eisner ended the Disney's America project now, the company would upset countless Virginia politicians, including Governor Allen, who had fought on its behalf. The various groups of Piedmont residents who had supported Disney and were counting on the park to provide jobs and tax revenues would be upset as well. Giving up now would mean that Disney had lost a very public, hard-fought campaign. Eisner had said publicly that Disney would not give in, so ending now would risk going back on his word. But were these previous commitments worth the costs of keeping them in light of the vocal opposition and the risk to Disney's reputation?

Eisner considered the options. He had reached the point where he needed to make a decision regarding Disney's America so he could focus more closely on other business concerns.

CASE QUESTIONS

1. What are the key issues that Eisner must consider in this situation from a government relations perspective?
2. Where is Disney most vulnerable, from a communications standpoint?
3. How could Disney have better anticipated the opposition to its new theme park proposal?
4. What advice would you give Eisner?

CHAPTER TEN

Crisis Communication

Unlike many of the other topics covered in this book, a crisis is something *everyone* can relate to. The death of a close relative, the theft of one's car, or even a broken heart—all can become crises in one's personal life. Organizations face crises as well. A 2011 study of companies around the globe found that 59 percent of businesses have experienced a crisis. Tokyo Electric Power Co. (Tepco), for example, experienced a crisis when the earthquake and tsunami that hit Japan in March 2011 disabled its Fukushima Daiichi nuclear power plant. This led to nuclear leaks and mandatory evacuation of the surrounding area. Several universities, charitable organizations, and private investment groups also experienced a crisis in December 2008 when they suddenly learned that they had lost all of the money that they had invested with Bernie Madoff, who had been operating a Ponzi scheme.[1]

Thirty years ago, such events would have received some national attention but would more likely have been confined to the local and regional areas where the events occurred. Today, because of the increasingly digital makeup of the media and the ever present social networking community, a breaking corporate crisis is likely to be reported within minutes by interested individuals via social networks such as Twitter and Facebook. The news will then be reported by bloggers and, of course, by traditional media organizations on their websites as well. Thus, a more sophisticated media environment, as well as a new emphasis on technology in business, has created the need for a more sophisticated *response* to crises.

This chapter first defines what constitutes a crisis. It turns next to a discussion of several prominent crises of the last quarter century. Once we define what crises are all about, the focus shifts to how organizations can prepare for such events. Finally, the chapter offers approaches for organizations to follow when crises do occur.

What Is a Crisis?

Imagine for a moment that you are sleeping in bed on a warm evening in southern California. Suddenly, you feel the bed shaking, the light fixtures swaying, and the house trembling. If you are from California, you know that you are in the middle of an earthquake; if you are from New England, you might think that the world is coming to an end. Now picture yourself on a friend's boat, out for a leisurely sail on a sunny afternoon. Two hours later you discover that you have been having

[1] "2011 Crisis Preparedness Study," Burson-Marsteller and Penn Schoen Berland, http://www.burson-marsteller.com/Innovation_and_insights/Thought_Leadership/default_view.aspx?ID=27.

such a good time that you didn't notice yourself moving farther and farther away from shore into open ocean. Storm clouds are gathering on the horizon, and the sun seems to be mysteriously setting a bit early.

All of us would agree that, in these situations, we as individuals would be facing crises. If the earthquake turns out to be "the big one," or if your friend is a novice sailor and you are in fact drifting into a severe storm, these scenarios could become life-threatening situations.

Organizations also face crises that occur naturally: a hurricane rips through a town, leveling the local waste management company's primary facility; the earthquake we imagined earlier turns the three biggest supermarkets in the area into piles of rubble; a tsunami devastates a coastal area, crippling the local tourism industry for months if not years in its aftermath; a ship is battered at sea by a storm and sinks with a load of cargo destined for a foreign port. Although all of these incidents create havoc and most can't be predicted, they all can be planned for to some degree.

Natural disasters cannot be avoided, but there are many crises—those caused by human error, negligence, or, in some cases, malicious intent—that planning could have prevented in the first place. In fact, most of the crises described later in this chapter—such as the infamous crises that beset Tylenol, Perrier, Pepsi, and several online retailers and banks—were *human-induced crises* rather than natural disasters. Such crises can be more devastating than natural disasters in terms of the costs they entail for companies, in both dollars and reputation.

All human-induced crises cannot be lumped together, however. One type includes cases in which the company is clearly at fault, such as cases of negligence. A perfect example of this is the devastating 2010 oil spill in the Gulf of Mexico that occurred when BP pumped 4.9 million barrels of oil into the water during three long months, putting over 400 species of animals at risk and causing lasting damage to the local fishing and tourism industries. This is a crisis that could have been avoided. Financial or accounting frauds constitute another example of human-induced crises—a type increasingly exposed under the scrutiny required by the Sarbanes-Oxley Act. A devalued stock price and significant legal expenses are not the only aftereffects that a company must address following a human-induced crisis; often the most serious challenge is the damage to the company's reputation and the subsequent loss of trust with key constituencies. Following Toyota's 2010 recall of 9 million vehicles in three separate but related events, its reputation took a major hit. "Toyota is a great company and they'll go on, but that historic concept of superior quality is probably gone forever," said David E. Cole, chairman for Automotive Research in Ann Arbor, Michigan.[2]

A second type of human-induced crisis includes cases in which a company becomes a victim. Examples include Barclays, Citibank, eBay, and other major corporations, which were targeted by online information theft attempts, discussed later in this chapter. The company falls prey to circumstances in these situations, just as when natural disasters unexpectedly hit. A company's role as either the perpetrator or the victim in a crisis is the distinction upon which public perception often

---

[2] Frank Ahrens, "Toyota's Shares Slide as Its Reputation Loses Steam," *The Washington Post*, February 4, 2010.

hinges. The general public's attitude toward a company in crisis is more likely to be negative for crises that could have been avoided, such as the 2008 financial disaster caused by failures in regulation and oversight and fueled by dishonesty and greed, as opposed to crises that organizations had no control over, such as the devastating 2010 earthquake in Haiti that destroyed more than 30,000 commercial properties. Whatever the cause, in all crisis situations, constituencies will look to the organization's *response* to the crisis before making a final judgment. If a company responds well, some crises, such as the Tylenol tragedy, end up actually increasing the overall credibility of the organization involved.

Thus, to define *crisis* for organizations today is a bit more complicated than to simply say that it is an unpredictable, horrible event. For the purposes of this chapter, a crisis will be defined as follows:

> A crisis is a major catastrophe that may occur either naturally or as a result of human error, intervention, or even malicious intent. It can include tangible devastation, such as the destruction of lives or assets, or intangible devastation, such as the loss of an organization's credibility or other reputational damage. The latter outcomes may be the result of management's response to tangible devastation or the result of human error. A crisis usually has significant actual or potential financial impact on a company, and it usually affects multiple constituencies in more than one market.

Crisis Characteristics

Although all crises are unique, they do share some common characteristics, according to Ray O'Rourke,[3] the former managing director for Global Corporate Affairs at the investment bank Morgan Stanley. These include (1) *the element of surprise*—such as Philip Morris finding carcinogens in its filters or Pepsi learning of reports of a syringe found in a Diet Pepsi can; (2) *insufficient information*—the company doesn't have all the facts right away, but very quickly finds itself in a position of having to do a lot of explaining (the Perrier example later in this chapter is instructive here, in that it took the company over a week to figure out what was going on after reports of benzene contamination surfaced); (3) *the quick pace of events*—things escalate very rapidly (even before Exxon's crisis center was up and running after the *Valdez* incident, the state of Alaska and several environmental groups were mobilized); (4) *intense scrutiny*—executives are often unprepared for the media spotlight, which is instantaneous, as answers and results normally take time. Think of how much air time Tony Hayward, then CEO of BP, received in 2010 during the oil spill in the Gulf of Mexico.

What makes crises difficult for executives is that the element of surprise can lead to a loss of control. It's hard to think strategically when overwhelmed by unexpected outside events. In addition, the blogosphere and media frenzy that typically surround a crisis can prompt a siege mentality to ensue, causing management to adopt a short-term focus. Attention shifts from the business as a whole to the crisis alone, forcing all decision making into the shortest time frame. This has

---

[3] Ray O'Rourke, presentation to Corporate Reputation Conference, New York University, January 1997. At the time of this presentation, O'Rourke was with public relations firm Burson-Marsteller.

been true in crisis situations even before the Internet dramatically changed the pace of reporting. For example, in the early 1990s, the public relations firm Burson-Marsteller was hired six days after the Perrier benzene scare began, and already it had to undo three different explanations from the company—none of which were true. Perrier's uncoordinated and off-the-cuff statements only increased the likelihood that the crisis would escalate. When panic sets in, this is typically what happens in organizations.

Part of the problem in dealing with crises is that organizations tend to not understand or acknowledge how vulnerable they are until *after* a major crisis occurs. Lack of preparation can make crises even more severe and prolonged when they do happen. Let's take a closer look at some major crises from the past 25 years to bring our definition to life.

Crises from the Past 25 Years

For baby boomers, the defining crisis of their time was the assassination of President John F. Kennedy. Virtually everyone who was alive at that time can remember what he or she was doing when the news was announced that President Kennedy had been shot. Generation Xers in the United States today probably feel the same way about the explosion of the space shuttle *Challenger* in January 1986. People near the United States and even internationally will remember the terror attacks of September 11, 2001, in New York City as a defining moment for the new millennium. These events have become etched in the public consciousness for a variety of reasons.

First, people tend to remember and be moved by negative news more than positive news. Americans in particular seem to have a preoccupation with such negative news. Network and cable news broadcasts underscore this point. Viewers rarely see "good" news stories because they just don't sell to an audience that has become accustomed to the more dramatic events that come out of the prime-time fare on television.

Second, the human tragedy associated with a crisis strikes a psychological chord with most everyone. Two high-speed trains collide and fall off of a bridge, killing 35 and injuring hundreds in eastern China; Hurricane Katrina devastates New Orleans and the U.S. government bungles its response; a deranged shooter injures 19, 6 fatally, at a 2011 open meeting held in Tucson in a grocery store parking lot by U.S. Representative Gabrielle Giffords—such events make us realize how vulnerable we all are and how quickly events can make innocent victims out of ordinary people.

Third, crises associated with major corporations stick in the public's mind because many large organizations lack credibility in the first place. A public predisposed to distrust big oil companies could not be completely surprised by what happened during BP's Deep Horizon oil spill or by the nationwide sex discrimination class action lawsuit against Walmart, which was the largest civil rights class action ever certified against a private employer. Indeed, these events validated the public's suspicions, and the public appears to have taken as much pleasure in the turmoil that these corporations faced in the aftermath. In other cases, crises have such a significant impact on the public because they take it by surprise. Consider the investment bank Lehman Brothers. Before filing for Chapter 11 bankruptcy, it was one of the four largest investment banks in the world. After the U.S. Government denied aid to Lehman and to other banks during the early days of the subprime mortgage crisis, Lehman filed for bankruptcy on September 15, 2008. At the time, it held assets of $691 billion. It was the largest bankruptcy in U.S. history and affected thousands of jobs and life savings of many. As we look at other major crises, we will start to see more clearly why these events linger in the public psyche.

1982: Johnson & Johnson's Tylenol Recall

Johnson & Johnson's (J&J's) Tylenol recall in the early 1980s is held by many as "the gold standard" of product-recall crisis management. Although over 30 years have passed since the crisis, the lessons to be learned from it are still relevant. Johnson & Johnson's handling of the crisis was characterized by a swift and coordinated response and a demonstration of concern for the public that only strengthened its reputation as "the caring company."

In late September and early October of 1982, seven people died after taking Tylenol capsules that had been laced with cyanide. At the time, Tylenol had close to 40 percent of the over-the-counter market for pain relievers. Within days of the first report of these poisonings, sales had dropped by close to 90 percent.

Certainly the irony of something that is supposed to relieve pain turning into a killer made this episode one of the most memorable in the history of corporate crises. However, many experts on crisis communication, marketing, and psychology have praised Johnson & Johnson for its swift and caring response that was primarily responsible for turning this disaster into a triumph for the company. Despite losses exceeding $100 million, Tylenol came back from the crisis stronger than ever within a matter of years.

What did Johnson & Johnson do? First, it did not simply *react* to what was happening. Instead, it took the offensive and removed the potentially deadly product from shelves. (In the end, 31 million bottles of Tylenol were recalled.) Second, it leveraged the goodwill it had built up over the years with constituencies ranging from doctors to the media and decided to try to save the brand rather than come out with a new identity for the product. Third, the company reacted in a caring and humane way rather than simply looking at the incident from a purely legal or financial perspective. Thousands of J&J employees made over 1 million personal visits to hospitals, physicians, and pharmacists around the nation to restore faith in the Tylenol name.[4] Fourth, when J&J reintroduced Tylenol to the market, the product was packaged in triple-seal tamper resistant packaging.

Why did the company go to these lengths? Despite its decentralized structure, Johnson & Johnson's management is bound together by a document known as the "Credo." The Credo is a 308-word companywide code of ethics that was created in 1935 to boost morale during the Depression, and it is carved in stone at company

---

[4] Harold J. Leavitt, "Hot Groups," *Harvard Business Review*, July 1, 1995, p. 109.

headquarters in New Brunswick, New Jersey, today. It acknowledges: "We believe our first responsibility is to the doctors, nurses, and patients, to mothers and all others who use our products and services." Then-CEO James Burke made sure that the principles of the Credo guided the company's actions during the Tylenol crisis, helping J&J react to tragedy without losing focus on what was most important.

What is most amazing is not that J&J handled this crisis so formidably but that the perception of the company was actually *strengthened* by what happened. As Burke—who was brought in early as the lead person handling the crisis—explained, "We had to put our money where our mouth was. We'd committed to putting the public first, and everybody in the company was looking to see if we'd live up to our pretensions."[5] J&J management did, and the public rewarded them for it. Within three months of the crisis, the company regained 95 percent of its previous market share.[6] More than two decades later, Johnson & Johnson ranks consistently on Interbrand's annual list of the 100 Top Global Brands, with a brand portfolio valued at 4.1 billion.[7]

1990: The Perrier Benzene Scare

Another classic crisis in business history is the 1990 Perrier benzene scare. Perrier Sparkling Water faced a contamination crisis of its own nearly 10 years after the Tylenol episode. Although Perrier's contamination crisis did not lead to any deaths, or even reported illnesses, it still demanded resolution and an explanation from the public and the media. Perrier's actions during the 1990 benzene scare provide as many lessons in how *not* to handle a crisis as J&J's did of how to handle one effectively.

In February 1990, Perrier issued the following press release:

> The Perrier Group of America, Inc. is voluntarily recalling all Perrier Sparkling Water (regular and flavored) in the United States. Testing by the Food and Drug Administration and the State of North Carolina showed the presence of the chemical benzene at levels above proposed federal standards in isolated samples of product produced between June 1989 and January 1990.[8]

This press release marked the beginning of the end of Perrier's reign over the sparkling water industry. In 1989 Perrier, one of the most distinguished names in bottled water, sold 1 billion bottles of sparkling water, riding high on the wave of 1980s health consciousness. Then in January 1990, a technician in the Mecklenberg County Environmental Protection Department in Charlotte, North Carolina, discovered a minute amount of benzene, 12.3 to 19.9 parts per billion (less than what is contained in a non–freeze-dried cup of coffee), in the water.[9] After receiving confirmation from both the state and federal officials, Mecklenberg briefed Perrier Group of America about the contamination.

[5] Brian O'Reilly, "Managing: J&J Is on a Roll," *Fortune*, December 26, 1994, p. 109.

[6] Ibid.

[7] "Special Report: The Best Global Brands," *BusinessWeek*, July 25, 2005.

[8] Perrier press release, The Perrier Group, February 10, 1990.

[9] "When the Bubble Burst," *Economist*, August 3, 1991, p. 67.

Two full days after the crisis broke, after recalling over 70 million bottles from North America (but before identifying the source of the contamination), Perrier America president Ronald Davis confidently announced that the problem was limited to North America. Officials had reported a cleaning fluid containing benzene had been mistakenly used on a production line machine.[10] The real cause of the contamination—defective filters at its spring[11]—was discovered less than three days later, and contrary to what Ronald Davis had previously announced, six months' worth of production would be affected, covering Perrier's entire global market.[12] The firm was forced to change its story.

Without an official crisis plan of its own, Perrier relied on the media to communicate its story during the crisis, which proved to be a fatal decision. The press only served to expose the lack of internal communication and the lack of global coordination within the company. At a news conference in Paris, when Perrier-France announced that it was also issuing a recall due to the presence of benzene, the president of Perrier's international division, Frederik Zimmer, offered the explanation that "Perrier water naturally contains several gases, including benzene."[13] From the contradictory messages released to the press, it was clear that the U.S. operations were not communicating well—if at all—with their European counterparts. Moreover, yet another story emerged to explain the presence of benzene, and it contradicted the previous explanations: According to Perrier officials, "the benzene entered the water because of a dirty pipe filter at an underground spring at Vergeze in southern France."[14] All of this hurt the company's credibility.

The cost of the recall and eventual relaunch of the product—ushered in by a pricey advertising campaign—meant that customers found the new 750-mL bottles selling at the same price as the old one-liter bottles. Perrier's pre-crisis 1989 market share of 44.8 percent plummeted to 20.7 percent by 1991, and the brand has continued to lose ground in the imported mineral water category.

The Perrier benzene crisis illustrates not only the consequences of having a *reactive* strategy to deal with crises but also the problems of not having a coordinated and fact-based approach to crisis communication.

1993: Pepsi-Cola's Syringe Crisis

A third classic case in crisis management is the 1993 Pepsi-Cola syringe contamination scare. In 1993, Pepsi-Cola faced a highly publicized contamination crisis of its own shortly after Perrier's benzene episode. Pepsi's handling of the syringe hoax of 1993 starkly contrasts with the Perrier example. In addition to showing concern for the public and demonstrating resoluteness in getting to the bottom of the problem, Pepsi also skillfully leveraged two other critical constituencies—the government and, most important, the media—to help it combat the bogus tampering claims and win back the public's trust.

[10] Ibid.

[11] "Handling Corporate Crises; Total Recall," *Economist* 335 (June 3, 1995), p. 61.

[12] Ibid.

[13] "Poor Perrier, It's Gone to Water," *Sydney Morning Herald*, February 15, 1990, p. 34.

[14] Ibid.

In June 1993, a man in Washington State reported that, after drinking half a can of Diet Pepsi the night before, he had discovered a syringe in the can the following morning when he shook out the rest of the contents into the sink.[15] This claim was the beginning of a major crisis for Pepsi-Cola.

The CEO of Pepsi-Cola North America, Craig E. Weatherup, did not let the surprise of the crisis overwhelm him when he was contacted at home by Food and Drug Administration (FDA) Commissioner David Kessler and informed of the situation. His first action was to engage Pepsi-Cola's four-person crisis management team—made up of "experienced crisis managers from public affairs, regulatory affairs, consumer relations, and operations"[16]—to swiftly deal with the unfolding situation, including opening lines of communication with FDA regulatory officials, the media, and consumers. Internally, Pepsi prevented organizational chaos by updating employees with daily advisories to over 400 Pepsi facilities nationwide.[17] Unlike Johnson & Johnson's immediate recall of Tylenol from the shelves, by the next morning, Weatherup had decided *not* to recall the product—despite a flood of new reports to the FDA of dangerous objects found in Pepsi cans.

When television networks contacted the company looking for a response or any formal statements, Weatherup realized that the crisis was rooted in the disturbing imagery of syringes in cans—and decided to supply the media with an equally "visual" response. Weatherup had his staff prepare video footage of the canning process at Pepsi that showed how it would be virtually impossible to insert a syringe into the cans. Additionally, Pepsi later distributed a grocery-store surveillance tape of a woman stealthily dropping a syringe into her Pepsi can. After the footage appeared as the lead story on three major networks, no new reports of syringes were made.[18]

Weatherup also made several television appearances throughout the day, on *"The MacNeil/Lehrer News Hour"* and *"Larry King Live."* In his last appearance, FDA Commissioner David Kessler accompanied him. Both men stressed the implausibility of the claims and the criminality of making false statements (five years in prison and up to $250,000 in fines).

Pepsi's highly visible work with the FDA in investigating the crisis boosted its credibility in the public eye. Additionally, without an investigative reporting team of its own, Pepsi found that the government agency was invaluable to the company during the crisis. The FDA established a center in 1989 to provide the agency with a team of forensic science experts who can respond immediately to all tampering incidents and provide expert advice and scientific evidence to FDA officials. It was an FDA investigation that provided the evidence used to convict a tamperer who had falsely claimed to find a mouse inside a Pepsi can when she opened it. Several days later, the FBI arrested four individuals for making false claims, and the contamination scare appeared even more like the hoax it turned out to be. In the end, 20 arrests were made, and the crisis was resolved.

---

[15] David Birkland, "Couple Say They Found Used Needle in Pepsi," *Seattle Times*, June 11, 1993, p. 18.

[16] Sandi Sonnenfeld, "Media Policy—What Media Policy?" *Harvard Business Review*, July 1, 1994, p. 18.

[17] Ibid.

[18] Glenn Kessler and Theodore Spencer, "How the Media Put the Fizz into the Pepsi Scare Story," *Newsday*, June 20, 1993, p. 69.

Pepsi-Cola did not stop there, however. To ensure that consumers knew that the tampering claims were false, Weatherup took out an ad to address the concerns of employees and customers. As he explained, "On Monday, Pepsi will run full-page advertisements in 200 newspapers around the country, including the *Washington Post*. The ad reads: 'Pepsi is pleased to announce . . . nothing. As America now knows, those stories about Diet Pepsi were a hoax. Plain and simple, not true.' It ends with an invitation: 'Drink All The Diet Pepsi You Want. Uh Huh.'"[19] Pepsi-Cola remains one of America's leading soft drinks, with a 29.3 percent market share in 2010,[20] demonstrating the negative publicity generated in crisis situations can be overcome when the crisis is successfully handled.

The New Millennium: The Online Face of Crises—Data Theft and Beyond

With personal computers, the Internet, and smart phones now integral parts of the fabric of business, organizations face new challenges and the potential for crises that they have not dealt with before. Mobile apps have created a new playground for cyber-thieves. Since Apple's App Store opened in July of 2008, Apple has lost $450 million in piracy.[21] The Conficker Virus, targeting the Microsoft Windows operating system, was released in 2009 and cost businesses across a range of industries $9.1 billion.[22] Companies of all kinds are also grappling with information security issues involving the theft or attempted theft of company and customer data.

Although all businesses need to be on guard against these threats, Internet-based businesses in particular are on the front lines of the information security battle. Operation Aurora was a cyber attack launched in 2009 seeking source code from Google, Adobe, and other high-profile Internet-based companies. The hackers used tactics such as encryption, stealth programming, and an unknown hole in Internet Explorer, according to the anti-virus firm McAfee.[23] "We have never, outside of the defense industry, seen commercial industrial companies come under that level of sophisticated attack," said Dmitri Alperovitch, vice president of threat research for McAfee, "It's totally changing the threat model."[24]

Hacking into Reputations

Today, the majority of online thieves are opting for more surreptitious tactics to steal confidential information. Viruses are now more commonly used to plant "Trojans" in personal or office computers. Trojans are malicious software programs that steal sensitive information stored in a computer and relay it back to the criminal. "Phishing" is another popular tactic used by scammers who send spoof (but

---

[19] John Schwartz, "Pepsi Punches Back with PR Blitz; Crisis Team Worked around the Clock," *Washington Post*, June 19, 1993, p. C1.

[20] "Top 10 CSD Results for 2010," *Beverage Digest*, March 17, 2011, http://www.beverage-digest.com/pdf/top-10_2011.pdf.

[21] Garrett W. McIntyre with Phil PacDonald, "Apple App Store has Lost $450 million to Piracy," 24/7 Wall Street, January 13, 2010.

[22] Dancho Danchev, "Conficker's Estimated Economic Costs? $9.1 billion," ZDNET.

[23] Kin Zetter, "Threat Level: Google's Hack Attack was Ultra Sophisticated, New Details Show," *WIRED*, January 14, 2010.

[24] Ibid.

often legitimate-looking) e-mails to customers, posing as well-known and trusted companies and requesting personal information such as account passwords and social security numbers under the auspices of updating the company's online records.

The proliferation of such online security threats has resulted in crisis situations for countless companies worldwide that now must redouble their efforts to defend themselves against attacks. They must successfully protect themselves if they wish to maintain the confidence and trust of their customers. The battle is not an easy one, especially as technological advances enable cyber-criminals to become more creative. The software industry missed out on more than $51 billion in 2009 as a result of software piracy, says a study released by the IDC and the Business Software Alliance (BSA).[25] According to a report by Javelin Strategy and Research, 8.1 million people became victims of identity theft in 2010, and overall losses from fraud that year totaled $37 billion.

Why has this problem reached crisis level for so many companies? Besides the reparations and damages businesses must cover for affected customers, doubts about online security have cast a shadow on many online retailer and banks' corporate reputations. And in the online arena, reputation may indeed be everything. A survey commissioned by KindSight in 2010 found that 65 percent of American consumers are concerned about identity theft as it relates to having their bank account, credit card, or other personal information stolen online. In addition, 34 percent identified phishing as one of their top concerns, with other common concerns being the use of Trojan viruses to hijack personal computers and using them to distribute spam, infections, or child pornography. The survey found that 81 percent of respondents had already been infected at some point in their computing lives.[27] Even marketing can be negatively affected—More than likely, businesses' legitimate marketing e-mails are lost to consumers as a result of their personal defense mechanisms.[28]

What are companies doing to battle back? Some are responding more effectively than others. In May 2003, despite the heightened sensitivity of online consumers, Wachovia sent online banking customers an e-mail asking them to update their user names and passwords by clicking on a link—a widely recognized phishing tactic customers have been trained to ward against. Although the request was legitimate as Wachovia was migrating customers onto a new system, a quarter of its customers questioned the e-mail and flooded Wachovia's call centers with calls.[29] More and more, businesses must understand and function within today's context of increased customer suspicion and make concerted efforts to quell those fears.

[25] Lance Whitney, "Piracy Costs Software Industry $51 billion in '09," CNET, May 12, 2010.

[26] Michelle Singletary, "Identity Theft Statistics Look Better, but You Still Don't Want to Be One," *The Washington Post*, February 9, 2011.

[27] KindSight, "Kindsight Survey Reveals Identity Theft Continues to Be a Major Concern for Consumers," August 24, 2010.

[28] Ibid.

[29] Alice Dragoon, "Fighting Phish, Fakes and Frauds," *CIO*, September 1, 2004.

The most effective reactions have focused on clear, consistent communications disseminated to customers prior to and in the immediate wake of an online attack. Most important, communications should concentrate on consumer education. Citibank, for example, highlights ways to ward off phishing on its website, including a "Spot a Spoof" chart that outlines ways to identify fraudulent e-mails.[30] The company also has taken its preventative measures a step further: In May 2005, Citigroup announced its collaboration with the National District Attorneys Association (NDAA) to work with prosecutors nationwide to develop new strategies for the arrest and prosecution of identity thieves.[31] Companies also have begun banding together and increasing intercorporate dialogues to communicate best practices and experiences and better combat online security incidents. For example, in 2003, the Anti-Phishing Working Group was founded in the United States, with members from more than 400 companies.[32]

Internet service provider EarthLink has led the charge in increasing customer awareness since it became a phishing target in early 2003. In fact, EarthLink makes its consumer education products—including a "ScamBlocker" toolbar—available to all Internet users, not merely EarthLink subscribers, to promote a "better, safer online experience" for all.[33] ScamBlocker displays a rating for each website visited, which alerts users before they enter a page included on EarthLink's blacklist of fraudulent sites. For the greater good of the customer, EarthLink has gone so far as to share its blacklist with eBay to use in its own security toolbar.[34]

Perhaps the most noteworthy example of a data security breach, subsequent crisis, and following reputation-rebuilding process in recent years is that of ChoicePoint. In October 2004, the data broker realized it had been tricked into turning over personal data of hundreds of thousands of its customers to an organized crime ring. The police were informed of the security breach, but ChoicePoint executives made the mistake of failing to tell affected individuals until February 2005; to further exacerbate the crisis, it only told California residents because doing so was required by state law. It wasn't until Congress intervened that the company contacted nearly 130,000 more individuals. A $10 million fine to the FTC and $5 million to victims followed suit.[35]

If this example seems like a textbook example of what not to do in a similar event, it is—at least, in the early stages of the crisis. Since then, ChoicePoint executives have made huge strides in reversing the company's negative reputation and reestablishing it as a leader in privacy and information security. ChoicePoint established an Office of Privacy, Ethics and Compliance intranet site to keep employees informed about all updates and new policies. A new consumer advocacy department works directly with consumers to handle their concerns; a risk and

---

[30] Jeanette Borzo, "E-Commerce—Something's Phishy," *The Wall Street Journal*, November 15, 2004, p. R8.

[31] Citigroup press release, "Citigroup Teams with State and Local Prosecutors to Lock Up ID Thieves," May 5, 2005.

[32] Borzo, "E-Commerce."

[33] EarthLink website, http://www.earthlink.net/software/nmfree.

[34] Dragoon, "Fighting Phish, Fakes and Frauds."

[35] Richard Levick, *Stop the Presses: The Crisis and Litigation PR Desk Reference*, 2nd edition (Ann Arbor, MI: Watershed Press, 2008).

compliance framework that extends across the entire business cinches the company's ability to target messages to internal and external stakeholders. Demonstrating genuine efforts to educate consumers and safeguard against threats is an important first step for a company that needs to rebuild the trust that many of its customers have lost in recent years, troubled by the potential threats of online transactions.

Online Opinions: Louder Than Ever

Data theft is only one type of threat companies need to guard against online. Another dimension of the new face of crisis is how the Internet can be used to create anticorporate, antibrand "communities" in which people can share information, opinions, and grievances about companies. One of the earliest examples of the influence of such sites is the crisis Dunkin' Donuts experienced in the summer of 1999, when a dissatisfied customer used the Internet to share his own bad experience at a Dunkin' Donuts store. When Dunkin' Donuts advertised coffee "your way," this customer was displeased to learn that it did not offer his choice of skim milk. Because the company did not have a corporate website where he could formally lodge a complaint, he created his own, writing: "Dunkin' Donuts sucks. Here's Why."[36]

Although the site started out as a small section of this individual's personal web page, it was not long before Yahoo! picked up the page in its consumer opinion section. Soon, it was generating 1,000 hits a day. Because Dunkin' Donuts had no official forum for customer suggestions or complaints, this fledgling site—out of the company's control—effectively became that forum. The complainant eventually purchased new web space and the domain name www.dunkindonuts.org, moving the discussion to a place with a seemingly official name.[37]

It was a full two years after the site was launched that Dunkin' Donuts purchased it (after first writing a letter to the individual who created it, politely requesting that he close it, and then threatening him with a lawsuit) and built its own corporate website around it. Customers now have a wide variety of options for contacting specific franchise managers or company headquarters via e-mail or toll-free numbers to share feedback.

In the end, Dunkin' Donuts learned the value of offering its own web-based forum for customer feedback and complaints, but it could have mitigated the crisis by acting sooner to take control of the situation. This example demonstrates the power of the Internet to make the voice of one individual louder than that of a major corporation, and it also highlights how search engines are a cost-free means to further raise the visibility of anticorporate sites, however small and "home grown" they may be at first.

Since Dunkin' Donuts' online debacle, outlets for people to share anticorporate ideals have mushroomed. Not only are there websites such as unitedpackages mashers.com, a site for unhappy customers to rant about their experience with UPS and post pictures of damaged parcels, but something as simple as a tweet

---

[36] Joanna Weiss, "Dunkin' Donuts Complaint-Site Saga Shows Business Power of Internet," *Boston Globe*, August 25, 1999, Online Lexis-Nexis Academic, April 2002.

[37] Ibid.

(especially if posted by someone who has many followers) has the ability to inform and sway opinions in an instant. It's important for companies to be present in the major social media hubs to interact directly with their customers. Data collected in 2011 by Kantar Media's Compete says that 66.1 percent of Twitter users say retailer's tweets and accounts have influenced their purchases.[38]

In many ways, advocates and consumers alike now use technology to rally together and fuel or escalate a crisis—posing additional challenges for the corporation in question. Consider the thousands of text messages/phone calls the people of Egypt used to coordinate, stay informed, and communicate with the outside world while forcing out then President Hosni Bubarak.[39] They were able to share pictures and videos and get the rest of the world involved as the revolution unfolded. In this way, the proliferation of online blogs and social networking sites has greatly increased the visibility and reach of all current events, not excluding large corporate disasters. In 2010 alone, 21.4 million websites were created, over 58,000 per day, enabling information to spread like wildfire.[40] These electronic diaries often push a very specific agenda, and this can be an agenda that can tarnish a company's reputation if read and shared by any of the estimated 174 million adults who go online. In fact, a May 2011 Pew Research Center survey revealed that 78 percent of U.S. adults are now online, up from 68 percent in May/June 2005, and 47 percent in May/June 2000.[41] Because postings tend to remain online for long periods of time—and are often not removed at all—blogs also can have a much longer-lasting impact than those transmitted through traditional vehicles such as print media, which is often, recycled to the curb the next day.[42] With data preservation services such as Wayback Machine and Google Cache offering an archive of Internet pages dating back nine years, online information may never disappear entirely.[43]

In 2009, Domino's Pizza experienced the wrath of social media when two employees posted a video on YouTube of themselves preparing food with nasal mucus and ingredients that one of them had stuck up his nose. Within a few days the video had over a million hits, the former employees had felony charges, and Domino's was facing a public relations crisis. Executives decided not to respond aggressively at first, hoping that it would pass over. However, "when you think it's not going to spread, that's when is gets bigger," said Scott Hoffman, chief marketing officer of the social media firm Lotame. As many of the customers were responding to the situation and asking for more information via Twitter, Domino's created a Twitter account to address its customers directly.[44] And then there is

---

[38] Lauren Dugan, "Two of Three Twitter Users say Retailer Tweets Influence Purchase," Media Bistro, June 24, 2011.

[39] Zack Brisson, "Egypt: From Revolutions to Institutions; The Role of Technology in the Egyptian Revolution," *REBOOT*, March 18, 2011.269 10/31/08

[40] Royal Pingdom: Internet 2010 in Numbers, http://royal.pingdom.com/2011/01/12/internet-2010-in-numbers/.

[41] http://www.centerformediaresearch.com/cfmr_brief.cfm?fnl=060621/ http://www.centerformediaresearch.com/cfmr_brief.cfm?fnl=060621.

[42] Caspar van Vark, "Your Reputation Is Online," *Revolution*, March 4, 2004, p. 42.

[43] David Kesmodel, "Not Fade Away—Lawyers' Delight: Old Web Material Doesn't Disappear," *The Wall Street Journal*, July 27, 2005, p. A1.

[44] Stephanie Clifford, "Video Prank at Domino's Taints Brand," *The New York Times*, April 15, 2009

the example of Dell, whose reputation tanked after influential blogger Jeff Jarvis launched his "Dell Hell" diatribe (see Chapter 6).

Although many companies do not yet have an official approach to dealing with bloggers, a good place to start is to identify the most vocal and visible bloggers covering industry topics and then to proactively supply them with accurate corporate information.[45] In 2005, General Motors launched its first official corporate blog, titled the GM "FastLane," written by GM vice president for global product development Bob Lutz.[46] "FastLane" took a bold approach, allowing consumers to post unfiltered feedback about GM, its products, or previous blog postings.[47] Many benefits resulted, including free insights about products that the GM marketing team could use and, even more important, a reputational boost for GM. Customers—particularly savvy Internet users who regularly research companies and products online—appreciate a company's solicitation for candid feedback. In times of crisis, this previously established credibility can be invaluable. The public will more likely give the company the benefit of the doubt, listen to the corporate response before casting judgment, or—at the very least—know where to turn to get the latest information if a crisis strikes. GM regularly uses its blog in noncrisis situations to manage its reputation and to defend itself against "media articles that we considered unfair, unbalanced or uniformed," explained Lutz.[48] With the power of social media and the Internet growing exponentially, companies have no choice but to join the fray and jump online to manage their reputations and to deflect potential crises.

In the "new economy," companies must recognize the increasing influence of the Internet on a growing number of their constituencies (see Chapters 6 and 7 for more on media and investor constituencies, respectively) and the increased consumer concern about online privacy and security. Companies must take their operating environment into account when planning for and handling crises.

These are just some of the other major crises that organizations have faced in the past 10 years:

- In November 2011, Pennsylvania State University football assistant coach Jerry Sandusky was arrested for 40 counts of sex crimes against young boys. The university's administration was accused of ignoring earlier allegations, and public outrage over the university's lack of concern for the victims forced the board of trustees to fire legendary head football coach Joe Paterno and university president Graham Spanier.
- In October 2011, Research In Motion's BlackBerry service was disrupted for four days, leaving millions of its smartphone customers around the globe without access to text messaging, e-mail and the Internet. The company, which prides itself on reliability and provides service to numerous governmental agencies, was criticized for both the outage and for its slow response.

[45] van Vark, "Your Reputation Is Online."

[46] Andrew Bernstein, "The Blogosphere: Separating the Hype from the Reality," *PR News*, July 20, 2005.

[47] Ibid.

[48] Kyle Wingfield, "Blogging for Business," *The Wall Street Journal Europe*, July 20, 2005, p. A9.

- In the third quarter of 2011, Netflix announced that it would be unbundling its DVD delivery and online streaming services and charging for them separately. This followed an unpopular price hike earlier in the summer. Netflix lost 800,000 subscribers in the third quarter and decided to return to the bundled offering. CEO Reed Hastings personally apologized to customers.

- In the fall of 2010, top toy company Mattel recalled 7 million of its Fisher-Price brand tricycles when 10 young children reportedly injured themselves. This followed significant safety recalls in 2009 and 2007. The 2007 recalls were due to lead contamination from Chinese manufacturing.

- On April 20, 2010, an explosion at BP's oil rig, the Deepwater Horizon, killed 11 men working on site and caused the largest accidental marine oil spill in the history of the petroleum industry. Top management at BP was criticized for its offensive apologies and unwillingness to accept responsibility for the spill.[49]

- In 2010 and 2009, Toyota recalled an estimated 9 million vehicles in three separate but related recalls, due to drivers experiencing unintended acceleration. Prior to this crisis, Toyota had been widely admired for its quality in engineering.

- In late 2008, the financial crisis began and resulted in the collapse of major financial institutions, the downturn of the stock market, and the bailout of the largest U.S. banks by the government—all affecting thousands of businesses and families across the globe.

- On April 8, 2008, American Airlines began grounding its M-80 planes due to complications arising from mandatory wiring inspections, which led to more than 3,000 flight cancellations and widespread customer complaints about the company's handling of the crisis.

- In early 2008, German customers boycotted Nokia after 1,000 employees were unexpectedly laid off from its Bochum plant so that business could be outsourced to cheaper countries in Eastern Europe. Many of the employees learned of their layoffs not from Nokia managers, but from radio reports of the company's plans to cut staff and outsource.

- On October 23, 2007, the Federal Emergency Management Agency (FEMA) hosted a press conference on wildfires that had been raging in California. The press conference turned out to have been staged; FEMA employees acted as journalists and pitched easy questions to then-Deputy Director Harvey Johnson.

- In 2007, a wave of pet food recalls due to contaminated ingredients left Menu Foods and ChemNutra in the middle of an image crisis—and an FDA investigation.

- During a snowstorm in mid-February 2007, hundreds of JetBlue passengers were stranded on runways in the New York metro area for up to 10 hours without being deplaned. The company ultimately had to invest millions in reparations and drafted a constitution of passengers' rights, designed to protect JetBlue customers in the event of future weather delays (See chapter 1).

[49] Jeremy Warner, "The Gulf of Mexico Oil Spill Is Bad, but BP's PR Is Even Worse—Telegraph." Telegraph.co.uk—Telegraph Online, Daily Telegraph and Sunday Telegraph—Telegraph, June 18 2010, http://www.telegraph.co.uk/finance/newsbysector/energy/oilandgas/7839136/The-Gulf-of-Mexico-oil-spill-is-bad-but-BPs-PR-is-even-worse.html.

- In 2007, Whole Foods CEO John Mackey was caught denouncing his competitors—including one that he was in discussions to purchase—on Yahoo! financial message boards. The messages, which he posted anonymously, dated back seven years; his reputation (and, in turn, that of the company) suffered from his lack of transparency.

- In 2006, the University of Michigan suspended the sale of all Coca-Cola products on its campus after environmental concerns in India and labor issues in Colombia became public.

- On August 29, 2005, Hurricane Katrina made landfall in New Orleans and the city was devastated. The U.S. government, in particular its Federal Emergency Management Agency (FEMA), was widely criticized for its poor handling of the crisis.

- In June 2005, credit-processing firm CardSystems was hacked, exposing 40 million credit card account numbers from Visa, MasterCard, American Express, and Discover Financial—constituting one of the biggest breaches of consumer data security in history.

- In 2005, Boeing Co.'s board of directors forced out CEO Harry C. Stonecipher over his consensual extramarital affair with an employee.

- In September 2004, Merck & Co. recalled pain medication Vioxx in response to a study revealing that the risk of heart attacks and strokes tripled in individuals who had taken the drug for periods of more than 18 months.

- In 2002, accounting giant Arthur Andersen was convicted of obstruction of justice for shredding tons of documents related to long-time client Enron Corp. Although the Supreme Court overturned the ruling three years later in May 2005, Arthur Andersen's reputation was irreparably tarnished, and its workforce plummeted from 85,000 to fewer than 200.[50] The firm no longer exists.

How to Prepare for Crises

The first step in preparing for a crisis is to understand that any organization, no matter what industry or location it is in, can find itself involved in the kinds of crises discussed in the previous section. The 2011 Crisis Preparedness Study conducted by Burson-Marsteller and Penn Schoen Berland found that 79 percent of business leaders believe that their companies are only 12 months away from a potential crisis. Although those crises listed in this chapter may be some of the most noteworthy ones from recent history, those left out were likely just as devastating to the companies involved. Obviously, some industries—the chemical industry, Big Pharma, consumer packaged goods, mining, forest products, energy-related industries such as oil and gas and electric utilities, and online retailers—are more crisis-prone than others, but today, every organization is at risk.

The terror attacks of September 11, 2001, proved to be an important test of many companies' crisis plans. For other companies, the attacks underlined the impor-

[50] Diya Gullapalli, "Andersen Decision Is Bittersweet for Ex-Workers," *The Wall Street Journal*, June 1, 2005, p. A6.

tance of having a plan in place. A survey of nearly 200 CEOs conducted by Burson-Marsteller and *PR Week* magazine in late 2001 revealed that a full 21 percent of CEOs surveyed "had no crisis plan and were caught unprepared" by the events of September 11. Fifty-three percent acknowledged that their plan was good but "not totally adequate for such events." In response to the question of whether they had readdressed their crisis communication plan since the September 11 disaster, 63 percent indicated that they intended to do so.[51] According to a recent survey, crisis management is one of the top priorities among most *Fortune* 1000 companies' senior managers. According to respondents, this is mostly due to a recent crisis in their own company or those witnessed in the media, and an increased sense of vulnerability to natural disasters.[52]

Many companies located in the World Trade Center also had been tenants of the Twin Towers at the time of a previous terrorist attack. In 1993, an explosion blew out three of the underground floors of the World Trade Center, forcing the evacuation of more than 30,000 employees and thousands of visitors from the entire complex and a rescue operation lasting 12 hours.[53] After the 1993 bombing, many organizations developed or refined their evacuation plans from the Trade Center. When the second attack occurred in 2001, this preparation helped save many lives.

For example, the World Trade Center's largest tenant, Morgan Stanley Dean Witter, cited its own evacuation plan as critical to saving the lives of all but 6 of its 3,700 employees on September 11. A Morgan spokesman attributed the smooth evacuation to companywide familiarity with the plan: "Everybody knew about the contingency plan. We met constantly to talk about it."[54]

Communications managers must follow these examples and prepare company management for the worst by using anecdotal information about what has happened to unprepared organizations in earlier crises. There are so many to choose from that managers should not be hard pressed to find crisis examples in virtually every industry from experiences over the last 25 years. Once the groundwork is laid for management to accept the notion that a crisis is a possibility, real preparation should take the following form.

Assess the Risk for Your Organization

As mentioned earlier, some industries are more prone to crises than others. But how can organizations determine whether they are more or less likely to experience a crisis? Publicly traded companies are at risk because of the nature of their relationship with a key constituency—shareholders. If a major catastrophe hits a company that trades on one of the stock exchanges, the likelihood of a selloff in the stock is enormous. Such immediate financial consequences can threaten the

[51] Jonah Bloom, "CEOs: Leadership through Communication—The *PR Week* and Burson-Marsteller CEO Survey 2001 Finds U.S. Corporate Leaders Emulating the Strong, Open, Communicative Style of Rudy," *PR Week*, November 26, 2001, pp. 20–29.

[52] http://www.disaster-resource.com/articles/98nuggs.shtml.

[53] Carol Carey, "World Trade Center," *Access Control & Security Systems Integration*, July 1, 1997.

[54] Daren Fonda, "Girding against New Risks: Global Executives Are Working to Better Protect Their Employees and Businesses from Calamity," *Time*, October 8, 2001, p. B8.

organization's image as a stable ongoing operation in addition to the damage the crisis itself inflicts.

Although privately held companies do not have to worry about shareholders, they do have to worry about the loss of goodwill—which can affect sales—when a crisis hits. Often the owners of privately held companies become involved in communication during a crisis to lend their own credibility to the organization. So all organizations—public, private, and not-for-profit—are at some risk if a crisis actually occurs. The next section examines how a company can plan for the worst no matter what.

Plan for Crises

The person in charge of corporate communication should first call a *brainstorming session* that includes the most senior managers in the organization as well as representatives from the areas that are most likely to be affected by a crisis—for example, the head of manufacturing in some cases because of the potential for industrial accidents in the manufacturing process. It also might include the chief information officer because of the danger to computer systems when accidents happen. In the case of the explosion during the first World Trade Center attack in 1993, most of the organizations were service organizations. After the loss of lives, the loss of critical information was one of the worst outcomes of the explosion.

During the brainstorming session, participants should work together to develop ideas about potential crises. They should be encouraged to be as creative as possible during this stage. The facilitator should allow participants to share their ideas, no matter how outrageous, with the group and should encourage all participants to be open-minded as they think about possible crisis scenarios.

Once an inventory of possible crises exists, the facilitator should help the group to determine which of the ideas developed have the most potential to actually occur. It might be useful, for example, to ask the group to assign probabilities to the potential crises so that they can focus on the more likely scenarios rather than wasting time working through solutions to problems that have a very low probability of occurring. But even at this stage, participants must not rule out the worst-case scenario. The risk for an oil spill the size of the BP spill occurring was very low according to outside projections. Thus, neither the oil company nor governmental agencies prepared for the worst possible accident.

Determine Effect on Constituencies

Once the probability of risk has been assigned to potential crises, organizations need to determine *which constituencies would be most affected by the crisis*. Crisis communication experts spend too little time thinking about this question. Why is it so important? Because some constituencies are more important than others, organizations need to look at risk in terms of its effect on the most important constituencies.

When the World Trade Center came under attack on September 11, 2001, American Express CEO Ken Chenault phoned the company's headquarters across the street from the World Trade Center and instructed building security to evacuate employees immediately. As the day wore on, he contacted all his senior executives to check on their well-being.[55] Until Chenault was able to relocate the company's 3,000 employees to a new building across the river, AmEx's in-house communications staff worked from their homes to reach out to customers and let them know the company was open for business.[56] Two concerns guided Chenault in his actions following this crisis: employee safety and customer service.[57] Employees and customers, in this example, were the constituencies determined to be most affected by these events, and Chenault's actions reflected this determination.

Determining how to rank constituencies when a crisis actually happens is more difficult because so many other things are going on. But thinking about risk in terms of effect on constituencies in advance helps the organization further refine which potential crises it should spend the most time and money preparing for During the Tylenol crisis, for example, Johnson & Johnson could rely on its Credo to help the company set clear priorities and deal with its constituencies.

Set Communication Objectives for Potential Crises

Setting communication objectives for potential crises is different than figuring out how to deal with the crisis itself. Clearly, organizations must do both, but typically managers are more likely to focus on what kinds of things they will do during a crisis rather than what they will say and to whom. Communication takes on more importance than action when the crisis involves more intangible things such as the loss of reputation rather than the loss of lives.

Analyze Channel Choice

Once the ranking of constituencies is complete, the participants in a planning session should begin to think about what their communication objective will be for each constituency. Whether this objective will be achieved often depends on the effectiveness of the communication *channel* the company selects to convey the message.

Perhaps the mass distribution of a memo would be too impersonal for a message to employees in a time of crisis. The company might consider personal or group meetings or a "town hall" gathering instead. The choice of communication channel often can reflect how sensitive a company is to its constituencies' needs and emotions. What would be the most efficient and most sensitive way to communicate with consumers or their families during a crisis? Johnson & Johnson's caring and highly personalized reaction to the Tylenol crisis—involving a host of personal visits to hospitals and pharmacies nationwide—won the company significant goodwill. In a time of crisis, constituencies crave information and are often more sensitive than usual to how information is conveyed to them. In the case of the Kryptonite lock-picking debacle, when a customer posted a "how-to" video showing how to use a ballpoint pen to unlock a Kryptonite lock, Kryptonite issued a generic statement describing its current line of locks as serving as "deterrent to

[55] Bloom, "CEOs: Leadership through Communication."

[56] "Corporate America's Reaction," *PR Week*, September 24, 2001, p. 10.

[57] Bloom, "CEOs: Leadership through Communication."

theft" and noting that the new line of locks promised to be "tougher."[58] Hundreds of bloggers were unsatisfied with the empty answer and continued to write about the locks, prompting hundreds of thousands more to read about them, online and in print via *The New York Times* and Associated Press stories.[59] An estimated 1.8 million people read at least one blog posting about Kryptonite throughout the crisis, largely because Kryptonite failed to stage a swift, cohesive online response effort to battle the bloggers head on in their own forum.[60]

Assign a Different Team to Each Crisis

Another important part of planning for communicating in a crisis is determining in advance who will be on what team for each crisis. Different problems require different kinds of expertise, and planners should consider who is best suited to deal with one type of crisis versus another. For example, if the crisis is likely to have a financial focus, the chief financial officer may be the best person to lead a team dealing with such a problem. He or she also may be the best spokesperson when the problem develops. On the other hand, if the problem is more catastrophic, such as an airline crash, the CEO is probably the best person to put in charge of the team and to serve, at least initially, as head spokesperson for the crisis. In crises that result in loss of life, anyone other than the CEO will have less credibility with the general public and the media.

But managers should avoid putting senior-level executives in charge of communications for *all* crises. Sometimes the person closest to the crisis is the one people want to hear from. For example, the best spokesperson for a global company may be someone located in the country where the problem develops rather than a more senior manager from the head office due to considerations such as cultural issues, language differences, and local community concerns.

Assigning different teams to handle different crises helps the organization put the best people in charge of handling the crisis and communications. It also allows the organization to get a cross section of employees involved. The more involved managers are in planning and participating on a team in a crisis, the better equipped the organization will be as a whole.

Plan for Centralization

Although organizations can employ either a centralized or decentralized approach to corporate communication for general purposes (as we discussed in Chapter 3), when it comes to crisis, the approach must be completely centralized.

Conflicting stories from Perrier's U.S. and European divisions created problems in the company's handling of the benzene contamination scare, further compounding that crisis. Decentralized organizations often find it more difficult to communicate efficiently between divisions, especially if they have not given interdivisional communication full consideration in a crisis-planning phase. Planning for central-

[58] David Kirkpatrick, Daniel Roth, and Oliver Ryan, "Why There's No Escaping the Blog," *Fortune*, January 10, 2005, p. 44.
[59] Ibid.
[60] Ibid.

Pearson and Mitroff's Crisis Management Strategic Checklist

STRATEGIC ACTIONS

1. Integrate crisis management into strategic planning processes.
2. Integrate crisis management into statements of corporate excellence.
3. Include outsiders on the board and on crisis management teams.
4. Provide training and workshops in crisis management.
5. Expose organizational members to crisis simulations.
6. Create a diversity or portfolio of crisis management strategies.

TECHNICAL AND STRUCTURAL ACTIONS

1. Create a crisis management team.
2. Dedicate budget expenditures for crisis management.
3. Establish accountabilities for updating emergency policies/manuals.
4. Computerize inventories of crisis management resources (e.g., employee skills).
5. Designate an emergency command control room.
6. Assure technological redundancy in vital areas (e.g., computer systems).
7. Establish working relationship with outside experts in crisis management.

EVALUATION AND DIAGNOSTIC ACTIONS

1. Conduct legal and financial audits of threats and liabilities.
2. Modify insurance coverage to match crisis management contingencies.
3. Conduct environmental impact audits.
4. Prioritize activities necessary for daily operations.
5. Establish tracking system for early warning signals.
6. Establish tracking system to follow up past crises or near crises.

COMMUNICATION ACTIONS

1. Provide training for dealing with the media regarding crisis management.
2. Improve communication lines with local communities.
3. Improve communication with intervening stakeholders (e.g., police).

PSYCHOLOGICAL AND CULTURAL ACTIONS

1. Increase visibility of strong top management commitment to crisis management.
2. Improve relationships with activist groups.
3. Improve upward communication (including "whistle-blowers").
4. Improve downward communication regarding crisis management programs/accountabilities.
5. Provide training regarding human and emotional impacts of crises.
6. Provide psychological support services (e.g., stress/anxiety management).
7. Reinforce symbolic recall/corporate memory of past crises/dangers.

Source: Christine Pearson and Ian Mitroff, "From Crisis Prone to Crisis Prepared: A Framework for Crisis Management," *Academy of Management Executive* 7, no. 1 (1993), pp. 48–59.

ization can help strip away layers of bureaucracy, keep lines of communication open throughout the organization, and dissipate conflict, all of which are especially critical in a crisis.

What to Include in a Formal Plan

Every communications consultant will suggest that you develop a detailed plan for use in a crisis. These are formal in the sense that they are typically printed up and passed around to the appropriate managers, who may have to sign a statement swearing that they have read and agree to the plan. This step allows the organization to ensure that the plan has been acknowledged by the recipients and permits questions and clarifications to be discussed *in a noncrisis environment*. The last thing you want to happen is for a plant manager's first read of the plan to be when a real crisis occurs.

Research on crisis planning shows that the following information is almost always included in a crisis plan.

A List of Whom to Notify in an Emergency

This list should contain the names and numbers of everyone on the crisis team as well as numbers to call externally, such as the fire and police departments. The list should be kept updated as people leave the company or change responsibilities.

An Approach to Media Relations

Frank Corrado, the president of a firm that deals with crisis communications, suggests that the cardinal rule for communicating with all constituencies in a crisis should be "Tell It All, Tell It Fast!"[61] To a certain extent, this recommendation is true, but one should be extremely careful about applying such a rule too quickly to the media. Perhaps a friendly amendment to Corrado's rule might be, "Tell as much as you can, as soon as you can," so that you do not jeopardize the credibility of the organization. For example, Perrier's hasty communication with the media, in the absence of accurate information, was a crippling mistake.

If the organization has done a good job of building relations with the media when times are good, reporters will be more understanding when a crisis occurs. Having a reserve of goodwill with the media is what helped Johnson & Johnson during the Tylenol crisis. Generally, the person who has the best relationships with individual reporters is probably the right person to get involved with them during a crisis. By agreeing ahead of time that all crisis-related inquiries will go to a central location, organizations can avoid looking disorganized.

A Strategy for Notifying Employees

Employees should be seen as analogous to families in a personal crisis. Employees finding out from the media about something that affects the organization can be likened to a family member hearing about a personal problem from an outsider. An organization should take pains to ensure that a plan for employee notification is created with employee communication professionals in advance and is included in the overall crisis plan.

A Location to Serve as Crisis Headquarters

Although consultants and experts who have written about crises suggest that companies need to invest money in a special crisis center, all companies really need to do is identify ahead of time an area that can easily be converted to such an operation. A contingency location should be determined in the event of a natural disaster

---

[61] Frank Corrado, *Media for Managers* (New York: Prentice Hall, 1984), p. 101.

or terrorist attack affecting the safety or security of the chosen location. Gathering the appropriate technology (e.g., computers, fax machines, cell phones, hookups for media transmissions) as quickly as possible when a crisis hits is also important. This headquarters location should be shared ahead of time with all key internal and external constituencies. All information ideally should be centralized through this office. Other lines of communication should then flow through the headquarters for the duration of the crisis.

A Description of the Plan

Companies should have their crisis plans documented in writing. In addition to communication strategy, a crisis plan should address logistical details, for example, how and where the families of victims should be accommodated in the case of an airline crash.

Following the development of an overall plan, all managers should receive training about what to do if and when a crisis strikes. Several public relations firms and academic consultants now offer simulations that allow managers to test their crisis management skills in experiential exercises. Companies including MasterCard, Southwest Airlines, and General Motors use simulations to help their organizations work out the kinks before a real crisis hits.[62] Managers searching for the right training should be sure that the simulation or training session includes a heavy emphasis on communication in addition to management of the crisis itself.

Beyond managers, all employees should be versed in and trained regularly on the company's emergency procedures and plans. Involve all employees in continuity of business tests; although a genuine crisis cannot be simulated, test runs will help ensure familiarity with emergency plans throughout all levels of the organization. British Airways (BA) conducts a companywide crisis simulation exercise every 12 to 18 months. Guiding those trial runs is BA's crisis manual—200 pages outlining employee roles, responsibilities, and actions in the event of an emergency; third-party contact information; press release templates; as well as maps and key information on BA's fleet and partners.[63]

Communicating during the Crisis

All the planning that an organization can muster will only partially prepare it for an actual crisis. The true measure of success is how it deals with a problem when it occurs. If the plan is comprehensive enough, managers will at least start from a strong position. What follow are the most important steps to take when communicating during a crisis. Every crisis is different, which means that managers must adapt these suggestions to meet their needs, but crises have enough common elements for this prescription to be a starting point for all crisis management.

---

[62] "Crises: In-House, in Hand," *PR Week*, January 21, 2002, p. 13.

[63] Mary Cowlett, "Crisis Training: Prepared for Anything?" *PR Week*, May 6, 2005, p. 25.

Step 1: Get Control of the Situation

The first step is for the appropriate manager to get control of the situation as soon as possible. Such control involves defining the real problem with the use of reliable information and then setting measurable communication objectives for handling it. Failing to take this seemingly obvious, but crucial, first step can be devastating to crisis management efforts, as seen in the Perrier case. Perrier lacked sufficient information to *define* its benzene problem in the first place—though its spokespeople tried to convince the public otherwise—which only compromised its attempts to mitigate the crisis.

When a crisis erupts, everyone in the organization should know who needs to be contacted, but in large global organizations, this knowledge is often unrealistic. Therefore, the corporate communication department can initially serve as a clearinghouse. The vice president for corporate communication at the head office should know the composition of crisis teams and can then turn the situation over to the appropriate manager.

Step 2: Gather as Much Information as Possible

Understanding the problem at hand is the right place for communicators to begin dealing with a crisis. This understanding often involves managing information coming from many sources.

As information becomes available, someone should be assigned to mine that information: If it is an industrial accident, how serious is it? Were lives lost? Have families already been notified? If the incident involves an unfriendly takeover, what are the details of the offer? Was it absurdly low? Have any plans been made for the company to defend itself?

Many corporations have been criticized for reacting too slowly during a crisis because they were trying desperately to gather information about the incident. If it is going to take longer than a couple of hours to get the right information, a company spokesperson should communicate this delay to the media and other key constituencies right away to make it clear that the company is not stonewalling. No one will criticize an organization for trying to find out what is going on, but a company can face harsh treatment if its constituencies think that management is deliberately obstructing the flow of information.

Step 3: Set Up a Centralized Crisis Management Center

At the same time managers are getting in touch with the right people and gathering information, they also should be making arrangements for creating a crisis center as described earlier in this chapter. This location will serve as the platform for all communications during the crisis. Organizations also should provide a comfortable location for media to use during the crisis, including adequate computers or Internet hookups, phones, fax machines, and so on. All communications about the crisis should come from this one, centralized location.

Step 4: Communicate Early and Often

The organization's spokesperson needs to say whatever he or she can as soon as possible. Particularly if the crisis involves threat to lives and property, communicators should try to shield constituencies from panic by allaying some of the probable fears that people will have about the situation. Employees, the media, and other important constituencies should know that the crisis center will issue updates at regular intervals until further notice. Even if they retain public relations firms to assist them in handling a crisis, companies need to put good *inside* people on the front lines of crisis communication and should encourage managers to adopt a team approach with others involved.

Above all else, companies must avoid silence and delayed responses or their constituencies will fill any information voids with criticism and rumors, using powerful tools such as Twitter, Facebook, and blogs. Research In Motion (RIM), the maker of BlackBerry, discovered this lesson the hard way during its unprecedented global service outage that began on Monday, October 10, 2011, and affected millions of users. The company issued infrequent and vague statements as the outage continued and spread. One customer wrote on RIM's BlackBerry Facebook page on Wednesday, October 12, 2011: "totally appalled at the lack of communication from RIM. Love my Berry, but furious at the fact that no one can actually give a time frame of how long it's going to take to fix. Utterly disappointed!"[64] Countless BlackBerry users echoed these sentiments across the blogosphere. Many pointed out the irony that a high-tech communications company was doing such a poor job of communicating with its constituencies.

It was not until Thursday, October 13, 2011, a seeming eternity to those affected, that RIM CEO Mike Lazaridis, issued an apology and personal update via video posted on YouTube. This apology was viewed by many as "too little, too late" and was criticized as "insincere." Critics called for changes in RIM top management. It didn't help that the outage came at a time of eroding market share for BlackBerry and misfires by RIM, including a failed tablet launch. One British customer, upon viewing Lazaridis' apology, wrote:

> "In today's world of online news and social media, RIM's response was quite simply out of touch and inadequate. I will be changing my provider not because I'm petulant, or frustrated by the outage, but because the company looked rudderless and were clearly panicking over a period of not just 24 hours, but days. No one is still able to say with assurance what is wrong, no one appears to be issuing statements for market confidence."[65]

Communicating early and often is much easier said than done. As Larry Kamer, chairman of GCI Kamer Singer, notes that "nine and a half times out of 10 you have to communicate before the facts are in."[66] So companies need to communicate values, such as concern for public safety, and to show a commitment to coming to the aid of people affected by the crisis, even if they do not have all the details yet.

---

[64] Alastair Sharp and Georgina Prodhan, "RIM scrambles to End Global BlackBerry Outage," Reuters.com. October 12, 2011.

[65] "BlackBerry Outage: RIM Boss's YouTube Apology in Full, with Transcript," *The Guardian*, October 13, 2011, http://www.guardian.co.uk/technology/2011/oct/13/blackberry-outage-rim-apology-youtube.

[66] John Frank, "What Can We Learn from the Ford/Firestone Tire Recall? As John Frank Explains, Unlike the Tylenol Crisis, the Problem Is That They Just Can't Seem to Put a Lid on It," *PR Week*, October 9, 2000, p. 31.

Step 5: Understand the Media's Mission in a Crisis

Members of the media work in an extremely competitive environment, which explains why they all want to get the story first. They are also more accustomed to a crisis environment in their work. What they are looking for is a good story with victims, villains, and visuals.

The Pepsi syringe hoax had all of these sensational elements. As we have seen, CEO Craig Weatherup recognized the impact that visuals would have in reassuring the public that the tampering claims it was facing were simply impossible. The video footage of Pepsi canning procedures and the grocery-store surveillance tape, shown on television, and the full-page newspaper ad are all examples of Pepsi's using the media to help it beat a crisis.

Step 6: Communicate Directly with Affected Constituents

Using the media to get information out is critical for companies, but it's even more important that they communicate with their employees, sales staff, organized leadership, site security, operators, and receptionists, as these will be the media's best sources of information in the crisis. External constituencies need to be contacted as well. These include the other three key constituencies besides employees—customers, shareholders, and communities—as well as suppliers, emergency services, experts, and officials. All available technologies should be employed to communicate with them, including e-mail, voicemail, faxes, direct satellite broadcasts, and online services.

Several companies received praise from their customers for their direct communications surrounding the March 2011 hacking of mass e-mail service Epsilon's address files. Epsilon provided mass e-mail service for many respected companies, including Brookstone, Kroger, Marriott Rewards, and The College Board. Upon learning of the security breach, Best Buy, for example, quickly posted press releases and sent e-mails to customers informing them of the security breach. These e-mails contained clear details about what had happened; what data had and had not been compromised; the steps that Best Buy was taking to investigate the breach; and resources for further information.[67]

When designing the communication plan, companies also should consider which constituencies are the top priority, and what information is the most important for each constituency to receive first. Companies should also keep in mind the increasingly blurred lines between constituencies and should consider that any communication meant for one constituency may not be read exclusively by that constituency.

Step 7: Remember that Business Must Continue

To the managers involved, the crisis will most certainly be uppermost in their minds for the duration, but to others, the business must go on despite the crisis. In

---

[67] "Epsilon Hacking Exposes Customers of Best Buy, Capital One, Citi, JPMorgan Chase and Others," *Los Angeles Times*, April 4, 2011.=, http://latimesblogs.latimes.com/technology/2011/04/epsilon-cutsomer-files-email-addresses-breached-including-best-buy-jpmorgan-chase-us-bank-capital-on.html.

addition to finding suitable replacements ahead of time for those who are on the crisis team, managers must try to anticipate the effects of the crisis on other parts of the business. For example, if an advertising campaign is under way, should it be stopped during the crisis? Have financial officers stopped trading on the company's stock? Will it be necessary for the organization to move to a temporary location during the crisis? These and other questions related to the ongoing business need to be thought through by managers both on and off the crisis team as soon as possible.

Step 8: Make Plans to Avoid Another Crisis Immediately

After the crisis, corporate communications executives should work with other managers to ensure the organization will be even better prepared the next time it faces a crisis. Companies that have experienced crises are more likely to believe that such occurrences will happen again and also will recognize that preparation is key to handling crises successfully.

Johnson & Johnson's experience in 1982 helped the company to deal with another episode of Tylenol contamination four years later when a New Yorker died after taking cyanide-laced Tylenol capsules. There is no better time than the period immediately following a crisis to prepare for the next one, because motivation is high to learn from mistakes made the first time.

Conclusion *Webster's Dictionary* traces the word *crisis* back to the Greek word *krisis*—meaning a decision, from the verb *krinein*, to decide.[68] Today we know crises as pivotal times of instability during which leadership and decision making can determine the ultimate outcome of the situation—for better or for worse. As we've seen, sometimes companies can emerge even more respected in the wake of a well-handled crisis.

In this chapter, we explored some real-life examples of how companies across a number of industries dealt with crises of their own and saw that planning and preparation are key to effective crisis management and communication. As British author Aldous Huxley put it, "The amelioration of the world cannot be achieved by sacrifices in moments of crisis; it depends on the efforts made and constantly repeated during the humdrum, uninspiring periods, which separate one crisis from another, and of which normal lives mainly consist."[69]

[68] *Merriam-Webster Online Dictionary*, http://www.merriam-webster.com.

[69] Aldous Huxley, *Grey Eminence: A Study in Religion and Politics* (London: Chatto & Windus, 1941), chapter 10.

Case 10-1

Coca-Cola India

On August 20, 2003, Sanjiv Gupta, president and CEO of Coca-Cola India, sat in his office contemplating the events of the last two weeks and debating his next move. Sales had dropped by 30–40 percent[1] in only two weeks on the heels of a 75 percent five-year growth trajectory and 25–30 percent[2] year-to-date growth. Many leading clubs, retailers, restaurants, and college campuses across India had stopped selling Coca-Cola.[3] Only six weeks into his new role as CEO, Gupta was embroiled in a crisis that threatened the momentum gained from a highly successful two-year marketing campaign that had given Coca-Cola market leadership over Pepsi.

On August 5, the Center for Science and Environment (CSE), an activist group in India focused on environmental sustainability issues (specifically the effects of industrialization and economic growth), issued a press release stating: "12 major cold drink brands sold in and around Delhi contain a deadly cocktail of pesticide residues" (see Exhibit 10.1). According to tests conducted by the Pollution Monitoring Laboratory (PML) of the CSE from April to August, three samples of 12 PepsiCo and Coca-Cola brands from across the city were found to contain pesticide residues surpassing global standards by 30–36 times, including

lindane, DDT, malathion, and chlorpyrifos (see Exhibit 10.2). These four pesticides were known to cause cancer, damage to the nervous and reproductive systems, birth defects, and severe disruption of the immune system.[4]

In reaction to this report, the Indian government banned Coke and Pepsi products in Parliament, and state governments launched independent investigations, sending soft drink samples to labs for testing. The Coca-Cola Bottling Company (Coke) stock dipped by five dollars on the New York Stock Exchange from $55 to $50 in the six sessions following the August 5 disclosure, as did shares of Coca-Cola Enterprises (CCA).[5]

Pepsi and Coca-Cola called the CSE allegations "baseless" and questioned the method of testing, but the CSE claimed it had followed standard procedures documented by the U.S. Environmental Protection Agency, including gas chromatography and mass spectrometry. Pepsi's own tests conducted at an independent laboratory showed no detectable pesticides and led Pepsi to file a petition with the high court questioning the credibility of the CSE's claims,[6] while Coke's Gupta commented: "The allegation is serious and it has the potential to tarnish the image of our brands in the country. If this continues, we will consider legal recourse."[7]

Despite Coke and Pepsi's early responses denying the validity of the CSE's claims and threats of legal action, a survey conducted in Delhi a few days after the CSE announcement found that a

[1] "Toxic Effect: Coke Sales Fall by a Sharp 30–40%," *Economic Times*, August 13, 2003, p. 1.

[2] "Controversy-Ridden Year for Soft Drinks," *Business Line* (New Delhi), December 30, 2003, p. 6.

[3] "Toxic Effect."

Source: This case was prepared in 2005 by Jennifer Kaye, T'05, under the supervision of Professor Paul A. Argenti. The author wishes to thank Nymph Kaul for her research assistance and Rai University for its financial support in the development of this case, which was written with the cooperation of Coca-Cola India. © 2008 Trustees of Dartmouth College. All rights reserved. For permission to reprint, contact the Tuck School of Business at 603-646-3176.

[4] Center for Science and Environment, press release, "Hard Truths about Soft Drinks." August 5, 2003.

[5] "No Standards for World-Wide Pesticide Residues in Soft-Drinks," *Business Line* (New Delhi), October 3, 2003, p. 9.

[6] "Coke & Pepsi in India: Pesticides in Carbonated Beverages," http://www.vedpuriswar.org/articles/Indiancases.

[7] "Tests Show Pesticides in Soft Drinks, Claims CSE," *Economic Times*, August 6, 2003, p. 1.

EXHIBIT 10.1   Center for Science and Environment Press Release: Hard Truths about Soft Drinks

New Delhi, August 5, 2003: After bottled water, it's aerated water that has plugged the purity test. In another exposé, Down To Earth has found that 12 major cold drink brands sold in and around Delhi contain a deadly cocktail of pesticide residues. The results are based on tests conducted by the Pollution Monitoring Laboratory (PML) of the Centre for Science and Environment (CSE). In February this year, CSE had blasted the bottled water industry's claims of being 'pure' when its laboratory had found pesticide residues in bottled water sold in Delhi and Mumbai.

This time, it analysed the contents of 12 cold drink brands sold in and around the capital. They were tested for organochlorine and organophosphorus pesticides and synthetic pyrethroids—all commonly used in India as insecticides.

The test results were as shocking as those of bottled water.

All samples contained residues of four extremely toxic pesticides and insecticides: lindane, DDT, malathion and chlorpyrifos. In all samples, levels of pesticide residues far exceeded the maximum residue limit for pesticides in water used as 'food', set down by the European Economic Commission (EEC). Each sample had enough poison to cause—in the long term—cancer, damage to the nervous and reproductive systems, birth defects and severe disruption of the immune system.

WHAT WE FOUND

* Market leaders Coca-Cola and Pepsi had almost similar concentrations of pesticide residues. Total pesticides in all PepsiCo brands on an average were 0.0180 mg/l (milligramme per litre), 36 times higher than the EEC limit for total pesticides (0.0005 mg/l). Total pesticides in all Coca-Cola brands on an average were 0.0150 mg/l, 30 times higher than the EEC limit.
* While contaminants in the 'Dil mange more' Pepsi were 37 times higher than the EEC limit, they exceeded the norms by 45-times in the 'Thanda matlab Coca-Cola' product.
* Mirinda Lemon topped the chart among all the tested brand samples, with a total pesticide concentration of 0.0352 mg/l.

The cold drinks sector in India is a much bigger money-spinner than the bottled water segment. In 2001, Indians consumed over 6,500 million bottles of cold drinks. Its growing popularity means that children and teenagers, who glug these bottles, are drinking a toxic potion.

PML also tested two soft drink brands sold in the US, to see if they contained pesticides. They didn't.

The question, therefore, is: how can apparently quality-conscious multinationals market products unfit for human consumption?

CSE found that the regulations for the powerful and massive soft drinks industry are much weaker, indeed non-existent, as compared to those for the bottled water industry. The norms that exist to regulate the quality of cold drinks are a maze of meaningless definitions. This "food" sector is virtually unregulated.

The Prevention of Food Adulteration (PFA) Act of 1954, or the Fruit Products Order (FPO) of 1955—both mandatory acts aimed at regulating the quality of contents in beverages such as cold drinks—do not even provide any scope for regulating pesticides in soft drinks. The FPO, under which the industry gets its license to operate, has standards for lead and arsenic that are 50 times higher than those allowed for the bottled water industry.

What's more, the sector is also exempted from the provisions of industrial licensing under the Industries (Development and Regulation) Act, 1951. It gets a one-time license to operate from the ministry of food processing industries; this license includes a no-objection certificate from the local government as well as the state pollution control board, and a water analysis report. There are no environmental impact assessments, or citing regulations. The industry's use of water, therefore, is not regulated.

Source: CSE press release, "Hard Truths about Soft Drinks." August 5, 2003.

EXHIBIT 10.2 Pesticide Content in 12 Leading Soft Drink Brands

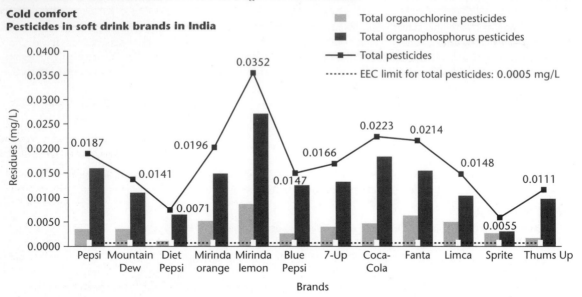

Cold comfort
Pesticides in soft drink brands in India

Source: CSE press release, "Hard Truths about Soft Drinks." August 5, 2003.

majority of consumers believed the findings were correct and agreed with Parliament's move to ban the sale of soft drinks.[8] The $1 billion Indian soft drink market[9] was at stake, and Gupta had to act.

HISTORY OF COKE

THE EARLY DAYS

Coca-Cola was created in 1886 by John Pemberton, a pharmacist in Atlanta, Georgia, who sold the syrup mixed with fountain water as a potion for mental and physical disorders. The formula changed hands three more times before Asa D. Candler added carbonation, and by 2003, Coca-Cola was the world's largest manufacturer, marketer, and distributor of nonalcoholic beverage concentrates and syrups, with more than 400 widely recognized beverage brands in its portfolio.

With the bubbles making the difference, Coca-Cola was registered as a trademark in

1887 and, by 1895, was being sold in every state and territory in the United States. In 1899, it franchised its bottling operations in the United States, growing quickly to reach 370 franchisees by 1910.[10] Headquartered in Atlanta with divisions and local operations in over 200 countries worldwide, Coca-Cola generated more than 70 percent of its income outside the United States by 2003 (see Exhibit 10.3).

INTERNATIONAL EXPANSION

Coke's first international bottling plants opened in 1906 in Canada, Cuba, and Panama.[11] By the end of the 1920s, Coca-Cola was bottled in 27 countries throughout the world and available in 51 more. In spite of this reach, volume was low, quality inconsistent, and effective advertising a challenge with language, culture,

[8] "Coke & Pepsi in India."

[9] http://www.indiastat.com.

[10] Nymph Kaul, "Coca-Cola India," Rai University, 2004; Coca-Cola Company website, http://www2.coca-cola.com/heritage/; Mark Pendergrast, *For God, Country and Coca-Cola* (New York: Charles Scribner's Sons, 1993).

[11] http://www2.coca-cola.com/ourcompany/aroundworld.html.

EXHIBIT 10.3   The Coca-Cola Company Income Statement

(in million $ except per share data)	2002	2001	2000
Net operating revenues	19,564	17,545	17,354
Cost of goods sold	7,105	6,044	6,204
Gross profit	12,459	11,501	11,150
Selling, general, and administrative expenses	7,001	6,149	6,016
Other operating changes	0	0	1,443
Operating income	5,458	5,352	3,691
Interest income	209	325	345
Interest expense	199	289	447
Equity income (loss)	384	152	(289)
Other income (loss)—net	(353)	39	99
Gains on issuances of stock by equity investee	0	91	0
Income before income taxes and cumulative effect of accounting change	5,499	5,670	3,399
Income taxes	1,523	1,691	1,222
Net Income before cumulative effect of accounting change	3,976	3,979	2,177
Cumulative effect of accounting change for SFAS No. 142 net of income taxes:			
Company operations	(367)	0	0
Equity investments	(559)	0	0
Cumulative effect of accounting change for SFAS No. 133 net of income taxes	0	(10)	0
Net income	3,050	3,969	2,177
Basic net income per share before accounting change	1.60	1.60	0.88
Cumulative effect of accounting change	(0.37)	0	0
	1.23	1.60	0.88
Diluted net income per share before accounting change	1.60	1.60	0.88
Cumulative effect of accounting change	(0.37)	0	0
	1.23	1.60	0.88
Average shares outstanding	2,478	2,487	2,477
Effect of dilutive securities	5	0	10
Average shares outstanding assuming dilution	2,483	2,487	2,487

and government regulation all serving as barriers. Former CEO Robert Woodruff's insistence that Coca-Cola wouldn't "suffer the stigma of being an intrusive American product," and instead would use local bottles, caps, machinery, trucks, and personnel, contributed to Coke's challenges, as did the lack of standard processes and training, which degraded quality.[12]

Coca-Cola continued working for over 80 years on Woodruff's goal: to make Coke available wherever and whenever consumers wanted it, "in arm's reach of desire."[13] The Second World War proved to be the stimulus Coca-Cola needed to build effective capabilities around the world and achieve dominant global market share. Woodruff's patriotic commitment "that every man in uniform gets

[12] Pendergrast, *For God, Country and Coca-Cola*, p. 172.

[13] Ibid.

EXHIBIT 10.4 Interbrand's Global Brand Scoreboard 2003

Rank	Company	2003 Brand Value ($Billion)	2002 Brand Value ($Billion)	Percent Change	Country of Ownership
1	Coca-Cola	70.45	69.64	+1%	U.S.
2	Microsoft	65.17	54.09	+2%	U.S.
3	IBM	51.77	51.19	+1%	U.S.
4	GE	42.34	41.31	+2%	U.S.
5	Intel	31.11	30.86	+1%	U.S.
6	Nokia	29.44	29.97	−2%	Finland
7	Disney	28.04	29.26	−4%	U.S.
8	McDonald's	24.70	26.38	−6%	U.S.
9	Marlboro	22.18	24.15	−8%	U.S.
10	Mercedes	21.37	21.01	+2%	Germany

a bottle of Coca-Cola for five cents, wherever he is and at whatever cost to our company"[14] was more than just great public relations. As a result of Coke's status as a military supplier, Coca-Cola was exempt from sugar rationing and also received government subsidies to build bottling plants around the world to serve WWII troops.[15]

TURN-OF-THE-CENTURY GROWTH IMPERATIVE

The 1990s brought a slowdown in sales growth for the carbonated soft drink (CSD) industry in the United States, achieving only 0.2 percent growth by 2000 (just under 10 billion cases) in contrast to the 5–7 percent annual growth experienced during the 1980s. While per capita consumption throughout the world was a fraction of the United States', major beverage companies clearly had to look elsewhere for the growth their shareholders demanded. The looming opportunity for the twenty-first century was in the world's developing markets with their rapidly growing middle-class populations.

THE WORLD'S MOST POWERFUL BRAND

Interbrand's Global Brand Scorecard for 2003 ranked Coca-Cola the #1 Brand in the World and estimated its brand value at $70.45 billion (see Exhibit 10.4).[16] The ranking's methodology determined a brand's valuation on the basis of how much it was likely to earn in the future, distilling the percentage of revenues that could be credited to the brand, and assessing the brand's strength to determine the risk of future earnings forecasts. Considerations included market leadership, stability, and global reach, incorporating the brand's ability to cross both geographical and cultural borders.[17]

From the beginning, Coke understood the importance of branding and the creation of a distinct personality.[18] Its catchy, well-liked slogans[19] ("It's the Real Thing" [1942, 1969], "Things go better with Coke" [1963], "Coke Is It" [1982], "Can't Beat the Feeling" [1987], and

[14] Ibid., p. 199.

[15] Ibid., pp. 200–201.

[16] "The Top 100 Brands: Interbrand's Global Brand Scorecard 2003," Interbrand Special Report, *BussinessWeek*, August 4, 2003.

[17] Ibid.

[18] Nicholas Kochan, ed., and Interbrand, *The World's Greatest Brands* (Washington, NY: New York University Press, 1997.

[19] Kevin Lane Keller, *Strategic Brand Management* (Upper Saddle River, NJ: Prentice Hall, 1998), p. 153.

a 1992 return to "Can't Beat the Real Thing")[20] linked that personality to the core values of each generation and established Coke as the authentic, relevant, and trusted refreshment of choice across the decades and around the globe.

INDIAN HISTORY

India is home to one of the most ancient cultures in the world, dating back over 5,000 years. At the beginning of the twenty-first century, 26 different languages were spoken across India, 30 percent of the population knew English, and greater than 40 percent were illiterate. At this time, the nation was in the midst of great transition, and the dichotomy between the old India and the new was stark. Remnants of the caste system existed alongside the world's top engineering schools and growing metropolises as the historically agricultural economy shifted into the services sector. In the process, India had created the world's largest middle class.

A British colony since 1769, when the East India Company gained control of all European trade in the nation, India gained its independence in 1947 under Mahatma Gandhi and his principles of nonviolence and self-reliance. In the decades that followed, self-reliance was taken to the extreme as many Indians believed that economic independence was necessary to be truly independent. As a result, the economy was increasingly regulated, and many sectors were restricted to the public sector. This movement reached its peak in 1977 when the Janta party government came to power and Coca-Cola was thrown out of the country. In 1991, the first generation of economic reforms was introduced and liberalization began.

COKE IN INDIA

Coca-Cola was the leading soft drink brand in India until 1977, when it left the country rather

than reveal its formula to the government and reduce its equity stake, as required under the Foreign Exchange Regulation Act (FERA), which governed the operations of foreign companies in India. After a 16-year absence, Coca-Cola returned to India in 1993, cementing its presence with a deal that gave Coca-Cola ownership of the nation's top soft-drink brands and bottling network. Coke's acquisition of local popular Indian brands, including Thums Up (the most trusted brand in India[21]), Limca, Maaza, Citra, and Gold Spot, provided not only physical manufacturing, bottling, and distribution assets but also achieved strong consumer preferences. This combination of local and global brands enabled Coca-Cola to exploit the benefits of global branding and global trends in tastes while also tapping into traditional domestic markets. Leading Indian brands joined the company's international family of brands, including Coca-Cola, Diet Coke, Sprite, and Fanta, plus the Schweppes product range. In 2000, the company launched the Kinley water brand, and in 2001, Shock energy drink and the powdered concentrate Sunfill hit the market.

From 1993 to 2003, Coca-Cola invested more than US$1 billion in India, making it one of the country's top international investors.[22] By 2003, Coca-Cola India had won the prestigious Woodruf Cup from among 22 divisions of the company, based on the three broad parameters of volume, profitability, and quality. Coca-Cola India achieved 39 percent volume growth in 2002 while the industry grew 23 percent nationally and the company reached break-even profitability in the region for the first time.[23] Encouraged by its 2002 performance, Coca-Cola India announced plans to double its capacity at an investment of $125 million (Rs. 750 crore) between September 2002 and March 2003.[24]

[20] http://www.portobello.com.au/portobello/reading/memorabilia_cocacola.htm.

[21] "Brands of Coca-Cola in India," Rai University, November 2004.

[22] http://www.coca-colaindia.com.

[23] Sanjiv Gupta biography, Rai University.

[24] "Coca-Cola India to Double Capacity," *Kolkata,* March 8, 2003.

Coca-Cola India produced its beverages with the help of 7,000 local employees at its 27 wholly owned bottling operations, supplemented by 17 franchisee-owned bottling operations and a network of 29 contract packers to manufacture a range of products for the company. The complete manufacturing process had a documented quality control and assurance program, including over 400 tests performed through out the process (see Exhibit 10.5).

The complexity of the consumer soft drink market demanded a distribution process to support 700,000 retail outlets serviced by a fleet that included 10-ton trucks, open-bay three wheelers, and trademarked tricycles and pushcarts that were used to navigate the narrow alleyways of the cities.[25] In addition to its own employees, Coke indirectly created employment for another 125,000 Indians through its procurement, supply, and distribution networks.

Sanjiv Gupta, president and CEO of Coca-Cola India, joined Coke in 1997 as vice president of marketing, and was instrumental to the company's success in developing a brand relevant to the Indian consumer and tapping India's vast rural market potential. Following his marketing responsibilities, Gupta served as head of operations for company-owned bottling operations and then as deputy president. Seen as the driving force behind recent successful forays into packaged drinking water, powdered drinks, and ready-to-serve tea and coffee, Gupta and his marketing prowess were critical to the continued growth of the company.[26]

THE INDIAN BEVERAGE MARKET[27]

India's 1 billion people, growing middle class, and low per capita consumption of soft drinks made it a highly contested prize in the global CSD market in the early twenty-first century. Ten percent of the country's population lived in urban areas or large cities and drank 10 bottles of soda per year, while the vast remainder lived in rural areas, villages, and small towns, where annual per capita consumption was less than four bottles. Coke and Pepsi dominated the market and together had a consolidated market share above 95 percent. Although soft drinks were once considered products only for the affluent, by 2003, 91 percent of sales were made to the lower, middle, and upper-middle classes. Soft drink sales in India grew 76 percent between 1998 and 2002, from 5,670 million bottles to over 10,000 million (see Exhibit 10.6), and were expected to grow at least 10 percent

EXHIBIT 10.5 Routine Tests Carried out by Bottling Operations and External Laboratories

	Process Parameter	No. of Tests
1.	Water	71
2.	Water treatment and auxiliary chemicals	68
3.	CO_2	50
4.	Sugar	13
5.	Syrup	17
6.	Packaging material	25
7.	Container washing	17
8.	Finished product	18
9.	Market samples	15
10.	External lab	147
	Total	441

Source: The Coca-Cola Company, http://www.myenjoyzone.com.

EXHIBIT 10.6 Soft Drink Sales in India

Fiscal Year	Million Bottles Sold
1998–1999	5,670
1999–2000	6,230
2000–2001	6,450
2001–2002	6,600
2002–2003	10,000

Source: "Soft Drink Sales Up 10.4%" *PTI*, September 29, 2004.

[25] http://www.coca-colaindia.com.

[26] Gupta biography.

[27] http://www.indiastat.com.

per year through 2012.[28] In spite of this growth, annual per capita consumption was only 6 bottles versus 17 in Pakistan, 73 in Thailand, 173 in the Philippines, and 800 in the United States.[29]

With its large population and low consumption, the rural market represented a significant opportunity for penetration and a critical battleground for market dominance. In 2001, Coca-Cola recognized that to compete with traditional refreshments, including lemon water, green coconut water, fruit juices, tea, and lassi, competitive pricing was essential. In response, Coke launched a smaller bottle priced at almost 50 percent the amount of the traditional package.

MARKETING COLA IN INDIA

The post-liberalization period in India saw the comeback of cola, but Pepsi had already beaten Coca-Cola to the punch, creatively entering the market in the 1980s in advance of liberalization by way of a joint venture (JV). As early as 1985, Pepsi tried to gain entry into India and finally succeeded with the Pepsi Foods Limited Project in 1988, as a JV among PepsiCo, the Punjab government–owned Punjab Agro Industrial Corporation (PAIC), and Voltas India Limited. Pepsi was marketed and sold as Lehar Pepsi until 1991, when the use of foreign brands was allowed under the new economic policy, after which Pepsi bought out its partners and became a fully owned subsidiary, ending the JV relationship in 1994.[30]

Although the joint venture was only marginally successful in its own right, it allowed Pepsi to gain precious early experience with the Indian market and served as an introduction of the Pepsi brand to the Indian consumer, such that it was well poised to reap the benefits when liberalization came. Although Coke benefited from Pepsi creating demand and developing

the market, Pepsi's head start gave Coke a disadvantage in the mind of the consumer. Pepsi's appeal focused on youth, and when Coke entered India in 1993 and approached the market selling an American way of life, it failed to resonate as expected.[31]

2001 MARKETING STRATEGY

Coca-Cola CEO Douglas Daft set the direction for the next generation of success for his global brand with a "Think local, act local" mantra. Recognizing that a single global strategy or single global campaign wouldn't work, locally relevant executions became an increasingly important element of supporting Coke's global brand strategy.

In 2001, after almost a decade of lagging rival Pepsi in the region, Coke India reexamined its approach in an attempt to gain leadership in the Indian market and capitalize on significant growth potential, particularly in rural markets. The foundation of the new strategy grounded brand positioning and marketing communications in consumer insights, acknowledging that urban versus rural India were two distinct markets on a variety of important dimensions. The soft drink category's role in people's lives, the degree of differentiation between consumer segments and their reasons for entering the category, and the degree to which brands in the category projected different perceptions to consumers were among the many important differences between how urban and rural consumers approached the market for refreshment.[32]

In rural markets, where both the soft drink category and individual brands were undeveloped, the task was to broaden the brand positioning, whereas in urban markets, with their higher category and brand development, the task was to narrow the brand positioning, focusing on differentiation through offering

---

[28] Ibid.

[29] Ibid.

[30] Kavaljit Singh, "Broken Commitments: The Case of Pepsi in India," *PIRG Update*, May 1997.

[31] Interview with Nymph Kaul, September 20, 2004.

[32] Coca-Cola India, "Marketing: Questioning Paradigms," internal marketing presentation.

unique and compelling value. This lens, informed by consumer insights, gave Coke direction on the trade-off between focus and breadth that a brand needed in a given market and made clear that to succeed in either segment, unique marketing strategies were required in urban versus rural India.

BRAND LOCALIZATION STRATEGY: THE TWO INDIAS
INDIA A: "LIFE HO TO AISI"

"India A," the designation Coca-Cola gave to the market segment including metropolitan areas and large towns, represented 4 percent of the country's population.[33] This segment sought social bonding as a need and responded to aspirational messages, celebrating the benefits of their increasing social and economic freedoms. "*Life ho to aisi*" (life as it should be) was the successful and relevant tagline that marked Coca-Cola's advertising to this audience.

INDIA B: "THANDA MATLAB COCA-COLA"

Coca-Cola India believed that the first brand to offer communication targeted to the smaller towns would own the rural market and therefore went after that objective with a comprehensive strategy. "India B" included small towns and rural areas, constituting the other 96 percent of the nation's population. This segment's primary need was out-of-home thirst-quenching; the soft drink category was undifferentiated in the minds of rural consumers. Additionally, with an average Coke costing Rs. 10 and an average day's wages around Rs. 100, Coke was perceived as a luxury that few could afford.[34]

In an effort to make the price point of Coke affordable for this high-potential market, Coca-Cola launched the Accessibility Campaign, introducing a new 200-mL bottle that was smaller than the traditional 300-mL bottle found in urban markets, and concurrently cutting the price in half to Rs. 5. This pricing strategy closed the gap between Coke and basic refreshments like lemonade and tea, making soft drinks truly accessible for the first time. At the same time, Coke invested in distribution infrastructure to serve a disbursed population effectively and doubled the number of retail outlets in rural areas from 80,000 in 2001 to 160,000 in 2003, increasing market penetration from 13 to 25 percent.[35]

Coke's advertising and promotion strategy pulled the marketing plan together using local language and idiomatic expressions. "Thanda," meaning cool/cold, is also generic for cold beverages and gave "Thanda Matlab Coca-Cola" delicious multiple meanings. Literally translated to "Coke means refreshment," the phrase directly addressed both the primary need of this segment for cold refreshment while positioning Coke as a "Thanda," or a generic cold beverage just like tea, lassi, or lemonade. As a result of the Thanda campaign, Coca-Cola won *Advertiser of the Year* and *Campaign of the Year* in 2003.

RURAL SUCCESS

Comprising 74 percent of the country's population, 41 percent of its middle class, and 58 percent of its disposable income, the rural market was an attractive target, and it delivered results. Coke experienced 37 percent growth in 2003 in this segment versus the 24 percent growth seen in urban areas. Driven by the launch of the new Rs. 5 product, per capita consumption doubled between 2001 and 2003. This market accounted for 80 percent of India's new Coke drinkers, 30 percent of 2002 volume, and was expected to account for 50 percent of the company's sales in 2003.[36]

---

[33] Ibid.

[34] Kaul, interview.

[35] Ibid.

[36] Ibid.

EXHIBIT 10.7   Coca-Cola Principles of Corporation Citizenship

Our reputation is built on trust. Through good citizenship we will nurture our relationships and continue to build that trust. That is the essence of our promise—The Coca-Cola Company exists to benefit and refresh everyone it touches.

Wherever Coca-Cola does business, we strive to be trusted partners and good citizens. We are committed to managing our business around the world with a consistent set of values that represent the highest standards of integrity and excellence. We share these values with our bottlers, making our system stronger.

These core values are essential to our long-term business success and will be reflected in all of our relationships and actions—in the marketplace, the workplace, the environment and the community.

MARKETPLACE

We will adhere to the highest ethical standards, knowing that the quality of our products, the integrity of our brands and the dedication of our people build trust and strengthen relationships. We will serve the people who enjoy our brands through innovation, superb customer service, and respect for the unique customs and cultures in the communities where we do business.

WORKPLACE

We will treat each other with dignity, fairness and respect. We will foster an inclusive environment that encourages all employees to develop and perform to their fullest potential, consistent with a commitment to human rights in our workplace. The Coca-Cola workplace will be a place where everyone's ideas and contributions are valued, and where responsibility and accountability are encouraged and rewarded.

ENVIRONMENT

We will conduct our business in ways that protect and preserve the environment. We will integrate principles of environmental stewardship and sustainable development into our business decisions and processes.

COMMUNITY

We will contribute our time, expertise and resources to help develop sustainable communities in partnership with local leaders. We will seek to improve the quality of life through locally-relevant initiatives wherever we do business.

Responsible corporate citizenship is at the heart of The Coca-Cola Promise. We believe that what is best for our employees, for the community and for the environment is also best for our business.

Source: Coca-Cola Company website.

CORPORATE SOCIAL RESPONSIBILITY

As one of the largest and most global companies in the world, Coca-Cola took seriously its ability and responsibility to affect the communities in which it operated. The company's mission statement, called the Coca-Cola Promise, stated: "The Coca-Cola Company exists to benefit and refresh everyone who is touched by our business." The company has made efforts toward good citizenship in the areas of community, by improving the quality of life in the communities in which it operates, and the environment, by addressing water, climate change, and waste management initiatives. Its activities also include The Coca-Cola Africa Foundation, created to combat the spread of HIV/AIDS through partnership with governments, UNAIDS, and other nongovernmental organizations (NGOs), and The Coca-Cola Foundation, focused on higher education as a vehicle to build strong communities and enhance individual opportunity (see Exhibit 10.7).[37]

Coca-Cola's footprint in India was significant as well. The company employed 7,000 citizens and believed that for every direct job, 30–40 more were created in the supply chain.[38]

[37] http://www.coca-colaindia.com.

[38] Ibid.

Like its parent, Coke India's corporate social responsibility (CSR) initiatives were both community- and environment-focused. Priorities included education, where primary education projects had been set up to benefit children in slums and villages; water conservation, where the company supported community-based rainwater harvesting projects to restore water levels and promote conservation education; and health, where Coke India partnered with NGOs and governments to provide medical access to poor people through regular health camps. In addition to outreach efforts, the company committed itself to environmental responsibility through its own business operations in India, including[39]

- Environmental due diligence before acquiring land or starting projects.
- Environmental impact assessment before commencing operations.
- Ground water and environmental surveys before selecting sites.
- Compliance with all regulatory environmental requirements.
- Ban on purchasing CFC-containing refrigeration equipment.
- Wastewater treatment facilities with trained personnel at all company-owned bottling operations.
- Energy conservation programs.
- 50 percent water savings in the last seven years of operations.

PREVIOUS COKE CRISES

Despite Coke's reputation as a socially responsible corporate citizen, the company has faced its share of controversy worldwide, surrounding both its products and its policies in the years preceding the Indian pesticide crisis.

INGRAM ET AL. V. THE COCA-COLA COMPANY—1999[40]

In the spring of 1999, four current and former Coca-Cola employees, led by information analyst Linda Ingram, filed bias charges against Coca-Cola in Atlanta federal court. The lawsuit charged the company with racial discrimination and stated: "This discrimination represents a company-wide pattern and practice, rather than a series of isolated incidents. Although Coca-Cola has carefully crafted African-American consumers of its product by public announcements, strategic alliances and specific marketing strategies, it has failed to place the same importance on its African-American employees."[41]

In the decades leading up to the suit, both internal and external warnings surrounding Coke's diversity practices were issued. In 1981, the Reverend Jesse Jackson, director of the Rainbow/PUSH Coalition, instigated a boycott against Coca-Cola, challenging the company to improve its business relationship significantly with the African-American community.[42]

The Ware report, written by Senior Vice President Carl Ware, an African-American executive at the company, cited a lack of diversity at the decision-making level, a basic lack of workplace diversity, a "ghettoization" among blacks who worked for Cola-Cola, and an overt lack of respect for cultural differences, as well as an implicit assumption that African-American employees lacked the intelligence to meet the challenges of the highest executive levels.[43]

Cyrus Mehri, one of the most visible and successful plaintiff advocates in the United

---

[39] Ibid.

[40] Nicola K. Graves and Randall L. Waller, "The Corporate Web Site as an Image Restoration Tool: The Case of Coca-Cola," *Proceedings of the 2004 Association for Business Communication 69th Annual Convention*, Cambridge, MA, October 25–29, 2004.

[41] H. Unger, "Coca-Cola Accused of a 'Companywide Pattern,'" *Atlanta Journal-Constitution*, April 24, 1999, p. H1

[42] C. L. Hays, *The Real Thing: Truth and Power at the Coca-Cola Company* (New York: Random House, 2004).

[43] Ibid.

States, represented the group and was skilled at leveraging the power of the media, creating a true crisis for the Coca-Cola Company and exerting tremendous pressure for settlement. In 2000, the lawsuit was settled for $192.5 million after the company had sent mixed messages and damaging statements regarding the merit of the suit for over a year. Analysts identified the bias suit as a prime reason for the $100 billion decrease in Coca-Cola's stock price between 1998 and 2000.[44]

BELGIUM—1999[45]

On June 8, 1999, 33 Belgian school children became ill after drinking Coke bottled at a local facility in Antwerp. A few days later, more Belgians complained of similar symptoms after drinking cans of Coke that had been bottled at a plant in Dunkirk, France, and 80 people in northern France were allegedly stricken by intestinal problems and nausea, bringing the total afflicted to over 250.

In the days following the first outbreak, 17 million cases of Coke from five European countries were recalled and destroyed. It was the largest product recall in Coke's history, and Belgian and French authorities banned the sale of Coca-Cola products for 10 days. Germany placed a temporary import ban on Coca-Cola produced in Belgium and the Netherlands, and Luxembourg banned all Coca-Cola products. Health ministers in Italy, Spain, and Switzerland warned people about consuming Coke products.

Coca-Cola sources explained that the contamination was due to defective carbon dioxide (CO_2) used at the Antwerp plant and that a wood preservative used on shipping pallets had concentrated on the outside of cans at the Dunkirk plant. The European Commission, however, believed production faults and contaminated pipes were more likely to be the cause of the problem.

Although CEO Ivester was in Paris when the news broke, he flew home to Atlanta and kept silent, waiting over a week to issue his first public statement on the crisis, citing that "Coke would do whatever necessary to ensure the safety of its products." A Netherlands-based toxicologist Coke had hired issued a report on June 29 exempting the company from blame for the CO_2 impurity in Antwerp and the fungicide at Dunkirk. Although the product ban was lifted, Coke had a tremendous amount of work to do to win back consumer confidence.

An aggressive PR campaign included vouchers and coupons for free product delivered to each of Belgium's 4.4 million homes; sponsored dances, beach parties, and summer fairs for teenagers; and significant television advertising reinforcing, "Today, more than ever, we thank you for your loyalty."

KINLEY BOTTLED WATER—2003

On February 4, 2003, the Center for Science and Environment (CSE) in India released a report based on tests conducted by the Pollution Monitoring Laboratory (PML) titled "Pure Water or Pure Peril?" Analyses of 17 packaged drinking water brands sold across the country revealed evidence of pesticide residues, including lindane, DDT, malathion, and chlorpyrifos. The CSE used European norms for maximum permissible limits for pesticides in packaged water, "because the standards set for pesticide residues by the Bureau of Indian Standards (BIS) are vague and undefined."[46] Coca-Cola's Kinley water brand had concentration levels 15 times higher than stipulated limits, top-seller Biserli had 79 times, and Aquaplus topped the list at 109 times.[47] In the wake of this statement, Coca-Cola remained largely silent, and the buzz went away.

---

[44] K. MacArthur and R. Linnett, "Coke Crisis: Equity Erodes as Brand Troubles Mount," *Advertising Age*, April 24, 2000, p. 3.

[45] "Coke & Pepsi in India," p. 8.

[46] "Pure Water or Pure Peril?" CSE press release, February 2003.

[47] Ibid.

EXHIBIT 10.8 Corporate Communication at Coca-Cola

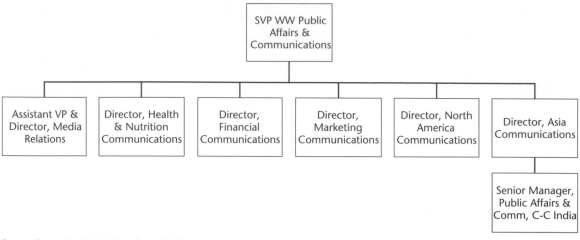

Source: Case writer derived from Coca-Cola Company website.

CORPORATE COMMUNICATION AT COCA-COLA

Corporate communication was a critical function at the Coca-Cola Company, given the number of constituencies both internal and external to the company. In addition, the complexity and global reach of the company's operations could not be centrally managed and instead demanded a matrixed team organization.

The senior communications position at the company, senior vice president, Worldwide Public Affairs & Communication, sat on the company's executive committee and reported to the chairman and CEO at the time of the crisis in India. Director-level corporate communication functions included media relations, nutrition communications, financial communications, and marketing communications, but the geographic diversity of the company's businesses required regionally based communication leaders in addition to the corporate resources in place. As a result, five regional communication directors serviced North America, Latin America, Asia, Europe, and Africa with their own teams of communications professionals (see Exhibit 10.8).

NGO ACTIVISM[48]

Nongovernmental organizations initially evolved to influence governments, but by the early twenty-first century, many realized that targeting corporations and key corporate constituents such as investors and customers could be an even more powerful way to effect change. Along with their ability to focus, gain attention, and act quickly was the high level of credibility NGOs had cultivated with many constituencies. This credibility stemmed in part from their emotional, rather than fact-based, appeals and the impassioned nature of their arguments.

The most common tactic of NGOs was to develop campaigns against business through which they garnered support from consumers and the media. These campaigns, such as Greenpeace's attack on Shell Oil following the company's decision to dump the Brent Spar oil rig in the ocean in the 1990s, typically focused on a single issue; targeted companies with successful and well-known brands such as McDonald's and Nike; and were augmented by market trends such as the homogenization created by

---

[48] Paul A. Argenti, "Collaborating with Activists: How Starbucks Works with NGOs," *California Management Review* 47, no. 1 (Fall 2004).

chains like Walmart and Starbucks. The NGOs realized that anticorporate campaigns could be far more powerful than antigovernment campaigns. Global Exchange's attack on Nike for sweatshop labor conditions in the 1990s, for example, was one of the most highly publicized and also one of the most successful antibusiness campaigns by an NGO.

CENTER FOR SCIENCE AND ENVIRONMENT

The CSE, an NGO, was established in India in 1980 by a group of engineers, scientists, journalists, and environmentalists to "catalyze the growth of public awareness on vital issues in science, technology, environment, and development."[49] Led by Sumita Narain, a former schoolmate of Coke India CEO Gupta, the CSE's efforts included communication for awareness, research and advocacy, education and training, documentation, and pollution monitoring.

Spurred by the February 2003 report on bottled water and questions such as, "if what we found in bottled water was correct, then what about soft drinks?" the CSE's August 2003 report claimed that soft drinks were extremely dangerous to Indian citizens, according to tests conducted at the Pollution Monitoring Laboratory (PML). All samples contained residues of lindane, DDT, malathion, and chlorpyrifos, toxic pesticides and insecticides known to cause serious long-term health issues. Total pesticides in all Coca-Cola brands averaged 0.0150 mg/l, or 30 times higher than the European Economic Commission (EEC) limit. PML also tested samples of Coke and Pepsi products sold in the United States to see if they contained pesticides; they did not.

The CSE report called on the government to put in place legally enforceable water standards and chastised the multinationals for taking advantage of the situation at the expense of consumer health and well-being.

INDIAN REGULATORY ENVIRONMENT[50]

The main law governing food safety in India was the 1954 Prevention of Food Alteration Act (PFA), which contained a rule regulating pesticides in foods but did not include beverages. The Food Processing Order (1955) required that the main ingredient used in soft drinks be "potable water," but the Bureau of Indian Standards (BIS) had no prescribed standards for pesticides in water. One BIS directive stated that pesticides must be absent and set a limit of 0.001 part per million, but the Health Secretary admitted, "There are lapses in PFA regarding carbonated drinks."[51]

Indian law enforcement was minimal, with virtually no convictions under PFA. In the absence of national standards, NGOs such as the CSE turned to the United States and the European Union for "international norms." The appropriateness and feasibility of these standards for developing nations, however, remained a question for many. Under EU food laws, for example, milk, fruit, and basic staples such as rice and wheat would need to be imported into India to satisfy safety standards.

THE INITIAL RESPONSE

The day after the CSE's announcement, Coke and Pepsi came together in a rare show of solidarity at a joint press conference. The companies attacked the credibility of the CSE and their lab results, citing regular testing at independent laboratories proving the safety of their products. They promised to provide these data to the public, threatened legal action against the CSE while seeking a gag order, and contacted the U.S. Embassy in India for assistance.

[49] http://www.cseindia.org.

[50] "Coke & Pepsi in India," p. 3.

[51] Supriya Bezbaruat and Malini Goyal, "The Gulp War," *India Today*, August 25, 2003, pp. 50–53.

EXHIBIT 10.9 Myths and Facts from Coca-Cola India Website

Since August 5, 2003 the quality and safety of Coca-Cola and PepsiCo products in India have been called into question by a local NGO, the Centre for Science and Environment (CSE). The basis of the allegations are [sic] tests conducted on products of Coca-Cola and PepsiCo by CSE's internal unaccredited laboratory, the Pollution Monitoring Laboratory.

In India, as in the rest of the world, our plants use a multiple barrier system to remove potential contaminants and unwanted natural substances including iron, sulfur, heavy metals as well as pesticides. Our products in India are safe and are tested regularly to ensure that they meet the same rigorous standards we maintain across the world.

The result of these allegations has been consumer confusion, significant impact on the sale of a safe and high-quality product, and the erosion of international investor confidence in the Indian business sector. This situation calls for the development of national sampling and testing protocols for soft drinks, an end to sensationalizing unsubstantiated allegations, and co-operation by all parties concerned in the interests of both Indian consumers and companies with significant investments in the Indian economy.

The facts versus the fiction False statements made in recent weeks have led to false perceptions by Indian consumers:

Myth   Coca-Cola products in India contain pesticide residues that are above EU norms.
Fact   Throughout all of our operations in India, stringent quality monitoring takes place covering both the source water we use as well as our finished product. We test for traces of pesticide in groundwater to the level of parts per billion. This is equivalent to one drop in a billion drops. For comparison's sake, this would also be equivalent to measuring one second in 32 years, or less than one person in the entire population in India. These tests require specialized equipment at accredited labs to have accurate results. Even at these stringent miniscule levels we are well within the internationally accepted safety norms.

Myth   Coca-Cola products sold in India are "toxic" and unfit for human consumption.
Fact   There is no contamination or toxicity in our beverage brands. Our high-quality beverages are—and have always been—safe and refreshing. In over 200 countries across the globe, more than a billion times every day, consumers choose our brands for refreshment because Coca-Cola is a symbol of quality.

Myth   Coca-Cola has dual standards in the production of its products, one high standard for western countries, another for India.
Fact   The soft drinks manufactured in India conform to the same high standards of quality as in the USA and Europe. Through our globally accepted and validated manufacturing processes and Quality Management systems, we ensure that our state-of-the-art manufacturing facilities are equipped to provide the consumer the highest quality beverage each time. We stringently test our soft drinks in India at independent, accredited and world-class laboratories both locally and internationally.

Myth   In India the soft drinks industry is virtually unregulated.
Fact   There are no standards for soft drinks in the US, the EU, or India. In India, water used for beverage manufacture must conform to drinking water standards. The water used by Coca-Cola conforms to both BIS and EU standards for drinking water and our production protocols ensure this through a focus on process control and testing of the water used in our manufacturing process and the final product quality.

Myth   Coca-Cola has put out results for Kinley water only and not for their soft drinks.
Fact   The results of product tests conducted by TNO Nutrition and Food Research Laboratory in the Netherlands is [sic] conclusive and is [sic] available on The Science Behind Our Quality web page.

Myth   International companies like Coca-Cola are "colonizing" India.
Fact   The Coca-Cola business in India is a local business. Our beverages in India are produced locally, we employ thousands of Indian citizens, our product range and marketing reflect Indian tastes and lifestyles, and we are deeply involved in the life of the local communities in which we operate. The Coca-Cola business system directly employs approximately 10,000 local people in India. In addition, independent studies have documented that, by providing opportunities for local enterprises, the Coca-Cola business also generates a significant employment "multiplier effect." In India, we indirectly create employment for more than 125,000 people in related industries through our vast procurement, supply and distribution system.

Myth   Farmers in India are using Coca-Cola and other soft drinks as pesticides by spraying them on their crops.
Fact   Soft drinks do not act in a similar way to pesticides when applied to the ground or crops. There is no scientific basis for this and the use of soft drinks for this purpose would be totally ineffective. In India, as in the rest of the world, our products are world class and safe and the treated water used to make our beverages there meets the highest international standards.

Source: Coca-Cola Company website.

Coca-Cola India's CEO Sanjiv Gupta published the following statement for the Indian public:[52]

You may have seen recently in the media some allegations about the quality standards of our products in India. We take these allegations extremely seriously. I want to reassure you that our products in India are safe and are tested regularly to ensure that they meet the same rigorous standards we maintain across the world.

Maintaining quality standards is the most important element of our business and we cannot stand by while misleading and unaccredited data is used to discredit trusted and world-class brands. Recent allegations have caused unnecessary panic among consumers in India and, if unchecked, would impair our business in India and impact the livelihoods of our thousands of employees across the country.

This site is about the truth behind the headlines. It provides some context and facts on these issues and we hope it helps you understand exactly why you can trust our beverage brands and continue to enjoy them as millions of Indians do each day.

Sanjiv Gupta, Division President, Coca-Cola India

In the following days, the Delhi High Court asked the government to convene an expert committee to test and report on the safety of soft drinks within three weeks and to revise existing standards to include pesticide norms. Coca-Cola and Pepsi launched independent campaigns to reassure the public, taking out full-page newspaper advertisements and directing consumers to their corporate websites to review test results and safety protocol in greater detail (see Exhibit 10.9). In spite of these actions, the public seemed to believe the CSE's claims, and the crisis was far from over for the beverage giants. With sales continuing to experience a precipitous drop, one Delhi medical student's sentiments appeared to be widespread: "For a person drinking at least one bottle a day, the report came as a rude shock. I haven't picked up a bottle today and most definitely will not consume soft drinks in the future. The reports of pesticides and other pollutants have made soft drinks a strict no-no and we will now stick to juices and plain drinking water."[53]

GUPTA'S DILEMMA

As he contemplated the crisis at hand, Sanjiv Gupta questioned what action, if any, was necessary. Coke India was well within the country's legal guidelines, and the crisis had not been widely reported outside of India. Gupta knew that the Indian public had a short attention span and had reason to think that it wouldn't be long before the CSE's report faded, just as the Kinley water issue had earlier in the year.

On the other hand, he wondered if the situation might offer the company an opportunity to display higher standards of social responsibility at a time when it needed to differentiate itself from the competition. Multinationals had slipped in numerous situations of late and were being blamed for not adhering to the same standards in developing countries as in industrialized nations. The additive effect of this negative press meant that the potential damage to Coke's reputation was even greater. Finally, an ineffective resolution would be a devastating blow to the momentum Coke had gained after three long years of work on the marketing front.

CASE QUESTIONS

1. What are the key problems that Gupta should focus on in the short term and in the long term?
2. How would you evaluate the crisis?
3. How well prepared was Coke India to deal with the CSE's allegations?
4. What is your recommendation for Coke's communication strategy? Who are the key constituents?

---

[52] http://www.coca-colaindia.com.

[53] "Shocked Delhites Stay Away from Soft Drinks," *The Hindu* (New Delhi), August 7, 2003, p. 1.

5. Could Coke India have avoided this crisis?
6. What should Gupta do now?

CASE BIBLIOGRAPHY

Argenti, Paul A. "Collaborating with Activists: How Starbucks Works with NGOs." *California Management Review* 47, no. 1 (Fall 2004).

Bhatia, Gauri. "Multinational Corporations: Pro or Con?" *Outlook India*, October 29, 2003.

Centre for Science and Environment (CSE). "Analysis of Pesticide Residues in Soft Drinks," August 5, 2003.

Coca-Cola India. "Marketing: Questioning Paradigms," internal company presentation.

"Coca-Cola, Philips Win Marketing Awards." http://www.financialexpress.com, October 7, 2004.

"Coke, Pepsi Challenge India Pesticide Claim." http://www.ajc.com/business/content/business/coke/0803/06pesticide.html.

"Coke, Pepsi India Deny Pesticides in Soft Drinks." http://www.forbes.com/home_europe/newswire/2003/08/05/rtr1049160.html.

Dawar, Niraj, and Nancy Dai. "Cola Wars in China: The Future Is Here." HBS Case, August 21, 2003.

Dey, Saikat. Interview on Indian History and Economic Liberalization. January 10, 2005.

Graves, Nicola K., and Randall L. Waller. "The Corporate Web Site as an Image Restoration Tool: The Case of Coca-Cola." *Proceedings of the 2004 Association for Business Communication 69th Annual Convention*, Cambridge, MA, October 25–29, 2000.

http://www.coca-cola.com/flashIndex1.html.

http://www2.coca-cola.com/presscenter/viewpoints_india_situation.html.

http://www.coca-colaindia.com/.

http://www.indiaresource.org/.

http://www.killercoke.org.

http://www.myenjoyzone.com/press1/truth.htm.

Kaul, Nymph. Interview of Sanjiv Gupta, president and CEO of Coca-Cola India, June 2004.

Kaul, Nymph. Rai University, multiple interviews.

Kaul, Nymph. Rai University, "Coca-Cola India." 2004.

Keller, Kevin Lane. *Strategic Brand Management*. Upper Saddle River, NJ: Prentice Hall, 1998.

Kochan, Nicholas, ed., and Interbrand. *The World's Greatest Brands*. Washington, NY: New York University Press, 1997.

Pendergrast, Mark. *For God, Country and Coca-Cola*. New York: Charles Scribner's Sons, 1993.

"People's Forum against Coca-Cola." Brochure.

Sanghvi, Rish. Interviews on cola in India before liberalization and marketing/advertising of Coke and Pepsi in India, November 2004.

Society for Environmental Communications. "Colonisation's Dirty Dozen: Deadly Pesticides Found in 12 Leading Brands of Soft Drinks," August 15, 2003.

"Soft Drink Sales Up 10.4%." PTI, September 29, 2004.

Srivastava, Amit. "Coke with a New Twist: Toxic Cola." India Resource Center, February 15, 2004.

"Things Aren't Going Better with Coke." *BusinessWeek Online*, June 28, 1999.

"The Top 100 Brands: Interbrand's Global Brand Scorecard 2003." Interbrand Special Report, *BusinessWeek*, August 4, 2003.

Yoffie, David B., and Richard Seet. "Internationalizing the Cola Wars: The Battle for China and Asian Markets." HBS Case, May 31, 1995.

Yoffie, David B., and Yusi Wang. "Cola Wars Continue: Coke versus Pepsi in the Twenty-First Century." HBS Case, January 11, 2002.

Aaker, David A. *Building Strong Brands*. New York: Free Press, 1996.

Adams, Walter, and James W. Brock. *The Bigness Complex: Industry, Labor, and Government in the American Economy*. New York: Pantheon Books, 1986.

Aguilar, Francis Joseph. *General Managers in Action: Policies and Strategies*, 2nd ed. Oxford: Oxford University Press, 1992.

Ailes, Roger, and Jon Kraushar. *You Are the Message*. Garden City, NJ: Currency Doubleday, 1995.

Angell, Marcia, M. D. *Science on Trial: The Clash of Medical Evidence and the Law in the Breast Implant Case*. New York: W.W. Norton, 1996.

———. "Crisis Communication: Lessons from 9/11." *Harvard Business Review*, December 2002.

———. "Keeping to the Fairway." Case commentary, *Harvard Business Review*, April 2003.

Argenti, Paul A. "Collaborating with Activists: How Starbucks Works with NGOs to Enhance Its Emphasis on Social Responsibility." *California Management Review*, Fall 2004. Argenti "The Good, the Bad and the Trustworthy," in Strategy + Business in Winter 2011.

———. "Communications and Business Value: Measuring the Link." *Journal of Business Strategy*, November 2006.

———, and Courtney Barnes. *Digital Strategies for Powerful Corporate Communications*. New York: McGraw-Hill, 2009.

———, and Janis Forman. *The Power of Corporate Communication: Crafting the Voice and Image of Your Business*. New York: McGraw-Hill, 2002.

——— and ———. "The Employee Care Revolution." *Leader to Leader*, Summer 2004.

———, Robert Howell, and Karen Beck. "The Strategic Communication Imperative." *Sloan Management Review*, Spring 2005.

Aristotle. *The Art of Rhetoric* . Cambridge, MA: Harvard University Press, 1975.

Barton, Laurence. *Crisis in Organizations II*. Cincinnati, OH: South-Western, 2000.

Brown, Michael. *Laying Waste: The Poisoning of America by Toxic Chemicals*. New York: Pocket Books, 1981.

Byrne, John A. *Informed Consent*. New York: McGraw-Hill, 1996.

Chajet, Clive, and Tom Shachtman. *Image by Design: From Corporate Vision to Business Reality*, 2nd ed. New York: McGraw-Hill, 1997.

Collins, James C., and Jerry I. Porras. *Built to Last: Successful Habits of Visionary Companies*. New York: Harper Business, 1994, 1997.

Corrado, Frank M. *Media for Managers*. Englewood Cliffs, NJ: Prentice Hall, 1997.

Cutlip, Scott M. *Public Relations History: From the 17th to the 20th Century*. Hillsdale, NJ: Lawrence Erlbaum, 1995.

D'Aveni, Richard A. *Hypercompetition: Managing the Dynamics of Strategic Maneuvering*. New York: Free Press, 1994.

DeBower, Herbert F. *Modern Business*, vol. 7, *Advertising Principles*. New York: Alexander Hamilton Institute, 1917.

Dozier, David M., Larissa A. Grunig, and James E. Grunig. *Manager's Guide to Excellence in Public Relations and Communication Management*. Mahwah, NJ: Lawrence Erlbaum, 1995.

Edsell, Thomas. *The New Politics of Inequality*. New York: Norton, 1984.

Eichenwald, Kurt. *Conspiracy of Fools*. New York: Broadway Books, 2005.

Eisner, Michael D. *Work in Progress*. New York: Random House, 1998.

Fombrun, Charles J. *Reputation: Realizing Value from the Corporate Image*. Boston: Harvard Business School Press, 1996.

Ford, Daniel F. *Three Mile Island: Thirty Minutes to Meltdown*. New York: Penguin, 1981.

Forty, Adrian. *Objects of Desire: Design and Society from Wedgewood to IBM*. New York: Pantheon, 1986.

Fritschler, Lee. *Smoking and Politics: Policymaking and the Federal Bureaucracy*. 3rd ed. Englewood Cliffs, NJ: Prentice Hall, 1983.

Garbett, Thomas F. *Corporate Advertising*. New York: McGraw-Hill, 1981.

———. *How to Build a Corporation's Identity and Project Its Image*. Lexington, MA: Lexington Books, 1988.

Garten, Jeffrey. *The Mind of the CEO*. New York: Basic Books, 2001.

Gibbs, Lois Marie. *Love Canal: The Story Continues*. New York: New Society, 1988.

Goodman, Michael B., ed. *Corporate Communication: Theory and Practice*. Albany: State University of New York Press, 1994.

Gottschalk, Jack, ed. *Crisis Response: Inside Stories on Managing Image under Siege*. Detroit, MI: Gale Research, 1993.

Handler, Edward, and John R. Mulkern. *Business in Politics*. Lexington, MA: Lexington Books, 1982.

Hattersley, Michael E., and Linda McJannet. *Management Communication: Principles and Practice*. New York: McGraw-Hill, 1997.

Heath, Jim F. *John F. Kennedy and the Business Community*. Chicago: University of Chicago Press, 1969.

Hoffman, Paul. *The Dealmakers*. Garden City, NJ: Doubleday, 1984.

Hsieh, Tony. *Delivering Happiness: A Path to Profits, Passion, and Purpose*. New York: Grand Central Publishing, 2010.

Hughes, Jonathan R. T. *The Governmental Habit: Economic Controls from Colonial Times to the Present*. New York: Basic Books, 1977.

Hutton, James G., and Francis J. Mulhem. *Marketing Communications: Integrated Theory, Strategy & Tactics*. West Patterson, NJ: Pentagram, January 2002.

Huxley, Aldous. *Grey Eminence: A Study in Religion and Politics*. London: Chatto & Windus, 1941.

Klein, Naomi. *No Logo: Taking Aim at the Brand Bullies*. New York: Picador USA, 1999.

Lasswell, Harold D. "The Structure and Function of Communication in Society." In Lyman Bryson, ed. *The Communication of Ideas: A Series of Addresses*. New York: Institute for Religious and Social Studies, 1948, pp. 203–243.

Levine, Adeline Gordon. *Love Canal: Science, Politics, and People*. Lexington, MA: Lexington Books, 1982.

Levitan, Sara A., and Martha R. Cooper. *Business Lobbies: The Public Good and the Bottom Line*. Baltimore, MD: Johns Hopkins University Press, 1984.

Lorenz, Christopher. *The Design Dimension: Product Strategy and the Challenge of Global Marketing*. New York: Blackwell, 1986.

Low, Jonathan, and Pam Cohen Kalafut. *Invisible Advantage: How Intangibles Are Driving Business Performance.* Cambridge, MA: Perseus Books, 2002.

Margolis, Joshua Daniel, and James Patrick Walsh. *People and Profits? The Search for a Link Between a Company's Social and Financial Performance.* London: Lawrence Erlbaum Associates, Publishers, 2001.

McLean, Bethany, and Peter Elkind. *The Smartest Guys in the Room.* New York: The Penguin Group, 2004.

McLuhan, Marshall, and Bruce R. Powers. *The Global Village: Transformations in World Life and Media in the 21st Century.* New York: Oxford University Press, 1989.

McQuaid, Kim. *Big Business and Presidential Power.* New York: Morrow, 1982.

Munter, Mary. *Guide to Managerial Communication,* 7th ed. Upper Saddle River, NJ: Prentice Hall, 2006.

Olins, Wally. *Corporate Identity: Making Business Strategy Visible through Design.* London: Thames and Hudson, 1989.

Peters, Thomas J., and Robert H. Waterman Jr. *In Search of Excellence: Lessons from America's Best-Run Companies.* New York: Harper & Row, 1982.

Poster, Mark. *The Second Media Age.* Cambridge: Polity Press, 1995.

Postman, Neil. *Amusing Ourselves to Death: Public Discourse in the Age of Show Business.* New York: Penguin, 1985.

Riley, Charles A., II. *Small Business, Big Politics: What Entrepreneurs Need to Know to Use Their Political Power.* Princeton, NJ: Peterson's/Pacesetter, 1995.

Schenkler, Irv, and Tony Herrling. *Guide to Media Relations.* Upper Saddle River, NJ: Pearson/ Prentice Hall, 2004.

Schultz, Howard, and Dori Yang. *Pour Your Heart into It: How Starbucks Built a Company One Cup at a Time.* New York: Hyperion, 1999.

Schultz, Majken, Mary Jo Hatch, and Mogens Holten Larsen, eds. *The Expressive Organization.* Oxford: Oxford University Press, 2000.

Shannon, Claude Elwood, and Warren Weaver. *The Mathematical Theory of Communication.* University of Illinois Press, 1964.

Slywotzky, Adrian. *Value Migration: How to Think Several Moves ahead of the Competition.* Boston: Harvard Business School Press, 1996.

ten Berge, Dieudonnee. *The First 24 Hours.* Cambridge, MA: Basil Blackwell, 1990.

van Riel, Cees B. M. *Principles of Corporate Communication.* London: Prentice Hall, 1995.

Vogel, David. *Fluctuating Fortunes: The Political Power of Business in America.* New York: Basic Books, 1989.

Wallis, Allen W. *An Over Governed Society.* New York: The Free Press, 1976.

Weidenbaum, Murray L. *Business, Government, and the Public.* Englewood Cliffs, NJ: Prentice Hall, 1990.

Welch, Jack, and John A. Byrne. *Jack: Straight from the Gut.* New York: Warner Business, 2001.

White, Jon, and Laura Mazur. *Strategic Communications Management : Making Public Relations Work.* New York: Addison-Wesley, 1995.

Wilson, Graham. *Interest Groups in the United States.* New York: Oxford University Press, 1981.

Yankelovich, Daniel. *Profit with Honor: The News Stage of Market Capitalism.* New Haven, CT: Yale University Press, 2006.

Index